Methods in Molecular Biology™

Series Editor
John M. Walker
School of Life Sciences
University of Hertfordshire
Hatfield, Hertfordshire, AL10 9AB, UK

For further volumes:
http://www.springer.com/series/7651

Tandem Repeats in Genes, Proteins, and Disease

Methods and Protocols

Edited by

Danny M. Hatters

Department of Biochemistry and Molecular Biology
Bio21 Molecular Science and Biotechnology Institute, University of Melbourne
Parkville, VIC, Australia

Anthony J. Hannan

Florey Institute of Neuroscience and Mental Health
Melbourne Brain Centre, University of Melbourne
Parkville, VIC, Australia

Humana Press

Editors
Danny M. Hatters
Department of Biochemistry
and Molecular Biology
Bio21 Molecular Science
and Biotechnology Institute
University of Melbourne
Parkville, VIC, Australia

Anthony J. Hannan
Florey Institute of Neuroscience
and Mental Health
Melbourne Brain Centre
University of Melbourne
Parkville, VIC, Australia

ISSN 1064-3745 ISSN 1940-6029 (electronic)
ISBN 978-1-4939-6300-3 ISBN 978-1-62703-438-8 (eBook)
DOI 10.1007/978-1-62703-438-8
Springer New York Heidelberg Dordrecht London

Printed on acid-free paper

Humana Press is a brand of Springer
Springer is part of Springer Science+Business Media (www.springer.com)

Preface

The genomes of humans, as well as many other species are interspersed with hundreds of thousands of tandem repeats of DNA sequences. Those tandem repeats located as codons within open reading frames encode amino acid runs, such as polyglutamine and polyalanine. Tandem repeats have been implicated not only in biological evolution, development, and function but also in a large collection of human disorders. These disorders include the polyglutamine diseases, the most common of which is Huntington's disease, a range of spinocerebellar ataxias, Friedreich ataxia, fragile X syndrome, myotonic dystrophy, and polyalanine disorders. This book will present a diverse range of methods chapters, covering the analysis of tandem repeats in DNA, RNA, and protein, in healthy and diseased states. This will include molecular genetics, molecular biology, biochemistry, proteomics, biophysics, cell biology, and molecular and cellular approaches to animal models of tandem repeat disorders.

We would like to thank all the authors for their contributions and in particular their insight and solutions to some unique methodological challenges that come from repetitive DNA and poly-amino acid sequences.

Parkville, VIC, Australia

Danny M. Hatters
Anthony J. Hannan

Contents

Contributors

CHARLES ANDERSON • *Department of Computer Science, Colorado State University, Fort Collins, CO, USA*

D. MICHAEL ANDO • *Gladstone Institute of Neurological Disease, University of California, San Francisco, USA; Department of Neurology, University of California, San Francisco, USA; Department of Physiology, University of California, San Francisco, USA*

BRETT A. BARBARO • *Department of Developmental and Cell Biology, University of California, Irvine, Irvine, CA, USA*

ASA BEN-HUR • *Department of Computer Science, Colorado State University, Fort Collins, CO, USA*

DOUG J. BORNEMANN • *Department of Developmental and Cell Biology, University of California, Irvine, Irvine, CA, USA*

STEPHEN P. BOTTOMLEY • *Department of Biochemistry and Molecular Biology, Monash University, Clayton, VIC, Australia*

JOHN BURKE • *Department of Developmental and Cell Biology, University of California, Irvine, Irvine, CA, USA*

KATHLEEN A. BURKE • *The C. Eugene Bennett Department of Chemistry, West Virginia University, Morgantown, WV, USA*

EMMA L. BURROWS • *Florey Institute of Neuroscience and Mental Health, Melbourne Brain Centre, University of Melbourne, Parkville, VIC, Australia*

LORI J. COOPER • *Department of Biochemistry and Molecular Biology, Thomas Jefferson University, Philadelphia, PA, USA*

MARTIN L. DUENNWALD • *Department of Pathology, Schulich School of Medicine and Dentistry, University of Western Ontario, Ontario, Canada*

STEVEN FINKBEINER • *Gladstone Institute of Neurological Disease, Taube-Koret Center for Huntington's Disease Research, University of California, San Francisco, USA; Department of Neurology, University of California, San Francisco, USA; Department of Physiology, University of California, San Francisco, USA*

JOZEF GÉCZ • *Neurogenetics, Department of Genetic and Molecular Pathology, SA Pathology, Adelaide, SA, Australia; Department of Paediatrics, University of Adelaide, Adelaide, SA, Australia*

ANTHONY J. HANNAN • *Florey Institute of Neuroscience and Mental Health, Melbourne Brain Centre, University of Melbourne, Parkville, VIC, Australia*

DANNY M. HATTERS • *Department of Biochemistry and Molecular Biology, Bio21 Molecular Science and Biotechnology Institute, University of Melbourne, Parkville, VIC, Australia*

MATS HOLMBERG • *European Research Institute for the Biology of Ageing, University Medical Centre Groningen, University of Groningen, Groningen, The Netherlands*

JAMES N. HUGHES • *School of Molecular and Biomedical Science, University of Adelaide, Adelaide, SA, Australia*

D.T. Humphreys • *The Victor Chang Cardiac Research Institute, Darlinghurst, NSW, Australia; Faculty of Medicine, University of New South Wales, Kensington, Australia*
Patrick Lajoie • *Department of Anatomy and Structural Biology, Albert Einstein College of Medicine of Yeshiva University, Bronx, NY, USA*
K.T. Lawlor • *ARC Special Research Centre for the Molecular Genetics of Development and Discipline of Genetics, School of Molecular and Biomedical Sciences, The University of Adelaide, Adelaide, SA, Australia*
Justin Legleiter • *The C. Eugene Bennett Department of Chemistry, West Virginia University, Morgantown, WV, USA; WVnano Initiative and the Center for Neurosciences, West Virginia University, Morgantown, WV, USA*
Gregor P. Lotz • *Novartis Institutes for BioMedical Research (NIBR), Basel, Switzerland*
Tamas Lukacsovich • *Department of Developmental and Cell Biology, University of California, Irvine, Irvine, CA, USA*
Kyle S. MacLea • *Department of Biochemistry and Molecular Biology, Colorado State University, Fort Collins, CO, USA*
J. Lawrence Marsh • *Department of Developmental and Cell Biology, University of California, Irvine, Irvine, CA, USA*
T.R. Mattiske • *Department of Paediatrics, University of Adelaide, Adelaide, SA, Australia*
Diane E. Merry • *Department of Biochemistry and Molecular Biology, Thomas Jefferson University, Philadelphia, PA, USA*
Yee-Foong Mok • *Department of Biochemistry and Molecular Biology, Molecular Science and Biotechnology Institute, University of Melbourne, Parkville, VIC, Australia*
Regina M. Murphy • *Chemical and Biological Engineering Department, University of Wisconsin-Madison, Madison, WI, USA*
Ellen A.A. Nollen • *European Research Institute for the Biology of Ageing, University Medical Centre Groningen, University of Groningen, Groningen, The Netherlands*
L.V. O'Keefe • *ARC Special Research Centre for the Molecular Genetics of Development and Discipline of Genetics, School of Molecular and Biomedical Sciences, The University of Adelaide, Adelaide, SA, Australia*
Saskia Polling • *Department of Biochemistry and Molecular Biology, Molecular Science and Biotechnology Institute, University of Melbourne, Parkville, VIC, Australia*
Judith Purcell • *Department of Developmental and Cell Biology, University of California, Irvine, Irvine, CA, USA*
Yasmin M. Ramdzan • *Department of Biochemistry and Molecular Biology, Monash University, Clayton, VIC, Australia*
R.I. Richards • *ARC Special Research Centre for the Molecular Genetics of Development and Discipline of Genetics, School of Molecular and Biomedical Sciences, The University of Adelaide, Adelaide, SA, Australia*
Amy L. Robertson • *Department of Biochemistry and Molecular Biology, Monash University, Clayton, VIC, Australia*
Eric D. Ross • *Department of Biochemistry, Colorado State University, Fort Collins, CO, USA; Department of Molecular Biology, Colorado State University, Fort Collins, CO, USA*
S.E. Samaraweera • *ARC Special Research Centre for the Molecular Genetics of Development and Discipline of Genetics, School of Molecular and Biomedical Sciences, The University of Adelaide, Adelaide, SA, Australia*

CHERYL SHOUBRIDGE • *Department of Paediatrics, University of Adelaide, Adelaide, SA, Australia*
MARIANNE R. SMITH • *Department of Developmental and Cell Biology, University of California, Irvine, Irvine, CA, USA*
ERIK LEE SNAPP • *Department of Anatomy and Structural Biology, Albert Einstein College of Medicine of Yeshiva University, Bronx, NY, USA*
WAN SONG • *Department of Developmental and Cell Biology, University of California, Irvine, Irvine, CA, USA*
C.M. SUTER • *The Victor Chang Cardiac Research Institute, Darlinghurst, NSW, Australia; Faculty of Medicine, University of New South Wales, Kensington, Australia*
ADEELA SYED • *Department of Developmental and Cell Biology, University of California, Irvine, Irvine, CA, USA*
MAY H. TAN • *Department of Paediatrics, University of Adelaide, Adelaide, SA, Australia*
PAUL Q. THOMAS • *School of Molecular and Biomedical Science, University of Adelaide, Adelaide, SA, Australia*
ANDREY S. TSVETKOV • *Gladstone Institute of Neurological Disease, San Francisco, USA; Taube-Koret Center for Huntington's Disease Research, San Francisco, USA*
C.L. VAN EYK • *ARC Special Research Centre for the Molecular Genetics of Development and Discipline of Genetics, School of Molecular and Biomedical Sciences, The University of Adelaide, Adelaide, SA, Australia*
ANDREAS WEISS • *Novartis Institutes for BioMedical Research (NIBR), Basel, Switzerland*
REBECCA WOOD • *Department of Biochemistry and Molecular Biology, Monash University, Clayton, VIC, Australia*

Chapter 1

Longitudinal Imaging and Analysis of Neurons Expressing Polyglutamine-Expanded Proteins

Andrey S. Tsvetkov, D. Michael Ando, and Steven Finkbeiner

Abstract

Misfolded proteins have been implicated in most of the major neurodegenerative diseases, and identifying drugs and pathways that protect neurons from the toxicity of misfolded proteins is of paramount importance. We invented a form of automated imaging and analysis called robotic microscopy that is well suited to the study of neurodegeneration. It enables the monitoring of large cohorts of individual neurons over their lifetimes as they undergo neurodegeneration. With automated analysis, multiple endpoints in neurons can be measured, including survival. Statistical approaches, typically reserved for engineering and clinical medicine, can be applied to these data in an unbiased fashion to discover whether factors contribute positively or negatively to neuronal fate and to quantify the importance of their contribution. Ultimately, multivariate dynamic models can be constructed from these data, which can provide a systems-level understanding of the neurodegenerative disease process and guide the rationale for the development of therapies.

Key words Huntington's disease, Neurodegeneration, Huntingtin, Survival analysis, High-throughput screening

1 Introduction

Huntington's disease (HD) is the most common inherited neurodegenerative disorder and is characterized by abnormal motor movements and cognitive decline. A polyglutamine expansion in the huntingtin (htt) protein, which causes the disease, leads to inclusion body (IB) formation and neuronal toxicity.

Studying the cellular and molecular mechanisms of HD in animal models has limitations. Phenotypes vary among the mouse models. Pharmacological or genetic manipulations in mice can be challenging and expensive. Drugs may have effects outside the CNS that confound the interpretation of results. The development of faithful cellular models of neurodegenerative disease is therefore critical to rigorously elucidate disease mechanisms, discover therapeutic targets, and identify potential therapies.

Danny M. Hatters and Anthony J. Hannan (eds.), *Tandem Repeats in Genes, Proteins, and Disease: Methods and Protocols*, Methods in Molecular Biology, vol. 1017, DOI 10.1007/978-1-62703-438-8_1, © Springer Science+Business Media New York 2013

Many cell models of HD are based on immortalized cell lines, primary murine cultures of neurons and glia, and cultured human neurons and glia differentiated from patient-derived induced pluripotent stem (iPS) cells. The main advantage of cell line-based models is their homogeneity and ease of use. However, the cellular physiology of immortalized cell lines differs in important respects from primary post-mitotic neurons, namely, in their capacity to divide and lack of synapses. Therefore, to identify mechanisms and therapeutic targets, primary neurons are likely to be a more relevant system [1–4]. The recent development of human-cell models of HD from patient-derived iPS cells should yield insights into the disease that have increased physiological relevance. This could be especially important if a major reason for the failure of clinical trials for HD therapies is because murine models of HD are not representative of the human condition [5].

In our laboratory, we use several HD models that are based on human neurons or primary murine neurons from the striatum as well as other brain regions less affected in HD [1, 5]. Cultured neurons are transfected with N-terminal fragments of wild-type (WT) or polyQ-expanded huntingtin (htt) (171, 480, 586 N-terminal amino acids or full-length htt) fused to the N-terminus of a fluorescent protein. These HD models recapitulate at least 18 molecular and cellular features observed in HD patients [2]. Moreover, with these HD neuron models, we predicted results that were later confirmed by observations in mouse models or HD patients, making these models highly physiologically relevant [6–13]. Recently, we developed a human neuron model of HD, based on iPS cells, which holds great potential for future HD research [5].

Conventional methods aimed at determining the prognostic significance of histological changes seen in fixed cultures or brain tissue have significant shortcomings. These methods rely on static histological "snapshots," in which only a fraction of neurons that degenerate are caught, while others have not yet begun to die or are degenerated (and consequently missed) completely. Therefore, inferences of cause-and-effect relationships from a time series of these snapshots can be incomplete and potentially misleading.

To overcome these limitations, we built an automated imaging system that can perform high-throughput longitudinal single-cell analysis [3, 14]. In our third-generation system, image acquisition is controlled by custom-built computer software and begins when a program instructs a robotic arm to load a plate of cells on the microscope stage. The computer instructs the microscope to focus itself and executes an algorithm that enables the microscope to position the plate in precise alignment to a reference position. The microscope then moves the plate to the center of the first well and collects fluorescence images at predetermined wavelengths, thereafter moving the stage to each adjacent field in a well. These steps are repeated until an entire well or a plate is imaged. The robotic arm then removes the plate and puts it back in the incubator until the plate is scheduled to be imaged again.

Custom programs then analyze the images and quantify features of interest off-line. First, the program organizes the images by well, plate, and date. Each image from the same well is then electronically "stitched" together into a montage. Montages from the same well are organized by date, and the program performs a further fine alignment on the stack of montages. The analysis program then identifies each neuron from the first montage, assigning it a unique identifying number, and then tracks that cell through the subsequent montages. Thus, we can follow a large group of neurons (thousands) over time, as neurodegeneration unfolds, and identify the changes (e.g., levels of diffuse htt, IB formation) in each neuron and link these changes to some future fate of this particular neuron (survival or death), establishing cause-and-effect relationships [2, 3, 14–16].

Such datasets—composed of repeated measures on individual members of a cohort—are amenable to a powerful suite of statistical tools called survival analysis. Despite its name, this tool can be used to measure differences in essentially any time-dependent event among cohorts, including survival, and has been used widely in engineering and medicine. A related statistical tool, the Cox proportional hazards (CPH) analysis, makes it possible to construct explanatory statistical models that delineate the factors that contribute positively or negatively to a particular fate for individuals in the cohort and quantifies their relative importance. This capability is likely to be critical for studies of neurodegenerative processes, which appear to be complex, highly intertwined, and dynamic. Indeed, we found that the performance of our robotic microscope system for hypothesis-driven and discovery research is extraordinary. Direct comparisons with conventional snapshot approaches revealed that robotic microscopy is about 100- to 1,000-fold more sensitive for detecting differences in responses or behavior between cohorts. As few as eight cells per well can be sufficient to predict the complete-well result with 90 % accuracy. These advantages suggest that the robotic microscope is a uniquely powerful tool to study precious or difficult-to-culture cells (e.g., iPS cells) and for performing high-throughput screens on primary cells.

The utility and power of the approach can be illustrated with some recent applications. Using this automated imaging platform, we discovered that mutant htt-forming IBs, rather than being a toxic species themselves, were in fact a coping response used by neurons to mitigate mutant htt. Neurons that had them survived longer [14]. In a separate study, we discovered that, even before IBs form, the ubiquitin-proteasome system (UPS) becomes overwhelmed in the presence of mutant htt. Interestingly, IB formation and the accompanying sequestration of mutant htt into an IB appear to restore protein homeostasis and clearance systems [15], offering an explanation for how IBs might help neurons to cope with mutant htt. Since IBs form asynchronously in only a subset of neurons with mutant htt, the relationship between IB formation and UPS function could only be elucidated by an approach that utilizes longitudinal single-cell analysis. Indeed, with our findings in mind, others reexamined the relationship of the UPS to IB formation in vivo and found a similar result [17].

Autophagy is a cellular mechanism that directs proteins to lysosomes for degradation and recycling and also turns over monomeric, misfolded proteins and protein aggregates. By observing striatal neurons transfected with mutant htt over time, we identified small-molecule autophagy inducers that lower the levels of mutant htt [4]. When we treated a cohort of neurons with this small-molecule autophagy inducer, we found that the levels of diffuse mutant htt dropped and, importantly, the survival of striatal neurons increased. We also observed that neurons produced fewer IBs, suggesting that the autophagy stimulator enhanced mutant htt degradation, thereby reducing the need for IBs to form as a coping response. These findings underscore how analyzing spatiotemporal changes in neurons with automated microscopy represents an unprecedented opportunity to study the mechanisms of neurodegeneration.

Although the sophistication of our third-generation robotic microscope—a product of many years of technology development and refinement—would be difficult to replicate, the basic power of longitudinal single-cell analysis can be achieved by most labs with lower-throughput manual approaches. In this chapter, we describe many of our methods and offer suggestions for how experiments similar in design and scope can be performed with simpler instrumentation.

2 Materials

2.1 Fluorescently Tagged Proteins

We co-transfect neurons with a fluorescently tagged N-terminal fragment of htt (WT or mutant) and a morphology/viability fluorescent marker. In some experiments, we perform a triple transfection by adding a third fluorophore [15].

2.1.1 An N-Terminal Fragment of htt Fused to the N-Terminus of a Fluorescent Protein

1. pGW1-CMV plasmid (British Biotechnology; Oxford, UK) that gives us the highest expression levels in striatal neurons encoding WT or mutant htt fused to GFP (*see* **Notes 1** and **2**).
2. QIAGEN plasmid isolation MAXI kit (QIAGEN; Valencia, CA). Purified plasmid is stored at 4 °C.

2.1.2 A Fluorescent Morphology and Viability Marker

1. pGW1-CMV plasmid (British Biotechnology) encoding a red fluorescent protein (*see* **Notes 3** and **4**) [14].
2. Plasmid purified with the QIAGEN plasmid isolation MAXI kit (stored at 4 °C).

2.2 Preparation of Neuronal Cultures

1. Poly-D-lysine solution (catalog number A-003-E, 1 mg/ml, Millipore; Billerica, MA).
2. Mouse laminin (catalog number 354232, 1 mg, BD Biosciences; San Jose, CA).
3. Plates: 96-well plate (catalog number 92696, Swiss TPP; Trasadingen, Switzerland) or 24-well plate (catalog number 3337, Corning) (*see* **Note 5**).
4. Trypsin inhibitor (catalog number T9253-5 G, Sigma, St. Louis, MO).
5. Papain (PAP 100 mg, Worthington; Lakewood, NJ).
6. L-Cysteine.
7. 10× KY stock solution: 10 mM kynurenic acid, 0.0025 % (w/v) phenol red (catalog number P-0290, Sigma), 5 mM HEPES, 100 mM $MgCl_2$, pH 7.4. Filter-sterilized and stored at 4 °C.
8. Dissociation medium (DM): 81.8 mM Na_2SO_4, 30 mM K_2SO_4, 5.8 mM $MgCl_2$, 0.25 mM $CaCl_2$, 1 mM HEPES, 20 mM glucose, 0.001 % (w/v) phenol red, 0.16 mM NaOH. Filter-sterilized and stored at 4 °C.
9. Opti-MEM-glucose: 4 ml of 2.5 M glucose per 500 ml of Opti-MEM (catalog number 31985, GIBCO).
10. Trypan blue.
11. For plating onto a 96-well plate: a 50-ml reservoir (catalog number 4870, Corning), a multichannel 250-μl pipettor and required tips.
12. Neurobasal medium (catalog number 21103, GIBCO).
13. 50× B27 vitamin supplement (catalog number 17504-044, GIBCO).
14. 100× GlutaMAX.
15. 100× pen/strep.
16. Dissection microscope (catalog number NI-MA-MMD31000, Nikon).
17. Tools: 2–3 pairs of forceps, one pair of scissors, a chemical spatula (catalog number 11295-10, 14060-09, 10099-15, Fine Science Tools).
18. 0.20-μm filters.
19. 30-ml syringes.
20. Cell-culture facilities including a humidified 37 °C incubator with 5 % (v/v) atmospheric CO_2, hemocytometer, and 37 °C water bath.
21. Alcohol solutions: 70 % (v/v) ethanol and 96 % (v/v) ethanol.
22. Source of primary neurons: timed-pregnant rat (or a mouse).

2.3 Transfection

1. Lipofectamine 2000 (catalog number 11668-027, Invitrogen) (*see* **Note 6**).
2. Transfection solution: 1× KY in neurobasal medium (catalog number 21103, GIBCO).
3. Opti-MEM (catalog number 31985, GIBCO).
4. Neurobasal medium (catalog number 21103, GIBCO).

2.4 Automated Microscopy

1. Digital images are obtained with an inverted microscope (a Nikon TE2000E-PFS microscope, a long-working-distance Nikon CFI S Plan Fluor 20× (NA 0.45) objective) and a 300 W Xenon Lambda LS illuminator (Sutter Instruments, Novato, CA) (*see* **Note 7**).
2. Stage movements to an adjacent well and focusing are performed with an MS-2000 XY stage (Applied Scientific Instrumentation, Eugene, OR).
3. Images are acquired with a CCD camera (Clara, Andor; Belfast, Northern Ireland) driven by Image-Pro Plus software (Media Cybernetics; Bethesda, MD).
4. Image-Pro Plus software (Media Cybernetics).

2.5 Image Analysis and Storage

1. StatView (Apple, Cupertino, CA).
2. Image J (rsbweb.nih.gov/ij/, the National Institutes of Health) (*see* **Note 8**).
3. Image-Pro Plus (Media Cybernetics).
4. Pipeline Pilot (Accelrys; San Diego, CA).
5. A terabyte server (Sentinel Small Office Storage Server, Amazon) (*see* **Note 9**).

2.6 Data Analysis

1. The R Project for Statistical Computing (http://www.r-project.org/, R Development Core Team).

3 Methods

3.1 Fluorescently Tagged Proteins

3.1.1 N-Terminal Fragment of htt Fused to the N-Terminus of a Fluorescent Protein

1. Isolate the plasmid DNA encoding GFP fused to WT or mutant htt (the exon 1 fragment or longer) with a plasmid purification kit (*see* **Notes 1** and **2**). Follow the manufacturer's instructions.
2. Adjust concentration to 0.5–1.5 μg/ml with water.
3. Store the plasmid at 4 °C for up to 6 months.

3.1.2 A Morphology and Viability Fluorescent Marker

1. Isolate the plasmid DNA encoding a red morphology and viability marker (such as mRFP or mCherry) with a plasmid purification kit (*see* **Notes 3** and **4**); follow manufacturer's instructions.

2. Adjust concentration to 0.5–1.5 μg/ml with water.
3. Store the plasmid at 4 °C for up to 6 months.

3.2 Preparation of Neuronal Cultures

1. One day before preparing the cultures, prepare the plates. Into 100 ml of sterile water, add 5 ml of poly-D-lysine solution and a half vial of mouse laminin (the remainder can be kept at −80 °C for later use). Mix well.
2. *To a 96-well plate*: add 150 μl of the coating mix (poly-D-lysine/laminin solution). *To a 24-well plate*: add 500 μl of the coating mix (*see* **Note 5**). Swirl the plates to ensure that the coating mix completely covers the bottom of the wells.
3. Leave the plates in a 37 °C/5 % CO_2 incubator overnight.
4. Wash the plates twice with sterile water (150 μl of water for a 96-well plate and 500 μl for a 24-well plate). Remove the final wash and leave the plates in a 37 °C/5 % CO_2 incubator.
5. Make neuronal growth medium: mix neurobasal medium with B27 vitamin supplement (50× stock), GlutaMAX (100× stock) (catalog number 35050, GIBCO), and penicillin/streptomycin (100× stock) to a 1× final concentration. Keep in a 37 °C water bath.
6. Make Opti-MEM-glucose (4 ml of 2.5 M glucose per 500 ml of Opti-MEM). Keep in a 37 °C water bath.
7. Prepare DM/KY solution (300–500 ml): dilute the 10× KY stock into an appropriate volume of DM. Keep on ice.
8. Prepare the trypsin inhibitor solution: add 150 mg of trypsin inhibitor to 10 ml of DM/KY. The solution will become yellow. Adjust the pH of the solution with 1 M NaOH until it is pink again (pH 7.5). Keep at room temperature.
9. Prepare the solution for papain: add 2–3 mg of L-cysteine to 10 ml of DM/KY. Adjust pH of the solution with 5 M NaOH until it is pink again (pH 7.5). Keep at room temperature.
10. Pour ice-cold DM/KY solution into two 10-cm culture dishes (one for the pup heads and one for the brains) and into one 6-cm dish (for the dissected striatum). Keep dishes on ice.
11. Sterilize the dissection tools with 70 % (v/v) ethanol.
12. Sacrifice a timed-pregnant rat (or a mouse) according to an animal protocol approved at your institution (*see* **Note 10**).
13. After an animal is sacrificed, clean the belly with 96 % (v/v) alcohol. Cut along the abdomen and remove the uterus. Place the pups into the large culture dish.
14. Cut off the pup heads with scissors and put them in a 10-cm dish with DM/KY on ice.
15. Under a dissection microscope, remove the skin from a skull. Hold the head with forceps through the eyes. Use the other

forceps to incise the skin towards the eyes and peel the skin and open the skull. Cut the optic nerves. Take the chemical spatula, dig underneath the brain, and scoop it out. Place the brain into a new 10-cm dish with DM/KY on ice.

16. Repeat the process with other heads.
17. Dissect the striatum. Orient yourself so that the brain is facing forward. For each hemisphere, use your forceps to dissect longitudinally down the hemisphere. Your goal is to expose the structures just underneath the superficial cortex. Pierce the ventral surface of the cortex of each hemisphere to access the lateral ventricle. The capillary network and the choroid plexus should be visible. At this developmental stage, the striatum is the structure of brain tissue bulging into the lateral ventricle from the midline. The lateral edge of the striatum forms a semicircle. Isolate the striatum and discard the remaining brain tissue (mostly cortex and hippocampus). Dissect away the remaining cortex from the striatum. Dissect the striatum and place it in a new 10-cm dish with DM/KY on ice (*see* **Note 11**).
18. Repeat the process with the other cortical hemisphere and other brains. Keep the striata all together in DM/KY on ice.
19. Prepare the papain solution: add 100 U of papain to the cysteine/DM/KY solution.
20. Filter the trypsin inhibitor solution and papain solution through a 0.20-μm filter and 30-ml syringe. Place both the trypsin inhibitor solution and the papain solution in the 37 °C water bath.
21. Clean up after the dissection (5 min).
22. Transfer the striata with a cut 500-μl tip by pipettor suction to a 15-ml conical tube. Let the tissue settle and remove the extra DM/KY solution.
23. Add 10 ml of warm papain solution to the striata and incubate at 37 °C for 10 min. Carefully swirl the tube after 5 min.
24. Carefully remove the papain solution.
25. Add warm trypsin inhibitor solution and incubate at 37 °C for 10 min. Carefully swirl the tube after 5 min.
26. Remove the trypsin inhibitor solution and wash the striata with 10 ml of warm Opti-MEM-glucose. Remove the Opti-MEM-glucose solution.
27. Under a sterile hood, add 5 ml of Opti-MEM-glucose. Triturate gently several times with a 5-ml pipette until the solution turns cloudy.
28. Allow the tissue to settle. Take the supernatant and transfer it to a 50-ml conical tube. Keep it under the sterile hood at room temperature.

29. Add 5 ml of new Opti-MEM-glucose to the striata and repeat trituration. Triturate 8–10 times for striatum collected from 10 to 15 rat brains or 5 times for striatum collected from 5 to 6 mouse brains.
30. Allow the debris to settle in the 50-ml conical tube. Take a 5-ml pipette and remove the debris.
31. Counting neurons: mix the neuron suspension in the 50-ml conical tube. Take a 10-μl aliquot and add it to a tube that contains 10 μl of Opti-MEM-glucose and 10 μl of trypan blue. Carefully mix and add 10 μl to the hemocytometer. Count the cells in the 16-box squares in the two opposite corners of the field. Average the two counts and multiply the average by 30,000 to get the number of cells per ml.
32. Plate cells to a final concentration of 100,000 per well of a 96-well plate (in 100–200 μl of plating Opti-MEM-glucose solution) or 0.6–0.75 million per well of a 24-well plate (in 500–1,500 μl of plating Opti-MEM-glucose solution). Dilute the cells with Opti-MEM-glucose solution before plating, if needed. Swirl the plate to make sure cells are evenly distributed.
33. Incubate the plates at 37 °C for 1 h.
34. Check the cells under the microscope to ensure that the neurons adhered to the well surface.
35. Replace the Opti-MEM-glucose solution with warm neurobasal medium (that contains B27 vitamin supplement, GlutaMAX, and antibiotics): 200 μl of neurobasal medium per well of a 96-well plate, 1 ml of neurobasal medium per well of a 24-well plate.

3.3 Transfection

1. Prepare neurobasal/KY solution: dilute the 10× KY stock into an appropriate volume of neurobasal medium. Keep it in the 37 °C water bath.
2. Warm Opti-MEM to room temperature.
3. For each transfection sample, prepare the DNA/Lipofectamine 2000 complexes according to the manufacturer's protocol (*see* **Note 12**). Keep the tubes at room temperature for 20 min to allow the DNA/Lipofectamine 2000 complexes to form.
4. Use 5 days in vitro (DIV) neuronal cultures for transfection.
5. Wash neuronal cultures twice with neurobasal/KY solution.
6. Add 150 μl of neurobasal/KY solution per well of a 96-well plate or 500 μl of neurobasal/KY solution per well of a 24-well plate.
7. Vortex the DNA/Lipofectamine 2000 complexes, add them to neuronal cultures (*see* **Note 12**), and swirl the plate.

8. Incubate the plates in the 37 °C/5 % CO_2 incubator for 0.5–2.5 h (*see* **Note 13**).
9. Wash neuronal cultures twice with neurobasal medium.
10. Add 200 μl of neurobasal medium per well of a 96-well plate or 1 ml of neurobasal medium per well of a 24-well plate.

3.4 Longitudinal Single-Cell Image Acquisition

Several commercially available systems are capable of performing automated image acquisition and some forms of automated image analysis [18]. High-content screening (HCS) systems, such as Opera (PerkinElmer) or IN Cell Analyzer (GE Healthcare), provide fast image acquisition and powerful data-processing capabilities. In general, these systems tend to be oriented to the analysis of snapshots of immunocytochemically stained, fixed cells rather than live-cell imaging, and their cost and the extent in which their proprietary algorithms allow for customization of each user's applications vary. However, the main difference between the commercial systems and those we described above is the ability to do longitudinal single-cell analysis over arbitrarily long periods of time. This feature allows our system to unravel complex cause-and-effect relationships and to detect the effects of manipulations with extraordinary sensitivity. For a more in-depth overview of automated image acquisition platforms with primary neurons, *see* ref. 19.

Although we have automated image acquisition to improve precision and throughput, we can generate longitudinal series of images of single cells and extract datasets that are amenable to analysis with powerful statistical tools described below. Here we present a general protocol to generate longitudinal images of single cells spanning arbitrary time intervals. The four main steps are registration, stage movement, focus, and acquisition:

1. *Registration*. In most of our experiments, cultured cells are returned to an incubator between imaging intervals, so many experiments can be run on the same microscope in parallel. To find the same neuron each time the plate is returned to the stage, a method to register the position of the plate on the microscope is critical. Most manufacturers stamp alphanumeric codes on the plate itself, a fiduciary mark that can be used to register the plate position. If the plate lacks such a mark, the user can make one with a marker or by etching the plastic. At the first time point, an image of the fiduciary mark on the plate is collected and stored as a reference image. Subsequently, each time the same plate is imaged, the fiduciary mark on the plate is aligned with the reference image. Alignment can be performed manually.
2. *Stage movements*. To begin image acquisition, the plate needs to be moved from the fiduciary mark to a well of interest.

This can be accomplished by using the computer to direct the automated stage to a specific *x–y* location. We typically begin with the center of the well located at the most far-left position in the top row of the plate. In cases where multiple fields from the same well are desired, the user must systematically collect images of adjacent microscope fields. We use the computer to instruct the stage to make a pattern of movements relative to the fiduciary mark. This can also be done manually by creating an ordered list of coordinates visited on the first time point and then controlling the automated stage to return to each coordinate on that list during subsequent time points. More conveniently, some acquisition programs provide built-in functionality for moving between wells of a standard microplate and for imaging an array within a well. We found that a 3 × 3 image array using a 20× objective often contains enough cells to powerfully measure relationships between expression levels, IB formation, neuronal morphology, and survival while maintaining high throughput. To ensure that the array of images from a single well have optimal image content, the physical dimensions of a microscope field should be determined and used to program the stage to collect images from contiguous but slightly overlapping fields. This can be done using a reference slide with grids of known distance to determine the physical dimensions of a pixel.

3. *Focus.* Accurate and consistent focus is a critical issue for measuring fluorescence-intensity levels and subcellular changes such as inclusion formation. This can be done manually using a brightfield image before fluorescence imaging, but it is tedious and reduces throughput. To increase throughput, we have used automated focusing algorithms that are image-based or based on the direct detection of the physical distance between the objective and the bottom of the plate. Although the image-based algorithms were often very good, the physical methods have proven to be more accurate, faster, and especially well suited to plates that contain ≥96 wells.

4. *Acquisition.* Acquisition of fluorescence images can be done manually or with an automated acquisition program, which should have functions for controlling the excitation and emission filters, the polychroic cube, and the exposure time. To limit bleed-through of fluorescence from one channel to another, we use fluorescent proteins whose emissions are relatively well separated with commercially available optical filters. A common choice is cyan, yellow, and red fluorescent proteins for imaging morphology or when creating reporter constructs. The length of illumination required to generate a quality image will depend significantly on the expression level and quantum efficiency of the fluorescent protein, the efficiency of

the optical path, and the sensitivity of the camera. With a standard CCD camera, exposure times from 100 ms to 1 s provide a sufficient balance between signal collection and throughput (*see* **Note 14**) although much lower exposure times are possible with EM-CCD cameras.

3.5 Image Analysis and Storage

We use survival analysis, a statistical tool, to quantify how the factors that were measured during the course of the experiment contribute to a fate that is being measured [11]. Despite its name, survival analysis is not limited to the study of survival, but in fact can be used with our system to study any endpoint that can be measured over time from collected images.

For these analyses, the survival time is measured for each neuron from the moment a transfected neuron is visualized until its death, indicated by the abrupt disappearance of the transfected marker. Survival time for each neuron in a cohort is tabulated, and a survival curve for the neuronal cohort is constructed with software (R or StatView). Neurons that survived the entire experiment are weighted differently to account for an indeterminate survival time. Survival functions are fit with Kaplan–Meier analysis, and differences between the cohorts are assessed with the log-rank test. The cumulative nature of the analysis makes it much more sensitive (100–1,000×) than approaches offered by commercial image analysis software, which depend on averaged responses measured at particular time points (*see* ref. 3).

A hazard function, which describes the instantaneous risk that a member of the cohort will reach the endpoint of interest, can be obtained from the survival function (StatView or R). This analysis can be useful for assessing whether a change of interest in the cells we are following is associated with an increased or decreased likelihood that the cohort will achieve the endpoint of interest. In simple terms, as it applies to neurodegeneration, this analysis helps us determine whether an intermediate change is likely to be incidental, part of the pathogenic process, or part of a coping response.

If two cohorts are being compared and the differences in hazard between the cohorts are proportional over the course of the study, the data can be analyzed further using a CPH analysis. CPH analysis makes it possible to quantify the contribution, if any, of specific factors or covariates to a given cell's fate. With CPH analysis, increasingly accurate quantitative multivariate models can be constructed that predict outcomes in terms of a series of measurable covariates. CPH analysis deduces a coefficient for each covariate whose magnitude and sign indicate its importance and whether that covariate increases or decreases the likelihood of a particular fate.

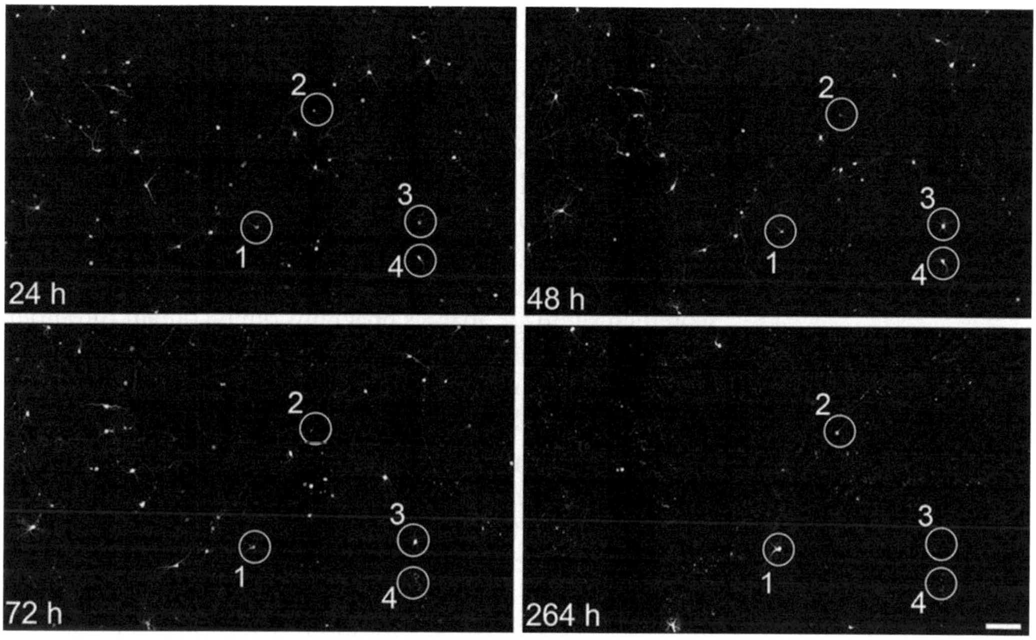

Fig. 1 Images collected at arbitrary intervals (i.e., 24, 48, 72, and 246 h after transfection) demonstrate the ability to return to the same field of neurons and track them over time. Every image is a montage of 20 nonoverlapping images captured in one well of a 96-well plate. Neurons express mCherry and can be identified using a custom-made algorithm (as an example, four neurons are circled and tracked). Neurons #1 and 2 survived throughout the experiment, neuron #3 died between 72 and 264 h, and neuron #4 died between 48 and 72 h. Images collected with a 20× objective. Scale bar is 300 μm

Below, we present a general protocol that might be used for analyzing images:

1. *Image organization*. Depending on the precision of the stage alignment, cells at the edge of a microscope field can appear in one image but also in a subsequent one. To avoid losing data at the edge of images and to facilitate the analysis of structures that can span multiple fields (e.g., axons, dendrites), we begin image analysis by electronically stitching images of all the microscope fields from the same well into a montage. Many image analysis programs offer this function, but it can also be performed manually. The key is to collect images of contiguous microscope fields that overlap slightly so as to ensure accurate alignment. One can then create a montage of the stitched images from one well at each time point and then combine those montages into a single image "stack" in which the *z*-axis represents time (Fig. 1). Since small errors occur during registration and stage movement, use of an alignment algorithm will aid in later analysis (*see* **Note 15**).
2. *Image segmentation*. This next step of the analysis is the most important and also the most difficult. First, we choose a threshold

intensity and identify pixels that exceed this threshold to consider for further analysis. The second step is a filter that analyzes contiguous groups of pixels to determine if the size of the group and the pattern of intensities within the border fit within our predetermined criteria for an object that we count as a neuronal soma. This is critical for distinguishing between positive pixels that belong to genuine cellular objects of interest versus those that represent debris, noise, or parts of cells. The algorithm that we developed performs with 80–90 % accuracy. Commercial image analysis programs often contain thresholding and spatial filter functions, which can be adapted for this purpose. For a more in-depth overview of image segmentation, *see* ref. 19 (Image Segmentation).

3. *Longitudinal tracking.* To get the benefits of longitudinal analysis, individual neurons must be tracked over the course of the experiment. Once image analysis has been performed on the first montage of a series for a particular experiment, each cell that has been identified is assigned a unique identifying number. The precise *x–y* location is recorded and used in images collected at subsequent time points to find the same cell or to determine that it has died and disappeared during the interval. As long as the movement of cells between images is relatively small and the "positive" cells are mostly well separated from each other, automated tracking is often possible. Some commercially available automated tracking programs do a reasonable job; alternatively, tracking can be performed manually.
4. *Analysis of image features.* Once a neuronal feature to analyze is selected (e.g., expression levels, IB formation, neurite length, or neurite branching), depending on the neuronal feature selected, open-source or commercially available software might exist with the functionality (or extensibility) to extract the desired information. If not, this may be done with manual intervention (*see* **Note 8**). Fluorescence data from the same cell in other channels can also be extracted. We binarize images of the cells generated from the fluorescence of the morphology marker and multiply those by the images of the same cell from the other fluorescence channels. The extracted information is then quantified and linked to that cell's unique identifier number.
5. *Survival analysis.* To perform survival analysis, a measureable endpoint must be chosen, and the time point when that endpoint is manifest must be determined. For survival analysis, the algorithm we developed to track neurons over time makes a determination of the time point at which death has occurred (*see* **Note 16**). Survival datasets are used to construct the

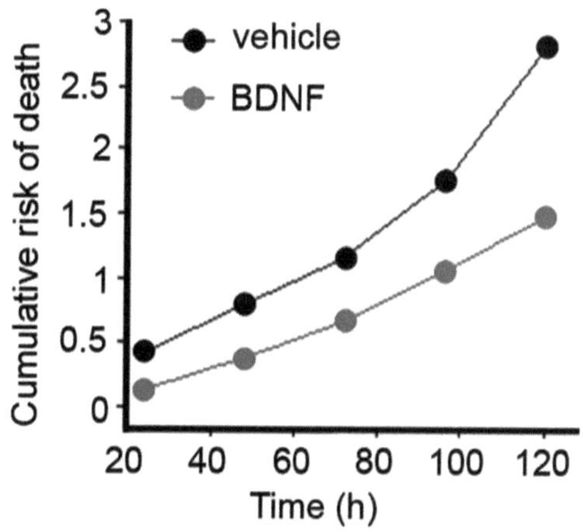

Fig. 2 An example of survival analysis. Striatal neurons transfected with mCherry and Htt^{ex1}_{72Q}-GFP were treated with 100 ng/ml BDNF or vehicle and followed with an automated microscope. Cumulative risk-of-death statistics were calculated from Kaplan–Meier curves. BDNF reduced the risk of death (i.e., improved survival) of neurons. * $p < 0.001$ (Mantel-Cox test)

survival curves (*see* **Notes 17** and **18**). Kaplan–Meier curves are used to estimate survival and hazard functions with R or StatView. Differences in Kaplan–Meier curves (e.g., differences between survival of different neuronal populations) are assessed with the log-rank test (Fig. 2).

6. *Cox proportional hazard analysis.* If intermediate measures were made of putative covariates (e.g., IB formation, expression levels), their contribution to fate can be estimated with CPH analysis. These types of analyses can be performed with commercially available or open-source programs (StatView or R).

4 Notes

1. We routinely use the exon 1 fragment of WT and mutant (containing either normal (Q_{17}) or disease-associated (Q_{47}, Q_{72}, Q_{103}) polyQ expansions) htt and longer N-terminal portions of htt (171, 480, 586 amino acids) [2–4, 14–16]. This approach to in vitro disease modeling can be applied broadly. In our lab, we developed a model of familial amyotrophic lateral sclerosis, based on expression of a fluorescently tagged version of the protein TDP-43 in cultured cortical neurons [20], and two models of Parkinson disease, based on the expression of

fluorescently tagged α-synuclein [21] and LRRK2 (personal communication, Steven Finkbeiner, Gaia Skibinski).

2. It is very important to use a truly monomeric version of a fluorescent protein, because dimerization or tetramerization can greatly affect the biochemistry of the protein to which it is fused. In our lab, we use a wide variety of monomeric fluorescent proteins [2–4, 14–16]. We prefer fluorescent proteins that have a high quantum yield and resist photobleaching. These characteristics allow us to use minimal illumination and avoid unwanted phototoxicity, and they simplify the analysis. Examples of fluorophore triads that can be resolved with most filter-based epifluorescence systems include monomeric (m) Apple or mCherry (red channel), Venus or mCitrine (yellow channel), and CFP or Cerulean (cyan channel).
3. Mutant htt forms IBs, which sequester diffuse mutant htt. After an IB has formed, it is difficult to accurately determine neuronal morphology using the fluorescence of mutant htt molecules. Therefore, to measure morphology and viability, it is important to use an inert fluorophore that is physically separate from the fluorophore fused to htt.
4. We routinely use mApple or mCherry (both are red fluorescent proteins) as a morphology and viability marker [3, 4]. Any red fluorescent protein can be used in a combination with htt fused to a yellow (mCitrine) or green (EGFP) fluorescent protein. Alternatively, htt could be labeled red and a morphology marker can be green.
5. The choice of plate type is based on several factors. 24-well plates (Corning) are cheaper and may work well for pilot experiments but allow for fewer conditions on one plate. In addition, each well requires more reagents (e.g., plasmids, transfection reagent, media, drugs). In our lab, we routinely use the Swiss TPP 96-well plates.
6. Ca^{2+} phosphate-based and lipid reagent-based (e.g., Lipofectamine 2000) methods can work comparably in the hands of an experienced user. The Ca^{2+}-phosphate method is very cheap but requires the detection of a Ca^{2+}-phosphate/DNA precipitate with a microscope, making this method less reproducible. Lipofectamine 2000 is expensive but gives more reproducible transfection, amenable to automated transfection, and easier for novices to perform. Other means of DNA delivery (e.g., viral delivery and electroporation, including nucleofection) produce a very high transfection efficiency, which leads to many transfected neurons with overlapping dendrites. This can make it difficult to identify the same individual neurons in a sequence of images.

7. Several commercial systems for image acquisition are now available, including Opera, Operetta, and UltraView (PerkinElmer); IN Cell Analyzer (GE Healthcare); IC200 (Vala Lifesciences); ImageXpress Micro, Ultra, and Velos (Molecular Devices); Cellomics ArrayScan (Thermo Scientific); Pathway 435 and 855 (BD); and Scan^R (Olympus) [18].
8. ImageJ was designed with open-source software and a variety of Java plugins. Some algorithms can be downloaded from the ImageJ website:

   http://rsb.info.nih.gov/ij/plugins/

   For additional reading, *see* ref. 22.

   An algorithm for ImageJ, NeuriteTracer, described in the following reference, measures neurite length and neuronal cell numbers in neuronal cultures:

   http://www.ncbi.nlm.nih.gov/pubmed/17936365

   Another algorithm for neurite tracing and quantification is described here:

   http://www.imagescience.org/meijering/software/neuronj/
9. For our typical scan of a 96-well plate, we collect images of nine microscope fields with fluorescence from two channels (a morphology/viability red fluorescent marker and green fluorescent protein-fused htt). This results in a collection of 18 images per well and requires 20 GB of storage space (image file size from a megapixel charged-coupled device (CCD) with 12–16 bit depth can be on the order of megabytes).
10. We prefer embryonic stage (E) 17 for rats for the striatal cultures, because the striatum is fairly well developed at this stage but much easier to dissect than at later gestational stages when the cortex is more fully developed. We also prefer rat E17 for isolating the cortex in comparative experiments (striatal cultures vs. cortical neurons), but E19–21 or even postnatal day 0 pups can be used, which will yield more neurons. For mouse cultures, we culture neurons from embryos at 18–20 days of gestation.
11. The striatum can also be isolated using coronal dissection. For a video *see* ref. 23.
12. Invitrogen's protocol can be found here:

    http://tools.invitrogen.com/content/sfs/manuals/lipofectamine2000_man.pdf

    We start any new transfection by testing the recommended concentrations of Lipofectamine 2000 reagent and the ratios of the transfected plasmids. The usual ratio is 2:1 for htt-GFP:mRFP.

13. Although the protocol suggests that "it is not necessary to remove complexes or change/add medium after transfection," we always remove complexes from neuronal cultures after 0.5–2.5 h. Longer incubations result in significant toxicity.
14. Transfected neurons can be imaged as early as 2 h after transfection. For imaging HD-model cells, we typically start observing neurons 12–24 h post-transfection and monitor them once daily for 14 days. Our longest experiment went 6 months. If an imaging system is not equipped with a controlled-environment chamber, use a strip of parafilm to seal the plate and limit the length of time the plate is outside the incubator.
15. Small errors occur during registration and stage movement; therefore, there is a portion of each field that might be missing from one image or another. Since we can only follow neurons that are in all the images, to avoid discarding the data, we montage the adjacent images. We use the Stitching 2D/3D plugin for ImageJ to montage our images [24]. For image alignment we use MultiStackReg, an adaption of StackReg [25].
16. We use a custom-made algorithm that identifies neuronal somata in a montaged image of microscope fields, tracks them from images collected at subsequent time points, and detects when an individual neuron died (Fig. 1). This custom algorithm records the tabulated survival times for each neuron in an Excel file along with information about fluorescence intensity and neuron morphology at each time point that it is alive.
17. To quantify neuronal survival, we tabulate survival time for each neuron in a cohort, and a survival curve for this cohort is constructed with StatView software. Neurons surviving the entire experiment are weighted differently (i.e., censored) to show that a precise survival time was not determined [3, 4].
18. For statistical analyses, we have adopted the convention of defining survival time as the imaging time point at which a neuron is last seen alive.

Acknowledgments

This work was supported by R01 2NS039746 and 2R01 NS045191 from the National Institute of Neurological Disease and Stroke, P01 2AG022074 from the National Institute on Aging, and the Gladstone Institutes (S.F.), the Milton Wexler Award, and a fellowship from the Hereditary Disease Foundation (A.T.). Gladstone Institutes received support from a National Center for Research Resources Grant RR18928-01. Kelley Nelson provided administrative assistance, and Gary C. Howard and Anna Lisa Lucido edited the manuscript.

References

1. Saudou F, Finkbeiner S, Devys D, Greenberg ME (1998) Huntingtin acts in the nucleus to induce apoptosis but death does not correlate with the formation of intranuclear inclusions. Cell 95:55–66
2. Miller J, Arrasate M, Shaby BA, Mitra S, Masliah E, Finkbeiner S (2010) Quantitative relationships between huntingtin levels, polyglutamine length, inclusion body formation, and neuronal death provide novel insight into huntington's disease molecular pathogenesis. J Neurosci 30:10541–10550
3. Arrasate M, Finkbeiner S (2005) Automated microscope system for determining factors that predict neuronal fate. Proc Natl Acad Sci USA 102:3840–3845
4. Tsvetkov AS, Miller J, Arrasate M, Wong JS, Pleiss MA, Finkbeiner S (2010) A small-molecule scaffold induces autophagy in primary neurons and protects against toxicity in a Huntington disease model. Proc Natl Acad Sci USA 107:16982–16987
5. Consortium, H. D. i (2012) Induced pluripotent stem cells from patients with Huntington's disease show CAG-repeat-expansion-associated phenotypes. Cell Stem Cell 11:264–278
6. Gutekunst CA, Li SH, Yi H, Mulroy JS, Kuemmerle S, Jones R, Rye D, Ferrante RJ, Hersch SM, Li XJ (1999) Nuclear and neuropil aggregates in Huntington's disease: relationship to neuropathology. J Neurosci 19:2522–2534
7. Schilling G, Savonenko AV, Klevytska A, Morton JL, Tucker SM, Poirier M, Gale A, Chan N, Gonzales V, Slunt HH, Coonfield ML, Jenkins NA, Copeland NG, Ross CA, Borchelt DR (2004) Nuclear-targeting of mutant huntingtin fragments produces Huntington's disease-like phenotypes in transgenic mice. Hum Mol Genet 13:1599–1610
8. Benn CL, Landles C, Li H, Strand AD, Woodman B, Sathasivam K, Li SH, Ghazi-Noori S, Hockly E, Faruque SM, Cha JH, Sharpe PT, Olson JM, Li XJ, Bates GP (2005) Contribution of nuclear and extranuclear polyQ to neurological phenotypes in mouse models of Huntington's disease. Hum Mol Genet 14: 3065–3078
9. Zala D, Bensadoun JC, Pereira de Almeida L, Leavitt BR, Gutekunst CA, Aebischer P, Hayden MR, Deglon N (2004) Long-term lentiviral-mediated expression of ciliary neurotrophic factor in the striatum of Huntington's disease transgenic mice. Exp Neurol 185:26–35
10. Kuemmerle S, Gutekunst CA, Klein AM, Li XJ, Li SH, Beal MF, Hersch SM, Ferrante RJ (1999) Huntington aggregates may not predict neuronal death in Huntington's disease. Ann Neurol 46:842–849
11. Slow EJ, Graham RK, Osmand AP, Devon RS, Lu G, Deng Y, Pearson J, Vaid K, Bissada N, Wetzel R, Leavitt BR, Hayden MR (2005) Absence of behavioral abnormalities and neurodegeneration in vivo despite widespread neuronal huntingtin inclusions. Proc Natl Acad Sci USA 102:11402–11407
12. Warby SC, Chan EY, Metzler M, Gan L, Singaraja RR, Crocker SF, Robertson HA, Hayden MR (2005) Huntingtin phosphorylation on serine 421 is significantly reduced in the striatum and by polyglutamine expansion in vivo. Hum Mol Genet 14:1569–1577
13. Graham RK, Slow EJ, Deng Y, Bissada N, Lu G, Pearson J, Shehadeh J, Leavitt BR, Raymond LA, Hayden MR (2006) Levels of mutant huntingtin influence the phenotypic severity of Huntington disease in YAC128 mouse models. Neurobiol Dis 21:444–455
14. Arrasate M, Mitra S, Schweitzer ES, Segal MR, Finkbeiner S (2004) Inclusion body formation reduces levels of mutant huntingtin and the risk of neuronal death. Nature 431:805–810
15. Mitra S, Tsvetkov AS, Finkbeiner S (2009) Single neuron ubiquitin-proteasome dynamics accompanying inclusion body formation in huntington disease. J Biol Chem 284: 4398–4403
16. Miller J, Arrasate M, Brooks E, Libeu CP, Legleiter J, Hatters D, Curtis J, Cheung K, Krishnan P, Mitra S, Widjaja K, Shaby BA, Lotz GP, Newhouse Y, Mitchell EJ, Osmand A, Gray M, Thulasiramin V, Saudou F, Segal M, Yang XW, Masliah E, Thompson LM, Muchowski PJ, Weisgraber KH, Finkbeiner S (2011) Identifying polyglutamine protein species in situ that best predict neurodegeneration. Nat Chem Biol 7:925–934
17. Ortega Z, Diaz-Hernandez M, Maynard CJ, Hernandez F, Dantuma NP, Lucas JJ (2010) Acute polyglutamine expression in inducible mouse model unravels ubiquitin/proteasome system impairment and permanent recovery attributable to aggregate formation. J Neurosci 30:3675–3688
18. Lang P, Yeow K, Nichols A, Scheer A (2006) Cellular imaging in drug discovery. Nat Rev Drug Discov 5:343–356
19. Sharma P, Ando DM, Daub A, Kaye JA, Finkbeiner S (2012) High-throughput screening in primary neurons. Methods Enzymol 506:331–360

20. Barmada SJ, Skibinski G, Korb E, Rao EJ, Wu JY, Finkbeiner S (2010) Cytoplasmic mislocalization of TDP-43 is toxic to neurons and enhanced by a mutation associated with familial amyotrophic lateral sclerosis. J Neurosci 30: 639–649
21. Nakamura K, Nemani VM, Azarbal F, Skibinski G, Levy JM, Egami K, Munishkina L, Zhang J, Gardner B, Wakabayashi J, Sesaki H, Cheng Y, Finkbeiner S, Nussbaum RL, Masliah E, Edwards RH (2011) Direct membrane association drives mitochondrial fission by the Parkinson disease-associated protein alpha-synuclein. J Biol Chem 286:20710–20726
22. Glory E, Murphy RF (2007) Automated subcellular location determination and high-throughput microscopy. Dev Cell 12:7–16
23. Chiu K, Lau WM, Lau HT, So KF, Chang RCC (2007) Micro-dissection of rat brain for RNA or protein extraction from specific brain region. J Vis Exp (7):e269. doi: 10.3791/269
24. Preibisch S, Saalfeld S, Tomancak P (2009) Globally optimal stitching of tiled 3D microscopic image acquisitions. Bioinformatics 25: 1463–1465
25. Thevenaz P, Ruttimann UE, Unser M (1998) A pyramid approach to subpixel registration based on intensity. IEEE Trans Image Process 7:27–41

Chapter 2

Atomic Force Microscopy Assays for Evaluating Polyglutamine Aggregation in Solution and on Surfaces

Kathleen A. Burke and Justin Legleiter

Abstract

Mutations which cause an expansion of CAG triplet repeats encoding polyglutamine (polyQ) are responsible for the subsequent misfolding of specific proteins that contribute directly to the pathogenesis of at least nine neurodegenerative disorders, including Huntington's disease (HD) and the spinocerebellar ataxias (SCAs). Expansion of polyQ tracts results in the aggregation of these proteins, potentially through a variety of precursor aggregates, into fibrillar structures. There may also be a variety of aggregates formed that are off-pathway to the formation of fibrils. Here, detailed protocols for analyzing the aggregation of mutant huntingtin (htt) fragments (associated with HD) and synthetic polyQ peptides with atomic force microscopy (AFM) are described. Ex situ AFM can be used to characterize htt aggregate formation and morphology. In situ AFM allows for tracking the formation and fate of individual polyQ peptide aggregates on surfaces. The interaction of htt with a variety of surfaces, including lipid bilayers, can also be investigated. Furthermore, the mechanical impact of htt on lipid surfaces can be studied using specialized AFM techniques. Methods to analyze AFM images of htt aggregates are also presented.

Key words Atomic force microscopy, Polyglutamine, Huntington's disease, Oligomers, Fibrils

1 Introduction

Polyglutamine (polyQ) expansions are the root cause for a number of inherited neurodegenerative disorders, such as Huntington's disease (HD) and the spinocerebellar ataxias [1, 2]. One clear consequence of polyQ expansion is the aggregation of the disease protein. The aggregation induced by expanded polyQ domains can proceed via a complex mechanism that involves the formation of a variety of different aggregate forms, including oligomers, annular aggregates, and fibrils (Fig. 1). Owing to the heterogeneous and dynamic aggregation process of polyQ-containing proteins and peptides, as well as the debate over which aggregate forms represent toxic species, there is a need for techniques capable of distinguishing and characterizing specific morphological aggregate features.

Danny M. Hatters and Anthony J. Hannan (eds.), *Tandem Repeats in Genes, Proteins, and Disease: Methods and Protocols*, Methods in Molecular Biology, vol. 1017, DOI 10.1007/978-1-62703-438-8_2, © Springer Science+Business Media New York 2013

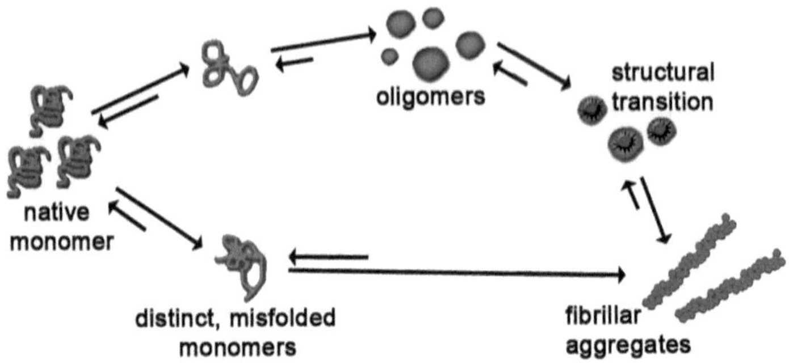

Fig. 1 A simplified aggregation model for polyQ-containing proteins. A native monomer can sample a variety of distinct conformations that may lead to different aggregation pathways to fibrillar aggregates. For example, aggregation into fibrils may occur via oligomeric intermediates, or monomers may nucleate fibril formation directly

Atomic force microscopy (AFM) has emerged as a powerful technique for studying disease-related protein aggregation [3–5]. The three-dimensional imaging capabilities of AFM enable the measurement of the height and volume of nanoscale objects. In contrast to other imaging techniques with comparable resolution, AFM is capable of not only operating in air (ex situ) but also in solution (in situ). This provides a unique ability to directly image important biological processes, such as protein aggregation, dynamically under near physiological conditions. AFM measures the vertical displacement of a flexible cantilever affixed with an ultrasharp probe tip as it physically interacts with a surface. A laser is reflected off the back of the cantilever and is focused onto a position-sensitive photo diode, resulting in an optical lever capable of measuring the vertical deflection of the cantilever with sub-angstrom resolution (Fig. 2). The three-dimensional images of the sample are generated by raster scanning the tip across the surface while maintaining constant deflection (contact mode) or amplitude (tapping or intermittent contact and noncontact) by means of a feedback loop.

Tapping mode is typically utilized for the study of soft biological materials, including aggregates of polyQ-containing proteins and peptides, due to its ability to greatly minimize the potentially damaging lateral forces associated with scanning. When operating in the tapping mode, the cantilever is oscillated harmonically near its resonance frequency, ω_0. When the probe tip is far from the surface, it oscillates with a free amplitude, A_0. When the probe tip is placed in the proximity of the surface, the oscillation amplitude decreases as the probe tip is allowed to intermittently contact (or tap) the surface at the bottom of each oscillation cycle, resulting in a tapping amplitude, A. An image is obtained by making adjustments

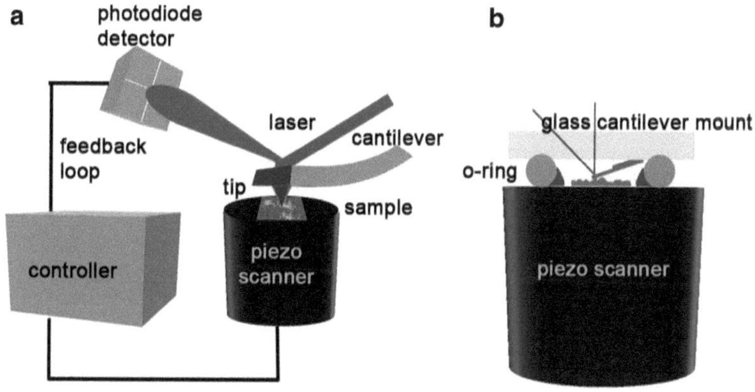

Fig. 2 The basic components of an atomic force microscope (AFM). (**a**) In AFM, a laser is focused onto the end of a cantilever with an ultrasharp tip and reflected onto a position-sensitive photodiode detector. The resulting optical lever is used to measure the bending (or deflection) of the cantilever. As the cantilever is scanned across a sample, perturbations in the deflection signal are monitored and used to construct topographical (or other types of information) images of the surface. (**b**) In situ operation often requires the use of a specialized fluid cell that allows for solution to be contained between two surfaces, one of which is the substrate on which the sample is deposited

to the vertical displacement of the scanner as the probe tip is raster scanned across a surface so that a constant set-point ratio $s = A/A_0$ is maintained.

2 Materials

1. Atomic force microscope system with the capability to image in fluids. There are numerous commercially available systems (Bruker, Asylum Research, JEOL, Agilent, Omicron, Pacific Scanning, Quesant Instrument Corporation, and Accurion Scientific Instruments) with a variety of features. The protocols described here make use of the tapping (or intermittent contact) imaging mode.
2. Fluid cell specific for your AFM system.
3. Cantilever probes. There are a wide array of AFM cantilever probes available (vendors include Veeco, Olympus, Asylum Research, BioForce, MikroMasch). The choice of cantilever can depend on several factors, e.g., the AFM system and imaging conditions. For ex situ tapping mode experiments described here, diving board-shaped cantilevers with a nominal spring constant of ~40 N/m and a typical resonance frequency ~300 kHz were used. For in situ AFM experiments, triangular silicon nitride cantilevers with nominal spring constant of ~0.5 N/m were used.

4. Protein sample. Huntingtin (htt) proteins can be expressed and purified in a variety of ways [6–8]. Synthetic polyQ peptides can be obtained from commercial sources (Keck Biotechnology Resource Lab).
5. Hexafluoroisopropanol (HFIP).
6. Trifluoroacetic acid (TFA).
7. Chemical fume hood.
8. Vacuum concentrator or similar to evaporate off organic solvents.
9. Bath sonicator.
10. Phosphate-buffered saline (PBS) buffer of approx pH 7.4 or other appropriate buffer.
11. Mica substrates. While the experiments described here all used mica as the underlying substrate, there are several other surfaces that can be used (e.g., graphite or silicon).
12. Porcine total brain lipid extract (TBLE) or other appropriate lipid system.
13. Chloroform.
14. Nitrogen gas.
15. Ultrapure water.
16. Double-sided tape or other adhesive to immobilize substrates for imaging.
17. Support for substrates. This is dependent on the AFM system used. For some systems this is simply a glass slide. Another commonly used support is metal pucks or discs that can be held in place on the AFM via a magnet.
18. For the scanning probe acceleration microscopy studies described in Subheading 3.4, access to the deflection signal of the AFM and a digital acquisition card are needed. The method of gaining access to this signal is AFM system dependent and often requires an additional signal access module.

3 Methods

In the described methods, htt exon1 proteins containing expanded polyQ domains and synthetic polyQ peptides will be used as model systems. Due to the variety of methods to express and purify polyQ-containing proteins (such as htt), protocols to obtain these samples are not discussed here (*see* **Note 1**). The methods described here can be used to investigate the aggregation of polyQ-containing peptides once they are obtained. The protocols to purify the htt exon1 proteins used here as examples are described elsewhere [8]. However, protocols for synthetic polyQ peptide sample preparation and handling will be discussed. Procedures for the deposition of polyQ-containing proteins and peptides on substrates for both ex situ and in situ AFM characterization are described. The preparation of supported lipid bilayers as substrates for AFM experiments with polyQ-containing proteins and peptides, as well as methods to measure the local mechanical properties of these lipid bilayers, is detailed. Finally, image analysis methods for analyzing AFM images of protein aggregates are presented.

3.1 Preparation of Synthetic PolyQ Peptides

The described protocols for preparing and handling polyQ-containing synthetic peptides are based on the extensive work performed in the laboratory of Ronald Wetzel [9–15]. To facilitate reproducibility, one of the key issues in preparing synthetic polyQ peptides is the removal of any preexisting aggregates that can act as seeds and nucleate fibril formation. The synthetic polyQ peptide used as an example in the described AFM experiments has 35 repeat glutamines flanked by two lysine residues on both the C- and N-terminus (*see* **Note 2**).

1. Lyophilized polyQ peptides are properly stored at −80 °C in sealed glass vials, and the sample should be allowed to equilibrate to room temperature (~30 min) before opening the vial to prevent condensation.
2. Once the vial has equilibrated to room temperature, the lyophilized polyQ peptide is treated with a 1:1 solution of HFIP and TFA. HFIP has been shown to remove β-sheet structure, disrupt hydrophobic forces, and promote α-helical secondary structure. As HFIP causes burns by all exposure routes, a chemical fume hood and appropriate safety precautions must be observed (refer to a material safety data sheet for HFIP for more information). Dissolve the polyQ peptide in the HFIP:TFA solvent to a final concentration of 0.5 mg/mL. The resulting solution can be distributed between several microcentrifuge tubes for later use, and the solvent is then evaporated off in the fume hood under a gentle stream of nitrogen. The clear peptide films that remain in the microcentrifuge tubes are further dried under vacuum to ensure complete removal of the HFIP and TFA. These peptide films can be stored desiccated for future use at −20 °C.
3. The resultant polyQ peptide films are dissolved in TFA-treated water at pH 3. As these stock solutions of polyQ peptides in TFA water will be further diluted into a physiological aqueous buffer, it is important to prepare these stocks with a high enough concentration of polyQ peptide (usually 100 μM but this can vary) so that final solution has a physiologically relevant pH as well as negligible amount of TFA, both which could influence polyQ peptide aggregation.

4. To further aid in the removal of preexisting aggregates, the peptide in the pH 3 aqueous TFA solution is centrifuged at 386,000×*g* for 3 h (or overnight if necessary) at 4 °C in an ultracentrifuge. Carefully remove the top two thirds of the solution and place on ice. As substantial amounts of peptide in the form of seeds may be removed during centrifugation, it is advised that the new concentration of the resulting solutions be determined. There are several methods to measure this concentration, including HPLC sedimentation assay or thioflavin T assays [16].
5. Solutions of polyQ peptides in pH 3 TFA water can be diluted into the desired physiological buffer (PBS, Tris, etc.). To ensure thorough mixing, the new solution should be gently vortexed for ~1 min. This mixing and change in pH initiates the aggregation process.
6. Incubation of these solutions (at room, elevated, or cooler temperatures) results in aggregation into higher ordered structures such as oligomers, protofibrils, and fibrils. The time needed to form these different structures depends on concentration, temperature, and other buffer conditions. Aliquots of these incubating samples can be taken at various time points, deposited on a substrate, and analyzed by ex situ AFM analysis (as described in Subheading 3.2).
7. For in situ AFM experiments aimed at observing the formation and fate of individual aggregates, freshly prepared solution should be injected into the fluid cell for AFM analysis (as described in Subheading 3.3) as soon as possible.

3.2 Ex Situ Studies of PolyQ-Containing Proteins and Peptides

Ex situ AFM experiments, i.e., imaging of dried protein samples in air, have rapidly become a common technique for analyzing protein aggregation state. A generic method for depositing and drying protein samples onto a surface for AFM analysis is provided. As this procedure consists of deposition, washing, and drying steps, caution must be exercised in preparing samples to assure reproducibility, avoid perturbation of fragile protein aggregates, and prevent contamination of the original sample. By sampling and depositing small aliquots of the same solution at various incubation times, it is possible to determine the transition between different aggregate forms and morphologies (as shown for aggregation of a htt exon1-51Q protein, as seen in Fig. 3). An advantage of drying samples for AFM analysis is that it allows one to take several images at a given time point to obtain a larger sampling of the potentially heterogeneous mixture of aggregates, allowing for a more robust statistical analysis of AFM images (*see* Subheading 3.5). For any given polyQ-containing protein or peptide, optimization of concentration, aliquot size, drying time, and other factors will be needed. To simplify analyzing images later, a bare substrate should be present between individual aggregates.

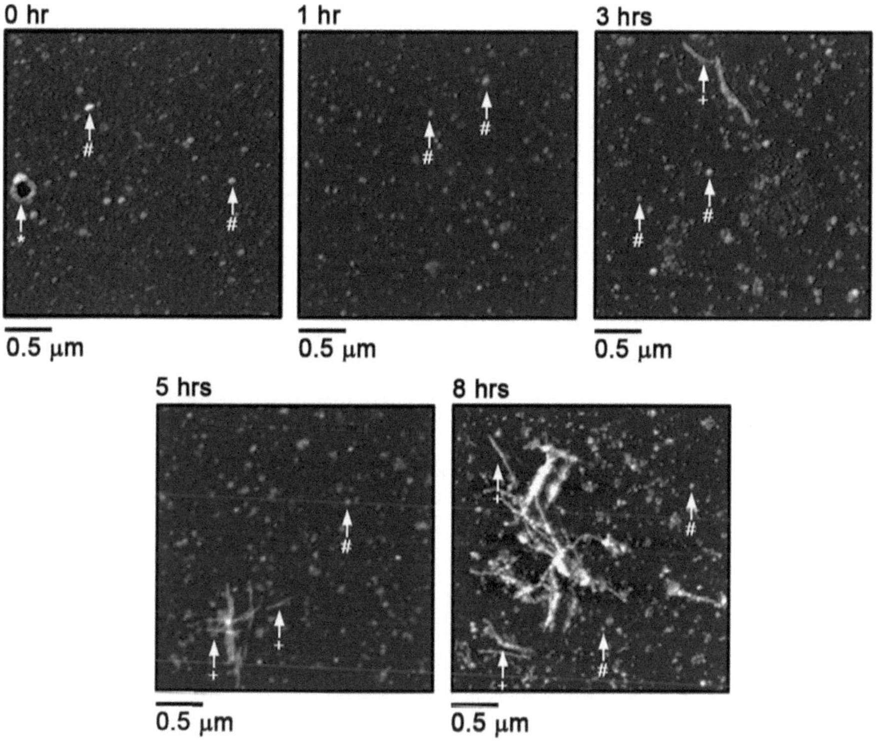

Fig. 3 Ex situ AFM images of aggregates formed by htt exon1-51Q. The htt exon1-51Q was incubated at a concentration of 40 µM and 37 °C. Aliquots were taken at different time points and deposited onto mica for AFM analysis. Heterogeneous mixtures of aggregates were observed at the different time points. Annular (*asterisk*), oligomeric (*hash*), and fibrillar (*plus*) aggregates are indicated by *arrows*

This provides an easily defined baseline for aggregate dimension measurements, such as height. The following brief protocol for deposition of a htt sample onto mica can be easily adapted to other proteins and surfaces.

1. Mark the backside of a rectangular piece of mica (~1 in. by 0.5 in.) using a marker pen with a small dot near one end of the substrate. This mark indicates where the sample will be deposited and aids in aligning the cantilever over the deposited sample. A brightly colored marker is recommended as it is easily distinguishable from other spurious features on the mica surface when observed from an overhead optical microscope during cantilever alignment. The mark should not be in the middle, but offset along the length of the rectangular mica piece. This will aid in washing the sample later.
2. Cleave the mica to expose a clean surface for deposition, making certain to remember which side was marked in **step 1** so that the sample can be deposited on the opposite side of the mica. Be careful not to touch the freshly exposed mica surface. It is recommended to handle the mica substrate from the sides and to wear latex gloves to avoid any contamination.

3. Using a pipettor or other appropriate liquid handling device, take an aliquot of 2–5 μL from the incubating polyQ-containing protein or peptide sample. Deposit this aliquot directly onto the mica surface directly above the dot (again making certain the marker dot is on the opposite side of your deposition). The droplet is left on the substrate for ~30 s to 2 min depending on the concentration sample and affinity of the protein for the substrate. This step may require some optimization to ensure a desirable aggregate coverage on the mica for the specific protein being investigated.
4. In order to remove excess salts and unbound protein, wash the sample with 200 μL of ultrapure water. To avoid potentially damaging fragile aggregate structures (such as fibrils), tilt the substrate ~45°, apply the wash to the top of the mica above the sample, and allow the wash to gently flow over the deposited sample. It is practical to place the bottom edge of the mica on a Kimwipe to absorb the excess wash as it flows downwards to prevent the wash from flowing back and contaminating your sample. The deposited sample should be closer to the top of the mica substrate where the wash is applied, which provides a longer piece of mica for the wash to flow away from the sample to avoid back contamination.
5. Dry the sample under a gentle stream of nitrogen applied directly to the deposited sample. Again, the mark on the substrate will provide a reference to where the sample is deposited.
6. The dry piece of mica can be mounted with double-sided side onto a substrate support for imaging. Excess mica can be trimmed away with scissors if the AFM system requires round metal pucks. Make sure that the side of mica on which the sample was deposited is exposed so that it can be accessed by the probe for imaging. The sample can now be imaged by ex situ tapping mode AFM.

3.3 In Situ AFM Studies

In situ tapping mode AFM provides the ability to observe the initial formation of aggregates on a surface and to track the fate of those aggregates as a function of time. An example of this capability is provided for a synthetic polyQ peptide with 35 repeat glutamines flanked by two lysine residues (Fig. 4).

3.3.1 In Situ PolyQ Studies on Solid Substrates

1. The choice of substrate is a major concern in using in situ tapping mode AFM to study the aggregation of polyQ-containing proteins and peptides. The surface chemistry associated with different substrates can exert a strong influence on the affinity of the protein for the surface, which will affect the rate of aggregation on the surface. Epitaxial effects can also have

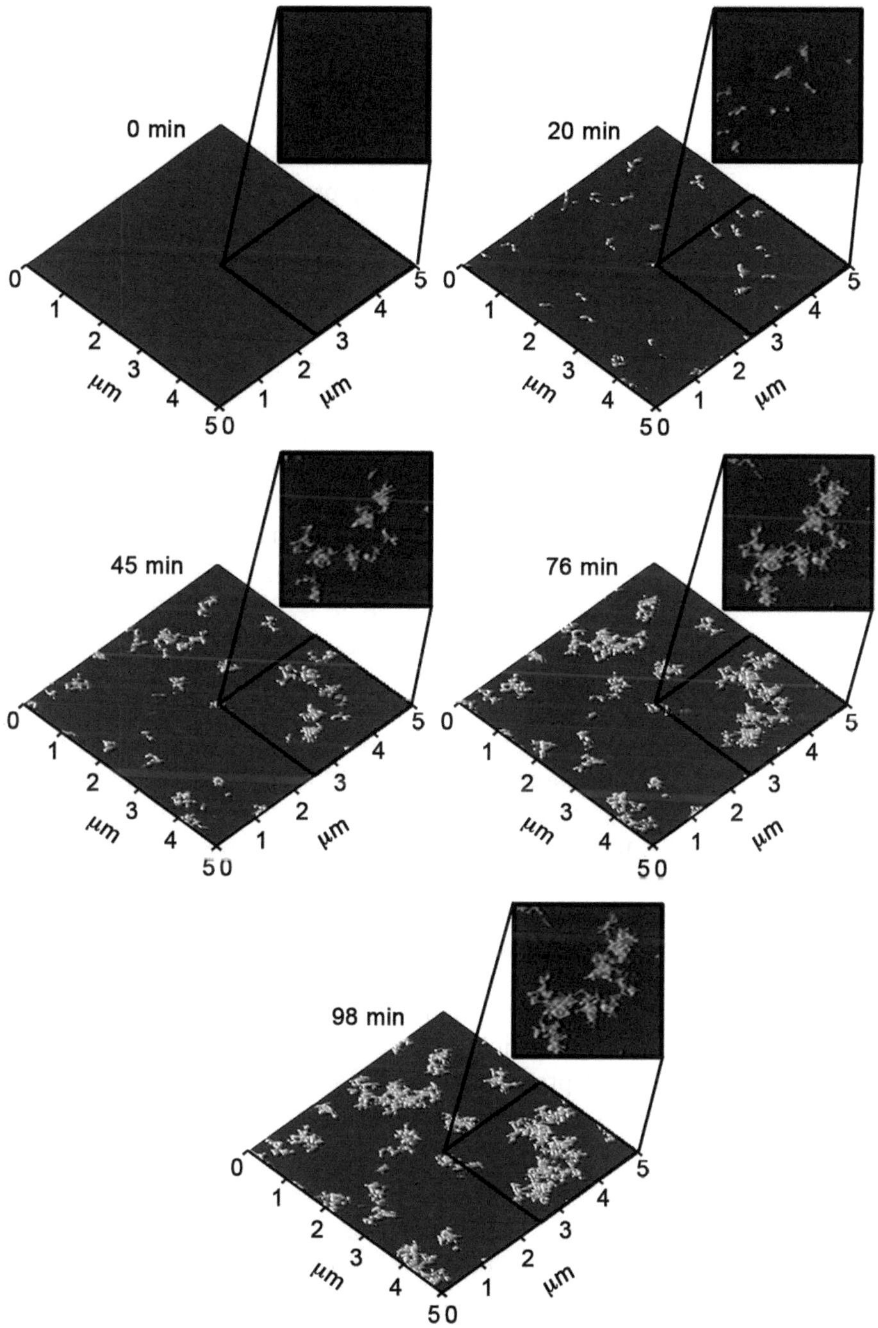

Fig. 4 Tracking the fate of discrete aggregates of a synthetic polyQ peptide using in situ AFM. A series of 3D and 2D in situ AFM images demonstrates the formation of fibrillar aggregates of KK-Q_{35}-KK peptides on a mica surface over 98 min

profound influence on the resulting aggregate morphology. Two commonly used model surfaces are mica (which is negatively charged in solution) and highly ordered pyrolytic graphite (which is hydrophobic).

2. When applying in situ AFM to study protein aggregation, it is advisable to obtain a background image of neat buffer to ensure the cleanliness of the AFM fluid cell and to avoid spurious artifacts.
3. Once cleanliness of the fluid cell is established, a freshly prepared polyQ-containing protein or peptide sample can be directly injected into the fluid cell. The exact method of accomplishing this depends on the AFM system being used. For synthetic polyQ peptide samples, preparation is carried out as described in Subheading 3.1.
4. To obtain time-lapse images of the aggregation process, the same area can be continuously imaged for the duration of the experiment. The concentration of the protein being studied may need to be optimized based on the affinity of the peptide for the surface and the resultant surface coverage of aggregates. Also, the imaging parameters need to be optimized to eliminate or greatly reduce any perturbations to the system due to the applied imaging force (*see* **Note 3**).
5. Pre-aggregated samples can be used if there is a desire to track the fate of higher order aggregates of polyQ-containing proteins. For example, the fate of preformed fibrils of a htt exon1-53Q protein exposed to various anti-htt antibodies was monitored via in situ AFM [17]. To accomplish this, the pre-aggregated sample can be injected directly into the fluid cell, and the aggregates are allowed to absorb onto the surface.

3.3.2 In Situ AFM Studies Using Supported Lipid Bilayers as Substrates

To enhance the physiological relevance of in situ AFM experiments, lipid bilayers supported on mica can be used as the substrate to study the aggregation of polyQ-containing proteins and peptides. Procedures on producing supported lipid bilayers comprised of total brain lipid extract (TBLE) and exposing it to htt exon1-51Q are presented (Figs. 5 and 6a). However, similar protocols would be applicable to preparing other lipid systems.

1. Lyophilized porcine total brain lipid extract (TBLE) is dissolved in chloroform. The chloroform is evaporated off, and the lipid is reconstituted in PBS buffer (or other appropriate buffer) to a final concentration of 1 mg/mL.
2. To promote the formation of lipid sheets, the lipid solution undergoes five freeze/thaw cycles using liquid nitrogen.
3. The solution is bath sonicated for 30 min to promote vesicle formation.
4. The resulting lipid vesicle solution can be injected directly into the AFM fluid cell to begin the process of vesicle fusion on the freshly cleaved mica substrate. It is advisable that a background

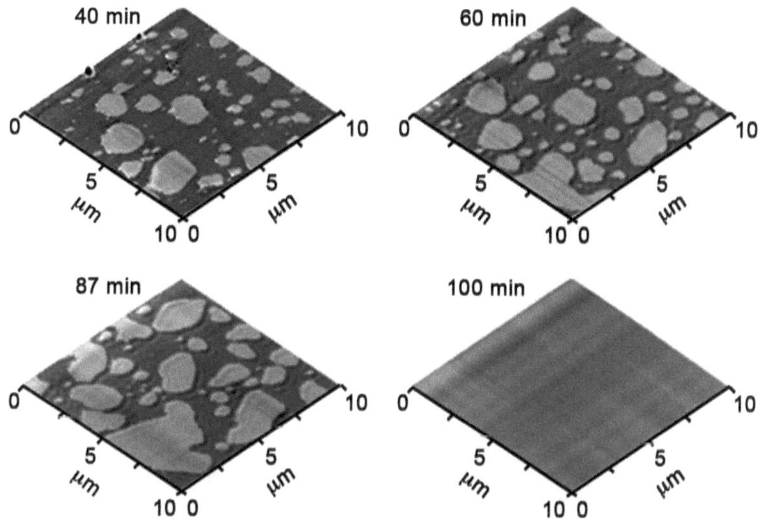

Fig. 5 The formation of a TBLE lipid bilayer supported on mica is demonstrated by a sequence of in situ AFM images

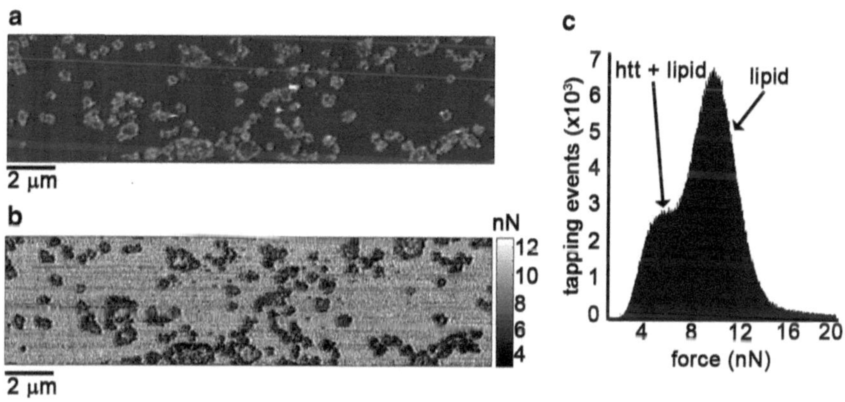

Fig. 6 The interaction of htt exon1-51Q with total brain lipid extract (TBLE) bilayers. (**a**) Topography and (**b**) maximum tapping force per oscillation cycle (F_{max}) AFM images are presented. Upon exposure to htt exon1-51Q, regions of increased roughness developed on the bilayer surface, as can be seen in the topography image. These rough regions are comprised of both htt and lipid. The lower F_{max} associated with the increased surface roughness indicates that the compression modulus is lower in these regions. This is further shown by (**c**) a histogram of F_{max} for every tapping event that occurred while capturing the image. Over 500,000 individual tapping events were measured while capturing this AFM image

image of the clean mica substrate is obtained prior to injecting the lipid vesicle solution into the fluid cell to ensure cleanliness.

5. Once the TBLE is directly injected into the fluid cell, the vesicles are allowed to deposit and fuse on the mica substrate in situ [18–21]. This process can be directly monitored by continuous imaging (Fig. 5), and the scanning process can aid in the formation of the lipid bilayer when large tapping forces are

used (*see* **Note 3**). Before the lipid bilayer is exposed to any polyQ-containing proteins or peptides, the resultant lipid bilayer should be relatively defect free, which can be observed by AFM. Once the desired lipid bilayer is formed, the imaging force should again be minimized.

6. Prior to the addition of any polyQ-containing protein or peptide, the fluid cell should be flushed with fresh buffer to remove any access vesicles still in solution. The exact procedure for this will depend on the AFM system being used. However, care must be taken that the flushing process does not damage the newly formed bilayer. After flushing, an additional image should be obtained to verify that the bilayer was not damaged.
7. Once a supported lipid bilayer is formed and the fluid cell is appropriately flushed, the polyQ-containing protein or peptide can be directly injected into the fluid cell as described in Subheading 3.3.1 to track the aggregation process on the lipid surface (Fig. 6a).

3.4 Measuring the Mechanical Impact of Proteins on Lipid Bilayers

As a consequence of the physical interaction between the sharp probe and the surface in tapping mode AFM, recent AFM technique development efforts have been focused on simultaneously obtaining measurements of physical properties of surfaces while imaging via tapping mode [22–29]. One method to accomplish this goal is to reconstruct the time-resolved force interaction between the tip and surface during tapping mode operation. The time-resolved tip/sample force during tapping contains information analogous to that obtained from the standard force curve experiment; however, until recently, there was no straightforward manner to obtain the value of this force in tapping mode imaging. One method for accomplishing this is scanning probe acceleration microscopy (SPAM) [24] (*see* **Note 4**). Here, a protocol to measure relative changes in the compression modulus of a lipid bilayer exposed to a polyQ-containing protein is presented (Fig. 2.6b, c).

1. The experiment is initially set up and performed as described in Subheading 3.3.2.
2. During the imaging process, the entire tip deflection trajectory is captured. The specifics on how to accomplish this are microscope dependent. The sampling rate of the deflection signal must allow for a minimum of 256 points per cycle and be captured with 14 bit vertical resolution (*see* **Note 5**). In order to easily correlate individual tapping events to features in the topography image, the initiation of capturing the deflection signal and the image needs to be synchronized.
3. Once the deflection signal is captured, a sliding Fourier transform-based filter is applied to the data to remove noise.

This filter design takes advantage of the fact that most of the information about the deflection trajectory is contained in the higher harmonics [30–32]. Therefore, it is possible to filter the signal by "comb" filtering, i.e., by taking its Fourier transform and inverting it while selectively retaining only the intensities at integer harmonic frequencies. Sliding this filter is necessary to not average out changes in local forces during the filtering process (*see* **Note 6**).

4. Once a deflection signal is appropriately filtered, the second derivative of the signal can be taken to acquire the tip acceleration. By appropriately scaling with the effective mass of the probe (*see* **Note 7**), the tip/sample force can be obtained (based on equation presented in **Note 4**) and used to reconstruct maps of specific force interactions (Fig. 6b).
5. The time-resolved tip/sample force interaction can be used to determine relative mechanical properties of the surface. For example, the maximum tapping force reflects the relative compression modulus of the surface (*see* **Note 8**).

3.5 Quantitative Analysis of AFM Images

Quantitative analysis of AFM images has been aided by the development of pattern recognition and image processing algorithms. Using these types of algorithms, routines to count particles and measure morphological features (such as height, diameter, and volume) of individual aggregates are easily constructed. Specific morphological features can be used to determine relative populations of specific aggregate forms in heterogeneous aggregation reactions (*see* **Note 9**). As the exact method to analyze AFM images will be dependent on the software package being used, below we will point out some potential issues that one should be aware of when processing and analyzing AFM images of polyQ aggregates.

1. Due to the finite size and shape of the probe, all observed features of aggregates in AFM images have some error associated with them. Aggregate features in the lateral direction will appear larger, and this error increases as the probe tip becomes more blunt. Irregular shaped probes can cause image artifacts such as double images. Using excessive imaging force can compress soft, compliant samples, resulting in reduced observed heights of aggregates. As a result, care should be taken when comparing AFM images taken with different probes. In time-lapse in situ AFM, the convolution of the image due to the tip may remain relatively constant, provided that the tip is not damaged and that peptide molecules are not significantly adhering onto the tip, changing its size and shape. There are methods to correct for some of these errors if necessary [33, 34].
2. Due to the mechanics of acquiring an AFM image, most AFM topography images contain curvature, which needs to be

corrected before quantitative image analysis can be completed. A standard technique for removing curvature from an image (often called flattening) is to fit each scan line with a polynomial and subtract this from the image (*see* **Note 10**).

4 Notes

1. AFM analysis has successfully been applied to a variety of polyQ peptide and proteins. This includes polyQ peptides [35, 36], htt exon1 and shorter fragments [7, 8, 17, 35–38], shortstop htt [6], ataxin-3 [39, 40], and the androgen receptor [41, 42] to name a few.
2. While many experimentalists use flanking lysine residues when synthesizing polyQ peptides to increase peptide solubility [12–14, 36, 43–45], simulations of polyQ dimer formation illustrated that the addition of flanking lysine residues strongly inhibit aggregation [46], which can potentially lead to erroneous conclusions when these residues are used. An experimental study also demonstrated that flanking lysine residues indeed have an impact on polyQ conformation, leading to a polyQ length-dependent reduction in peptide end-to-end distance [47].
3. Applying imaging forces to some protein systems can drive them to aggregate [48]. To avoid this issue, it is important to understand how to adjust the magnitude of the tapping forces and perform appropriate controls. The total tip/sample force per oscillation cycle can be approximated by:

$$F_{\text{total}} \approx 0.5ka_0 \frac{\Delta A}{A_0}, \tag{1}$$

where k is the spring constant, a_0 is the drive amplitude, ΔA is difference between the free amplitude and tapping amplitude, and A_0 is the free amplitude [49]. As is apparent from the equation, the imaging forces can be reduced by decreasing the drive, increasing the tapping amplitude (or set point) for a given free amplitude, reducing the free amplitude, or using a softer cantilever (smaller k). One way to verify that the AFM probe did not induce aggregation is to zoom out or to offset the probe position to a previously unscanned area. If the probe is not significantly altering the aggregation process, there should be no visible difference between the morphology and extent of aggregation in the previously unscanned portions of the surface.
4. The cantilever in a tapping mode AFM experiment can be modeled as a single degree of freedom damped driven harmonic oscillator [50–53]:

$$m_{\text{eff}}\ddot{z} + b\dot{z} + k\left[z - D_0 + a_0 \sin(\omega t)\right] = F_{\text{ext}}, \tag{2}$$

where m_{eff} is the effective mass of a cantilever, b is the damping coefficient, k is the cantilever spring constant, a_0 is the drive amplitude, ω is the drive frequency, D_0 is the resting position of the cantilever base, F_{ext} is the tip/sample force, and z is the position of the cantilever with respect to the surface. However, real AFM systems monitor the cantilever deflection signal (y) with respect to the base, rather than the position (z). The deflection signal (y) is related to position by:

$$y = z - D_0 + a_0 \sin(\omega t). \tag{3}$$

The second derivative of the deflection signal (or acceleration) is directly related to the tip/sample force, as can be shown by substituting Eq. 3 into Eq. 2 and rearranging to obtain:

$$\ddot{y} = \frac{1}{m_{\text{eff}}}\left[F_{\text{ext}} - b\dot{y} - ky + m_{\text{eff}}\omega^2 a_0 \sin(\omega t) - ba_0\omega \cos(\omega t)\right]. \tag{4}$$

The extra terms oscillate with the frequency ω and are easily removed from the resulting time-resolved tip/sample forces. SPAM utilizes the second derivative of the deflection signal to recover the tip acceleration trajectory and hence these tip/sample forces, which have been shown to be related to a variety of mechanical properties of the surface [24, 54].

5. To avoid errors associated with sampling the deflection signal, simulations suggest that a minimum of ~256 points per cantilever oscillation cycle must be captured. If the drive frequency for tapping mode in solution is in the range of 8–9 kHz, this would require a sampling rate of 2–2.5 million samples per second. As a result, 1×10^6 data points would be required to capture the deflection signal associated with imaging one line with a scan rate of 2 Hz. The total number of points would be determined by the number of scan lines captured to acquire the image.
6. For SPAM analysis, the deflection trajectory of the cantilever requires filtering. We typically use a Fourier transform-based harmonic comb filter. In short, the Fourier transform is taken of a portion of the deflection trajectory, and then an inverse Fourier transform is performed using only intensities that correspond to integer harmonic frequencies of the operating frequency, based on the following equation:

$$y_{\text{rec}}(t) = \hat{f}^{-1}\left[y(\omega)\sum_{k=1}^{N}\delta\left(\omega - k\omega_{\text{oper}}\right)\right] \tag{5}$$

where ω_{oper} is the operating frequency and δ is the Dirac's delta function. The summation is carried out up to N, which is the highest harmonic still distinguishable above the noise level, which is typically 20–25. To retain spatially resolved changes in the tip/sample force, the filter is typically applied to a five-cycle window. This window is then moved one cycle at a time.

7. To determine the effective mass of the cantilever, m_{eff}, the thermal tune method can be employed. This method obtains the spring constant, k, and resonance frequency (in Hz), f_{res}, which are related to m_{eff} by the equation:

$$f_{res} = \frac{1}{2\pi}\sqrt{\frac{k}{m_{eff}}}. \tag{6}$$

The thermal tune measures the displacement as a function of time of the cantilever due to thermal excitation. A Fourier transform coverts the data to the frequency domain. The spring constant can be determined by fitting the area under the resulting resonance peak with a Lorentzian function and relating it to the power of the cantilever displacement by the equation:

$$k = \frac{k_B T}{P} \tag{7}$$

where k_B is the Boltzmann constant, T is the temperature, and P is the area of the power spectrum due to thermal noise. The acquired m_{eff} is used to convert the tip acceleration into units of force.

8. Tapping mode AFM operates by keeping the total force between the tip and surface constant per oscillation cycle; however, specific features of the time-resolved tip/sample force can change within a cycle depending on surface properties. When tapping a rigid surface (high compression modulus), the contact time between the tip and surface will be short, and the maximum tapping force will be large. On a softer surface (low compression modulus), the contact time will be longer as the surface is compressed by the tip, and the resulting maximum tapping force will be smaller. By mapping maximum tapping force, a map of the relative rigidity of the surface can be obtained. Quantifying mechanical properties based on the tapping forces will depend on the model chosen. For example, modeling the tapping event with an elastic Hertz model results in a power law dependence of the maximum tapping force on the compression modulus.

9. It is possible to distinguish aggregates by morphological features. For example, oligomers and fibrils have vastly different aspect ratios. Some aggregate polymorphs can be distinguished by height [35, 36]. Additionally, correlation plots, such as height vs. diameter, may yield subtle differences in aggregate populations or subpopulations unidentified by other type of analyses [6].

10. When correcting for curvature in AFM images, it is often observed that "shadows" appear in the scan direction around discrete features (such as aggregates) in the topography image. These "shadows" arise from the inability of the polynomial fit to accommodate these discrete features when fitting the background curvature, and the flattening algorithm overcompensates for the background curvature in these regions. The error associated with this also affects the height of the feature itself, resulting in the absolute value of the pixels associated with the feature to appear shorter. Most AFM image processing software packages have the ability to avoid this error. Basically, the image can be flattened; then, features can be located to create a mask. Some software packages have the ability to locate features automatically; in other packages the features have to be manually located. The original flattening can then be undone, and a second flattening operation can be performed using the mask to remove discrete features from being fit by the polynomial. During imaging, it is often useful to use a real-time plane fit to allow one to observe the image as it is being collected. These real-time plane fit images often also have the "shadow" effect around features. If the image is saved with this real-time plane fit applied, it cannot be corrected later. Therefore, it is recommended that images are saved without this real-time plane fit.

Acknowledgements

Work in the author's laboratory is supported by the Brodie Entrepreneurial and Development Fund, the National Science Foundation (NSF#1054211), and the Alzheimer's Association.

References

1. Shao J, Diamond MI (2007) Polyglutamine diseases: emerging concepts in pathogenesis and therapy. Hum Mol Genet 16(R2):R115–R123. doi:10.1093/hmg/ddm213
2. Spada ARL, Wilson EM, Lubahn DB, Harding AE, Fischbeck KH (1991) Androgen receptor gene mutations in X-linked spinal and bulbar muscular atrophy. Nature 352(6330):77–79
3. Blackley HK, Patel N, Davies MC, Roberts CJ, Tendler SJ, Wilkinson MJ, Williams PM (1999) Morphological development of Aβ(1–40) amyloid fibrils. Exp Neurol 158(2):437–443
4. Blackley HKL, Sanders GHW, Davies MC, Roberts CJ, Tendler SJB, Wilkinson MJ (2000) *In-situ* atomic force microscopy study of β-amyloid fibrillization. J Mol Biol 298:833–840
5. Kowalewski T, Holtzman DM (1999) In situ atomic force microscopy study of Alzheimer's β-amyloid peptide on different substrates: new insights into mechanism of beta-sheet formation. Proc Natl Acad Sci USA 96(7):3688–3693
6. Nucifora LG, Burke KA, Feng X, Arbez N, Zhu S, Miller J, Yang G, Ratovitski T, Delannoy M, Muchowski PJ, Finkbeiner S, Legleiter J, Ross CA, Poirier MA (2012) Identification of novel potentially toxic oligomers formed in vitro from mammalian-derived expanded huntingtin exon-1 protein. J Biol Chem 287(19):16017–16028
7. Poirier MA, Li H, Macosko J, Cai S, Amzel M, Ross CA (2002) Huntingtin spheroids and

protofibrils as precursors in polyglutamine fibrillization. J Biol Chem 277(43):41032–41037. doi:10.1074/jbc.M205809200

8. Wacker JL, Zareie MH, Fong H, Sarikaya M, Muchowski PJ (2004) Hsp70 and Hsp40 attenuate formation of spherical and annular polyglutamine oligomers by partitioning monomer. Nat Struct Mol Biol 11: 1215–1222
9. Bhattacharyya A, Thakur AK, Chellgren VM, Thiagarajan G, Williams AD, Chellgren BW, Creamer TP, Wetzel R (2006) Oligoproline effects on polyglutamine conformation and aggregation. J Mol Biol 355(3):524–535
10. Bhattacharyya AM, Ashwani KT, Ronald W (2005) Polyglutamine aggregation nucleation: thermodynamics of a highly unfavorable protein folding reaction. Proc Natl Acad Sci USA 102(43):15400–15405
11. Chen S, Berthelier V, Hamilton JB, O'Nuallain B, Wetzel R (2002) Amyloid-like features of polyglutamine aggregates and their assembly kinetics. Biochemistry 41:7391–7399
12. Chen S, Berthelier V, Yang W, Wetzel R (2001) Polyglutamine aggregation behavior in vitro supports a recruitment mechanism of cytotoxicity. J Mol Biol 311(1):173–182
13. Chen S, Wetzel R (2001) Solubilization and disaggregation of polyglutamine peptides. Protein Sci 10:887–891
14. Thakur AK, Jayaraman M, Mishra R, Thakur M, Chellgren VM, Byeon I-J, Anjum DH, Kodali R, Creamer TP, Conway JF, Gronenborn AM, Wetzel R (2009) Polyglutamine disruption of the huntingtin exon 1 N terminus triggers a complex aggregation mechanism. Nat Struct Mol Biol 16(4):380–389
15. Yang W, Dunlap JR, Andrews RB, Wetzel R (2002) Aggregated polyglutamine peptides delivered to nuclei are toxic to mammalian cells. Hum Mol Genet 11(23):2905–2917
16. O'Nuallain B, Thakur AK, Williams AD, Bhattacharyya AM, Chen S, Thiagarajan G, Wetzel R (2006) Kinetics and thermodynamics of amyloid assembly using a high-performance liquid chromatography-based sedimentation assay. In: Indu K, Ronald W (eds) Methods in enzymology, vol 413. Academic, New York, pp 34–74. doi:10.1016/s0076-6879(06)13003-7
17. Legleiter J, Lotz GP, Miller J, Ko J, Ng C, Williams GL, Finkbeiner S, Patterson PH, Muchowski PJ (2009) Monoclonal antibodies recognize distinct conformational epitopes formed by polyglutamine in a mutant huntingtin fragment. J Biol Chem 284(32): 21647–21658
18. Legleiter J, Fryer JD, Holtzman DM, Kowalewski T (2011) The modulating effect of mechanical changes in lipid bilayers caused by ApoE-containing lipoproteins on a beta induced membrane disruption. ACS Chem Neurosci 2(10):588–599. doi:10.1021/cn2000475
19. Pifer PM, Yates EA, Legleiter J (2011) Point mutations in a beta result in the formation of distinct polymorphic aggregates in the presence of lipid bilayers. PLoS One 6(1):e16248. doi:10.1371/journal.pone.0016248
20. Yip CM, Elton EA, Darabie AA, Morrison MR, McLaurin J (2001) Cholesterol, a modulator of membrane-associated Aβ-fibrillogenesis and neurotoxicity. J Mol Biol 311(4): 723–734
21. Yip CM, McLaurin J (2001) Amyloid-β peptide assembly: a critical step in fibrillogenesis and membrane disruption. Biophys J 80: 1359–1371
22. Kumar B, Pifer PM, Giovengo A, Legleiter J (2010) The effect of set point ratio and surface Young's modulus on maximum tapping forces in fluid tapping mode atomic force microscopy. J Appl Phys 107:044508
23. Legleiter J (2009) The effect of drive frequency and set point amplitude on tapping forces in atomic force microscopy: simulation and experiment. Nanotechnology 20(24):245703
24. Legleiter J, Park M, Cusick B, Kowalewski T (2006) Scanning probe acceleration microscopy (SPAM) in fluids: mapping mechanical properties of surfaces at the nanoscale. Proc Natl Acad Sci USA 103(13):4813–4818
25. Sahin O (2007) Harnessing bifurcations in tapping-mode atomic force microscopy to calibrate time-varying tip-sample force measurements. Rev Sci Instrum 78(10)
26. Sahin O (2008) Accessing time-varying forces on the vibrating tip of the dynamic atomic force microscope to map material composition. Isr J Chem 48(2):55–63
27. Sahin O (2008) Time-varying tip-sample force measurements and steady-state dynamics in tapping-mode atomic force microscopy. Phys Rev B 77(11)
28. Sahin O, Erina N (2008) High-resolution and large dynamic range nanomechanical mapping in tapping-mode atomic force microscopy. Nanotechnology 19(44)
29. Sahin O, Magonov S, Su C, Quate CF, Solgaard O (2007) An atomic force microscope tip

designed to measure time-varying nanomechanical forces. Nat Nanotechnol 2(8):507–514
30. Hillenbrand R, Stark M, Guckenberger R (2000) Higher-harmonics generation in tapping-mode atomic-force microscopy: insights into the tip-sample interaction. Appl Phys Lett 76:3478–3480
31. Stark M, Stark RW, Heckl WM, Guckenberger R (2002) Inverting dynamic force microscopy: from signals to time-resolved interaction forces. Proc Natl Acad Sci USA 99:8473–8478
32. Stark RW, Heckl WM (2003) Higher harmonics imaging in tapping-mode atomic-force microscopy. Rev Sci Instrum 74:5111–5114
33. Legleiter J, Demattos R, Holtzman D, Kowalewski T (2004) In situ AFM studies of astrocyte-secreted apolipoprotein E and J-containing lipoproteins. J Colloid Interface Sci 278:96–106
34. Tian F, Qian X, Villarrubia JS (2008) Blind estimation of general tip shape in AFM imaging. Ultramicroscopy 109(1):44–53. doi:10.1016/j.ultramic.2008.08.002
35. Burke KA, Godbey J, Legleiter J (2011) Assessing mutant huntingtin fragment and polyglutamine aggregation by atomic force microscopy. Methods 53(3):275–284. doi:10.1016/j.ymeth.2010.12.028
36. Legleiter J, Mitchell E, Lotz GP, Sapp E, Ng C, DiFiglia M, Thompson LM, Muchowski PJ (2010) Mutant Huntingtin fragments form oligomers in a polyglutamine length-dependent manner in vitro and in vivo. J Biol Chem 285(19):14777–14790. doi:10.1074/jbc.M109.093708
37. Dahlgren PR, Karymov MA, Bankston J, Holden T, Thumfort P, Ingram VM, Lyubchenko YL (2005) Atomic force microscopy analysis of the Huntington protein nanofibril formation. Nanomedicine 1(1):52–57. doi:10.1016/j.nano.2004.11.004
38. Hands S, Sajjad MU, Newton MJ, Wyttenbach A (2011) In vitro and in vivo aggregation of a fragment of huntingtin protein directly causes free radical production. J Biol Chem 286(52):44512–44520. doi:10.1074/jbc.M111.307587
39. Masino L, Nicastro G, De Simone A, Calder L, Molloy J, Pastore A (2011) The Josephin domain determines the morphological and mechanical properties of ataxin-3 fibrils. Biophys J 100(8):2033–2042. doi:10.1016/j.bpj.2011.02.056
40. Natalello A, Frana AM, Relini A, Apicella A, Invernizzi G, Casari C, Gliozzi A, Doglia SM, Tortora P, Regonesi ME (2011) A major role for side-chain polyglutamine hydrogen bonding in irreversible ataxin-3 aggregation. PLoS One 6(4):e18789. doi:10.1371/journal.pone.0018789
41. Jochum T, Ritz ME, Schuster C, Funderburk SF, Jehle K, Schmitz K, Brinkmann F, Hirtz M, Moss D, Cato ACB (2012) Toxic and non-toxic aggregates from the SBMA and normal forms of androgen receptor have distinct oligomeric structures. Biochim Biophys Acta 1822(6):1070–1078. doi:10.1016/j.bbadis.2012.02.006
42. Li M, Chevalier-Larsen ES, Merry DE, Diamond MI (2007) Soluble androgen receptor oligomers underlie pathology in a mouse model of spinobulbar muscular atrophy. J Biol Chem 282(5):3157–3164. doi:10.1074/jbc.M609972200
43. Chen S, Frank AF, Ronald W (2002) Huntington's disease age-of-onset linked to polyglutamine aggregation nucleation. Proc Natl Acad Sci USA 99(18):11884–11889
44. Singh VR, Lapidus LJ (2008) The intrinsic stiffness of polyglutamine peptides. J Phys Chem B 112(42):13172–13176
45. Thakur AK, Ronald W (2002) Mutational analysis of the structural organization of polyglutamine aggregates. Proc Natl Acad Sci USA 99(26):17014–17019
46. Williamson TE, Vitalis A, Crick SL, Pappu RV (2010) Modulation of polyglutamine conformations and dimer formation by the N-terminus of huntingtin. J Mol Biol 396(5):1295–1309. doi:10.1016/j.jmb.2009.12.017
47. Walters RH, Murphy RM (2009) Examining polyglutamine peptide length: a connection between collapsed conformations and increased aggregation. J Mol Biol 393(4):978–992
48. Hwang W, Kim B-H, Dandu R, Cappello J, Ghandehari H, Seog J (2009) Surface induced nanofiber growth by self-assembly of a silk-elastin-like protein polymer. Langmuir 25(21):12682–12686. doi:10.1021/la9015993
49. Kowalewski T, Legleiter J (2006) Imaging stability and average tip-sample force in tapping mode atomic force microscopy. J Appl Phys 99:064903
50. Kuhle A, Sorensen AH, Bohr J (1997) Role of attractive forces in tapping tip force microscopy. J Appl Phys 81(10):6562–6569

51. Salapaka MV, Chen DJ, Cleveland JP (2000) Linearity of amplitude and phase in tapping-mode atomic force microscopy. Phys Rev B 61(2):1106
52. Song YX, Bhushan B (2008) Atomic force microscopy dynamic modes: modeling and applications. J Phys Condens Matter 20(22)
53. Burnham NA, Behrend OP, Oulevey F, Gremaud G, Gallo P-J, Gourdon D, Dupas E, Kulik AJ, Pollock HM, Briggs GAD (1997) How does a tip tap? Nanotechnology 8(2): 67–75
54. Kumar B, Pifer PM, Giovengo A, Legleiter J (2010) The effect of set point ratio and surface Young's modulus on maximum tapping forces in fluid tapping mode atomic force microscopy. J Appl Phys 107(4):044508. doi:10.1063/1.3309330

Chapter 3

Morphometric Analysis of Huntington's Disease Neurodegeneration in *Drosophila*

Wan Song, Marianne R. Smith, Adeela Syed, Tamas Lukacsovich, Brett A. Barbaro, Judith Purcell, Doug J. Bornemann, John Burke, and J. Lawrence Marsh

Abstract

Huntington's disease (HD) is an autosomal dominant neurodegenerative disorder. The HD gene encodes the huntingtin protein (HTT) that contains polyglutamine tracts of variable length. Expansions of the CAG repeat near the amino terminus to encode 40 or more glutamines (polyQ) lead to disease. At least eight other expanded polyQ diseases have been described [1]. HD can be faithfully modeled in *Drosophila* with the key features of the disease such as late onset, slowly progressing degeneration, formation of abnormal protein aggregates and the dependence on polyQ length being evident. Such invertebrate model organisms provide powerful platforms to explore neurodegenerative mechanisms and to productively speed the identification of targets and agents that are likely to be effective at treating diseases in humans. Here we describe an optical pseudopupil method that can be readily quantified to provide a fast and sensitive assay for assessing the degree of HD neurodegeneration *in vivo*. We discuss detailed crossing schemes as well as factors including different drivers, various constructs, the number of UAS sites, genetic background, and temperature that can influence the result of pseudopupil measurements.

Key words Huntington's disease, *Drosophila* model, Neurodegeneration, Polyglutamine disease, Pseudopupil assay, Ommatidium, Photoreceptor cell death

1 Introduction

Huntington's disease (HD) is an inherited genetic disorder, characterized by a combination of chorea, cognitive impairment, and affective changes. The behavioral symptoms of HD are precipitated by progressive neurodegeneration that is particularly acute in the striatum but also involves other regions, primarily the cerebral cortex. HD is the most prevalent autosomal dominant neurodegenerative disorder associated with the expansion of unstable CAG tracts [2]. The product of the HD gene, *huntingtin* (*Htt*), is a 350 kDa cytoplasmic protein with 67 exons [3]. The amino terminus

Danny M. Hatters and Anthony J. Hannan (eds.), *Tandem Repeats in Genes, Proteins, and Disease: Methods and Protocols*, Methods in Molecular Biology, vol. 1017, DOI 10.1007/978-1-62703-438-8_3, © Springer Science+Business Media New York 2013

of the HTT protein contains a CAG repeat beginning at codon 17 that encodes a glutamine repeat (polyQ) followed by two short stretches of prolines within exon 1. The CAG repeat gives rise to proteins containing polyglutamine tracts of varying size. Normal alleles have 35 or fewer glutamines while disease alleles range from 40 to over 150 glutamines [4]. The length of the CAG/polyglutamine repeat sequence is inversely correlated with the age of disease onset [5]. Despite the discovery of the HD gene almost 20 years ago, the pathogenic mechanisms of HD are still unknown. To understand degenerative mechanisms and discover methods of suppression, this dominant gain-of-function disease has been modeled by expressing mutant human huntingtin (*mHtt*) in *Drosophila melanogaster*. Multiple transgenic Drosophila models of HD disease have been generated [6–15]. Fly models recapitulate the core phenotypes observed in HD patients, including late onset, progressive cellular pathology as a function of polyQ repeat length, motor dysfunction, protein inclusions, transcriptional dysregulation, mitochondrial dysfunction, and shortened adult lifespan; hence, they provide powerful genetic systems for dissecting the neuronal degeneration induced by glutamine repeat-containing proteins.

As a measure of neurodegeneration, pseudopupil analysis detailed in this chapter allows characterization of the photoreceptor neurons by visualizing rhabdomeres (the light gathering organ of photoreceptor neurons) in the ommatidia (the individual eyes) of the compound eye [16]. Neurotoxicity in HD can be readily monitored by measuring the loss of visible photoreceptor neurons in the eye [11] and photoreceptor loss can be used as a quantitative marker for the *in vivo* assessment of neuronal loss [17].

2 Materials

1. Drosophila stocks.
2. Standard Drosophila food.
3. Incubators.
4. Glass vials.
5. Microscope slides.
6. CO_2 gas supply.
7. Coverslips.
8. Nail polish (clear).
9. Tweezers.
10. 26 G×5/8″ hypodermic needle.
11. 1 cm^3 syringe.

12. Dissecting microscope.
13. Light microscope, e.g., Nikon Optiphot-2, with a DPlan 50 0.09 oil lens, Labophot-2 with Plan50 0.85 oil lens, or a Zeiss Axioskop 2 Plus with a Plan100 1.40 oil lens. Important features are to have an oil objective with good depth of field and the ability to adjust the condenser diaphragm to the size of a pinhole.

3 Methods

3.1 Theory: Modeling HD Neurodegeneration in *Drosophila*

Nearly all of the current fly models of neurodegenerative diseases have been made using the GAL4/UAS system [18] which allows the ectopic expression of any transgene in a specific tissue or cell type (Fig. 1). In this system, two transgenic fly lines are used.

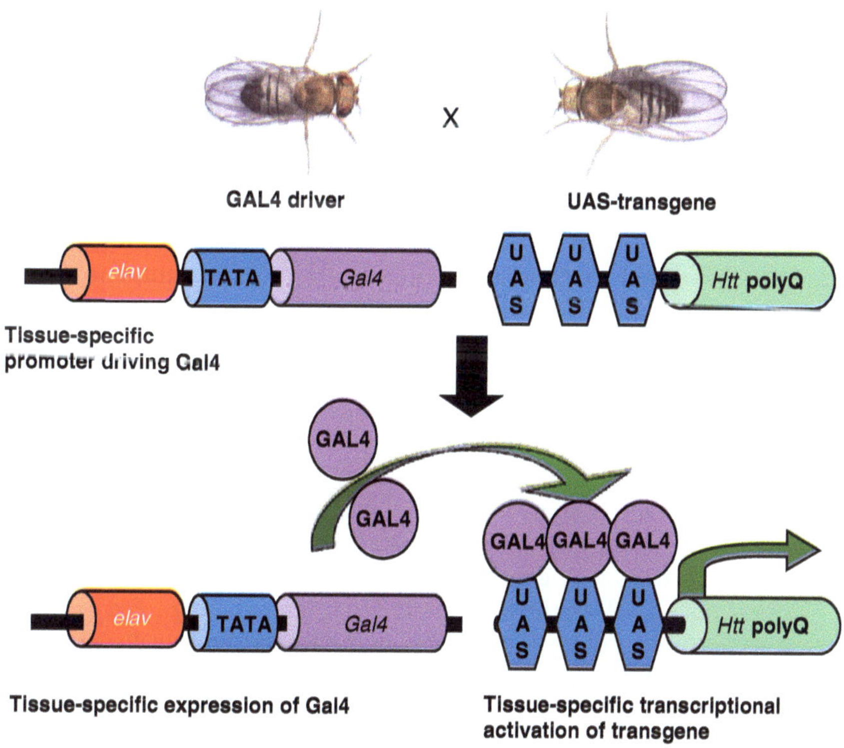

Fig. 1 GAL4-UAS system used to model Huntington's and other neurodegenerative diseases in *Drosophila melanogaster*. In this example, transgenic female flies that have a human *Htt* polyQ transgene inserted downstream of a yeast upstream activator sequence (UAS) are crossed with male transgenic flies containing the transcriptional activator, GAL4, under control of the *elav* promoter (*elav*-GAL4/Y; +; + (males) and *w*; +; UAS-*Htt* ex1p-93Q females). In the resulting female progeny, the *elav* promoter induces expression of GAL4 specifically in neuronal cells of the nervous system. GAL4 then binds to the UAS sequences upstream of the *Htt* polyQ gene, activating transcription of the *Htt* polyQ gene in these cells. Male progeny will not have the *elav* driver on the X chromosome and thus will not express the *Htt* polyQ transgene

In one, the human disease-related gene is placed downstream of a yeast upstream activating sequence (UAS) that typically contains several GAL4-binding sites ("UAS-transgene"). GAL4 is a yeast transcriptional activator. In the absence of ectopically expressed GAL4, the transgene is inactive. To activate the disease gene, the UAS-transgene flies are crossed to a "GAL4 driver" line that expresses GAL4 under control of a specific promoter. A wide array of "driver" lines that express in a variety of cell types is available at the Drosophila stock centers (e.g., http://flystocks.bio.indiana.edu/) and other sources (http://flybase.org/). Drivers that are particularly useful for neurodegeneration studies include the pan-neural driver *elav* (embryonic lethal, abnormal vision) or the eye-specific promoter *GMR* (*Glass Multimer Reporter*). In the progeny of a cross, the transgene will be activated in a specific cell or tissue type, depending on the "driver". This is especially important in studying neurodegenerative diseases, as questions regarding cell-type-specific death can be investigated.

3.2 Crossing Schemes to Analyze Neurodegeneration in Htt Challenged Flies

Flies expressing mutant human *Htt* exon 1 pathogenic fragments with the pan-neuronal *elav* driver replicate key HD features such as late onset and progressive neuronal dysfunction and degeneration, leading to a decline in motor performance and premature death [11, 19]. A convenient measure of neuronal degeneration is the pseudopupil method. In this protocol, an *elav*-GAL4 driver (the strongest is the one located on the X chromosome, C155, [20]) is used to drive expression of a UAS-Htt transgene, i.e., UAS-*Htt* ex1p-93Q with 93Qs contained in exon 1 of the *Htt* gene [11]. Generally, a cross is made between *elav*-GAL4/Y; +; + (males) and +; +; UAS-*Htt* ex1p-93Q (virgin females) to produce progeny that include non-expressing males +/Y; +/+; UAS-*Htt* ex1p-93Q/+ and Htt-expressing *elav*-GAL4/+; +/+; UAS-*Htt* ex1p-93Q/+ females. The males, that do not express the transgene, serve as control animals. Another method is to use a line with the transgene over a marked chromosome such as a balancer and compare the transgene-expressing *vs.* non-expressing flies.

With the most toxic *Htt* fragments (e.g., *Htt* ex1p-93Q) there is considerable lethality in the larval and pupal stages when driven with *elav*. To ensure that enough adult flies are recovered for analysis, crosses can be made at reduced temperature (e.g., 22.5 °C). Virgin females (*w*; +; *Htt* ex1p-93Q) are crossed to males (*elav*-GAL4/Y; +; +) in vials containing fresh Drosophila food and the vials are kept at 22.5 °C until they eclose (about 12 days at this temperature). After eclosion at 22.5 °C, the Htt-expressing virgins and non-expressing male control flies are separated, placed at 25 °C and transferred to fresh food daily, especially if drug containing food is being used. The progeny larvae or adult flies can be used for a variety of analyses including assessing degeneration of photoreceptor neurons by the pseudopupil method [19], survival of neurons in the larval ventral nerve cord [21], viability/eclosion rate (*see* **Note 1**), motor function by using the climbing assay, survival/longevity assay, etc.

The adult compound eye of Drosophila consists of a unit structure, the ommatidium, that is repeated nearly 800 times in a regular symmetrical array. Each hexagonally shaped ommatidium consists of approximately 20 cells: 8 photoreceptors and 12 non-neuronal accessory cells. Each of the photoreceptor neurons forms a highly fenestrated membrane structure, the rhabdomere (subcellular light-gathering structure), and these are arranged in a regular trapezoidal array [22]. The rhabdomeres of photoreceptors 7 and 8 are stacked one above the other such that only seven rhabdomeres are seen at any given plane of focus. By measuring the loss of visible photoreceptor neurons, one can readily monitor the degree of neurotoxicity caused by a particular mutant *Htt* [11]. Adult flies expressing mutant human *Htt* ex1p-93Q exhibit progressive degeneration of the photoreceptor neurons and this can be measured by counting the number of intact rhabdomeres (*see* Fig. 2 and **Note 2**). Since the neuropathology is progressive and the number of rhabdomeres in the eyes decreases as the flies age [7, 19], freshly eclosed animals typically show mild levels of degeneration (because they have just emerged from 5 days in the pupal case where degeneration was ongoing) while the level of degeneration is more extensive and lethality is more aggravated with increasing time after eclosion (Fig. 2). For a single time point comparison, one can perform the psudopupil assay on the 7th day post-eclosion when the degeneration has sufficiently progressed and yet death of the flies has not occurred.

3.3 Factors Influencing the Neurodegeneration Analysis

Various factors, such as drivers, constructs, number of UAS transgenes, genetic background, and temperature, can influence the level of neurodegeneration. It is helpful to consider the potential advantages and challenges these factors may pose before carrying out the crosses.

3.3.1 Different Transgenes and Chromosomal Position Effects

Different methods of creating transgenic flies have distinct advantages and disadvantages depending on the experimental objectives. In some cases, the experimental goal is to create an "allelic" series of transgenes with varied levels or patterns of expression, while in other cases the objective is to compare the pathophysiology of two different transgenes (e.g., a mutant *Htt* ex1p *vs.* one where a putative phosphorylation site has been altered (e.g., S>A)). To generate an allelic series of a particular transgene, it is convenient to generate transgenes by random, traditional P-element mediated transformation [23]. Depending on where the transgene is inserted in the chromosome and the resulting chromosomal environment, the expression level and/or patterns may be classified as strong, medium, or weak in phenotype (e.g., ref. 9).

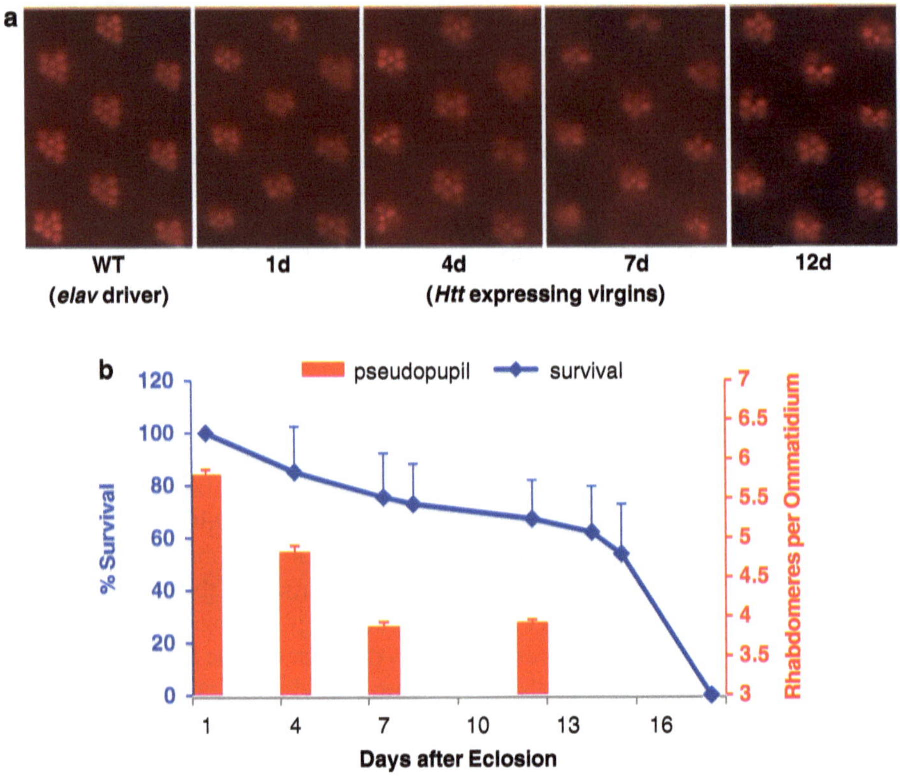

Fig. 2 Neurodegeneration is progressive. (**a**) Pseudopupil images of females from an *elav* driver line ("WT") and representative virgin females expressing *Htt* ex1p-93Q under control of the neuron-specific driver *elav*-GAL4 aged for 1, 4, 7, and 12 days after eclosion at 25 °C. Images were taken with a Zeiss microscope with 100× oil objective at the same settings. (**b**) Percent survival (*blue line, left vertical axis*) and number of rhabdomeres per ommatidia (*red columns, right vertical axis*) in 1, 4, 7, 12 day old Huntingtin-challenged flies. The percent survival is based on the loss of *Htt*-expressing virgins over time ($n = 15\sim25$ each in six vials). The different numbers of rhabdomeres per ommatidium in WT, *Htt*-expressing flies aged for 1, 4 and 7 days are significant ($p < 0.01$, *t*-test) while those between *Htt*-expressing flies aged for 7 and 12 days are not

On the other hand, if the intent is to compare the impact of two different transgenes, it is desirable that the two transgenes have the same level and pattern of expression and are in the same genetic background. For this purpose it is useful to utilize the targeted gene insertion approach that employs the phiC31 insertion system [24]. In these cases, the transgenes are all inserted into the same chromosomal location in the identical orientation and have similar levels of expression (barring expression differences occasioned by the design of the transgene itself). For example, two different *Htt* fragments of variable lengths inserted in the same chromosomal location (i.e., 51D) produce the same levels of RNA (Fig. 3).

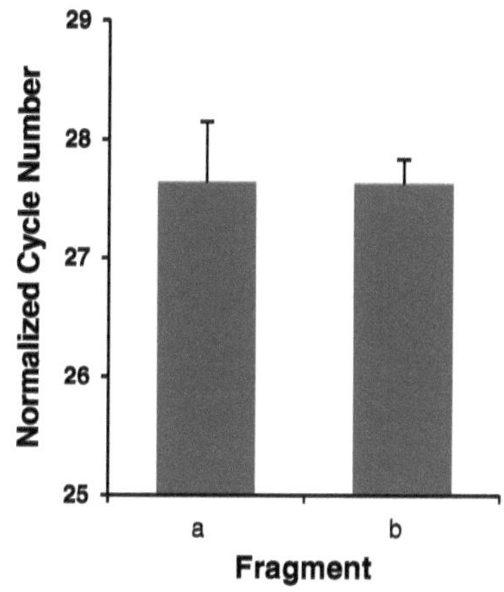

Fig. 3 Targeted inserts give similar expression levels. Expression of different lengths of human Huntingtin (represented by "a" and "b") inserted at cytological region 51D in Drosophila. RNA expression level is quantified by qRT-PCR and represented by the cycle number normalized to that of *rp49* ($n = 3$)

3.3.2 Temperature Effects

Induced expression of the transgene by the GAL4/UAS system is particularly temperature dependent. By rearing flies at temperatures ranging from 18 to 29 °C, a range of expression levels of any responder can be achieved [25]. We compared the temperature dependence of the expression of a GFP transgene and neurodegeneration in *Htt*-expressing flies at a set of temperatures ranging from 18 to 29 °C (Fig. 4). The GFP expression increases from 72.6 ± 12.7 (Mean ± SD) arbitrary units (at 18 °C) to 175.7 ± 15.5 at 29 °C, an ~2.4-fold difference. Concurrently, neurodegeneration becomes more severe with the increase of temperature, reflecting the augmented transgene protein production. The number of rhabdomeres per ommatidium in 7-day-old flies dropped from 6.1 ± 0.1 (at 18 °C) to 4.3 ± 0.1 (at 25 °C). Note also that higher temperature results in greater lethality (106 % eclosion rate at 18 °C vs. 0 % eclosion rate at 29 °C and 5 % at 25 °C).

3.3.3 Inducible GAL4

In some cases, the experimental objective is to monitor the progression of events from a defined starting time or in a particular developmental stage. Several inducible GAL4 systems are available for this purpose. One method, the TARGET technique, utilizes a temperature sensitive mutant of the yeast GAL80 protein (GAL80^{ts}) that specifically binds the transactivation domain of GAL4 to prevent transcription at low temperatures (i.e., 18 °C) while allowing expression of the transgene to be activated by shifting the flies to a higher temperature (i.e., 22.5 °C or greater) [26–28]. Another method uses temperature to activate a heat shock inducible FLP recombinase. The recombinase can then remove a terminator cassette flanked by FLP recognition target sequences that is placed in

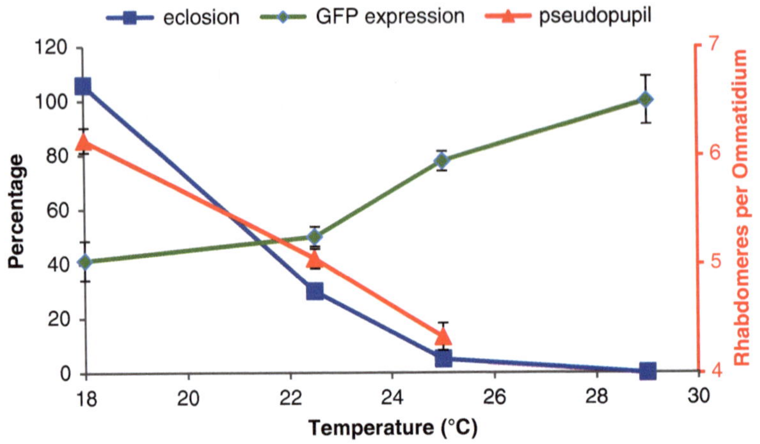

Fig. 4 Effect of temperature on UAS-GAL4 expression. The eclosion rate (*blue line, left vertical axis*), expression of transgene GFP (*green line, left vertical axis*) and neurodegeneration measured by pseudopupil assay (*red line, right vertical axis*) in flies expressing *Htt* ex1p-93Q and mCD8-GFP (mCD8 is a transmembrane protein and GFP labels the cell surface) under control of the neuron-specific driver *elav*-GAL4 (Bloomington stock #5146) raised at various temperatures. Percent eclosion is represented by the ratio of *Htt*-expressing females to non-expressing males that eclose ($n = 3\sim5$ vials). Percentage of GFP expression in eye discs is expressed relative to the highest expression at 29 °C ($n = 3\sim5$ discs) by measuring intensity across multiple lines drawn across each disc imaged on a Zeiss LSM780. The pseudopupil assay is performed on flies aged for 7 days after they eclose at different temperatures ($n = 5\sim7$)

front of GAL4, thus allowing expression of GAL4 and activation of the GAL4-UAS controlled transgene [27, 29, 30]. Other inducible drivers are controlled by specific ligands such as mifepristone/RU486 in a dose-dependent manner [27, 31, 32]. Where expression levels are critical, it is important to verify the relative levels in induced vs. uninduced systems.

3.3.4 UAS Dependence

A number of UAS-based vectors with various numbers of UAS binding sites are available [18, 33, 34]. In addition, because multiple UAS transgenes are often employed in a given experiment, the possibility of titration of GAL4 with additional UAS binding sites must be considered. In our experience, the effect of introducing a second UAS transgene on the expression of another is generally minimal. For example, in one case, we expressed homozygous or heterozygous Kaede (a fluorescent protein, [35]) with one or two copies of the *elav*-GAL4 driver and compared the Kaede intensities in eye discs (Fig. 5a, b). With one copy of *elav*-GAL4, doubling the UAS-Kaede transgene doubled the level of fluorescent protein while doubling the GAL4 driver with a fixed dose of UAS-Kaede did not change the level of expression (note that each

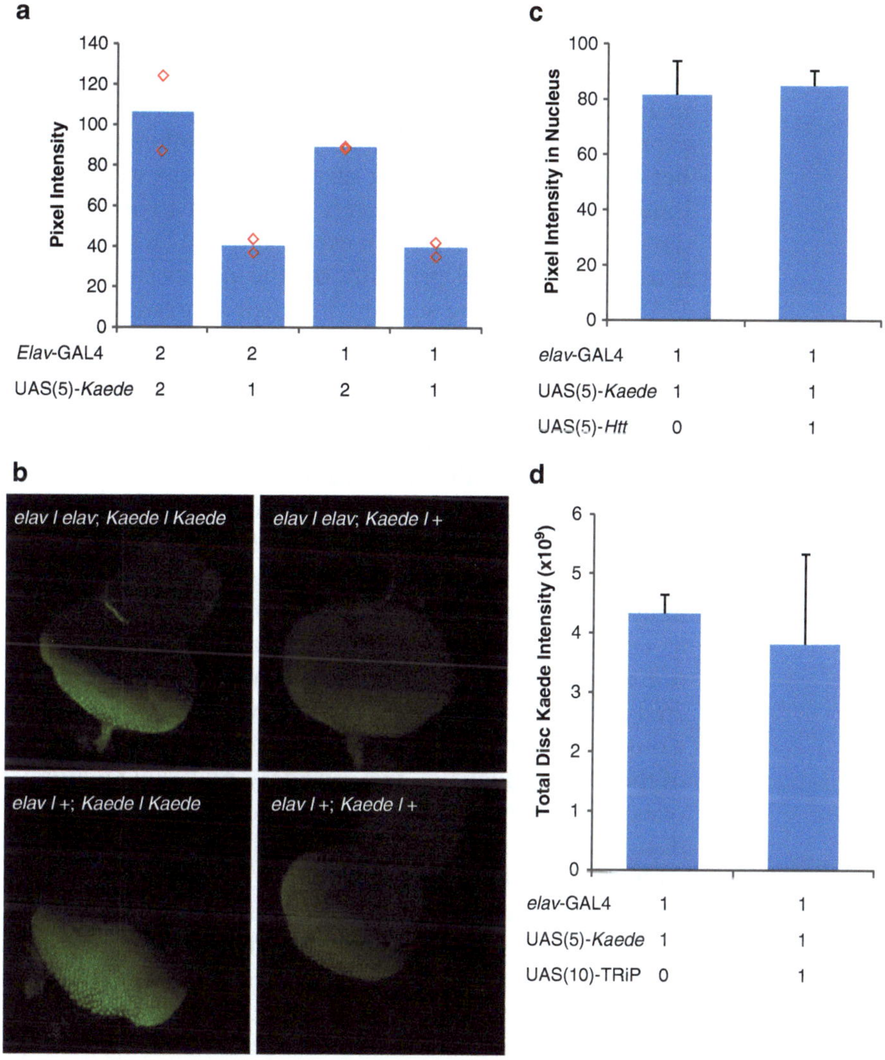

Fig. 5 GAL4 is not the limiting factor with multiple UASs. (**a**) Comparison of relative Kaede expression in eye discs of flies containing various numbers of *elav*-GAL4 and 5×UAS-*Kaede* constructs (*diamonds* represent high and low values from different discs). Pixel intensity is obtained by measuring intensity across multiple lines drawn across each disc ($n = 2$~3). (**b**) Representative compiled 3D confocal Z-stack images of fluorescent eye discs dissected from third instar larvae imaged on a Zeiss LSM780. (**c**) Comparison of relative Kaede expression in nuclei of salivary glands of *elav*-Kaede flies with or without the UAS-*Htt* transgene (containing 5×UASs). Pixel intensity is obtained by measuring intensity across the nucleus in each cell ($n = 5$ cells). (**d**) Comparison of Kaede expression in discs of *elav*-Kaede flies with or without another transgene (TRiP line containing 10×UASs). Total intensity of Kaede is obtained by summing up the pixel intensity of each Z stack slice of a disc ($n = 4$~5)

transgene has 5×UAS binding sites). In another experiment, we monitored the level of Kaede expression in salivary glands in the presence and absence of a UAS-*Htt* transgene and could detect no influence of the extra UAS-*Htt* on Kaede expression (Fig. 5c). In a third case, the levels of UAS-Kaede expression in discs with and without an unrelated TRiP RNAi construct [36] that has 10×UAS sites was monitored (Fig. 5d). Again, no major change in levels was observed although an increase in the variance from animal to animal was seen. Another independent study indicates that the level of expression from a single UAS-transgene is influenced by the number of GAL4 DNA-binding sites (e.g., vectors with 10 vs. 5×UAS binding sites boost GFP levels more than twofold) [37]. These data indicate that GAL4 is in excess and that levels of expression of a transgene depend primarily on the number of transgenes and UAS/GAL4 binding sites driving their expression rather than the number of GAL4 drivers.

3.3.5 Drivers: elav, GMR, Rh1, OK107, Da, Act, Arm

For many studies, and for neurodegenerative studies in particular, it is often advantageous to test the cell type specificity of an effect. Among the GAL4 drivers available for such studies, the *elav*-GAL4 construct is particularly useful for monitoring neurodegeneration by the pseudopupil method because the *elav* promoter is expressed in all neurons of the peripheral and central nervous system beginning in embryonic stages and continuing into adult life. Importantly, its expression in the eye is limited to neuronal cells only and it is not expressed in the support cells. In addition, expression is initiated only as the morphogenetic furrow passes by and commits previously uncommitted cells to a neuronal lineage [38]. In the case of the eye imaginal disc, this provides a powerful temporal gradient of exposure of neurons to the transgene in question (Fig. 6). In the disc, there are rows of developing photoreceptor neuron clusters with each row being ~2 h older and thus having been exposed to transgene expression ~2 h longer than the row anterior to it. This driver provides chronic expression of expanded polyQs in neurons. There are several *elav*-GAL4 drivers. The X chromosome *elav* driver is the strongest and is an enhancer trap insert into the endogenous *elav* locus while the other *elav* drivers (e.g., on chr II) are fusions of the *elav* promoter and GAL4, and are weaker [39].

The GMR driver is often used to express *Htt* and other degenerative genes in the eyes [9, 19, 40] but some considerations should be noted. Because the GMR promoter expresses GAL4 in all cells of the developing eye [41], and development of the eye involves extensive cell signaling [42], the *GMR*-GAL4 driver alone exhibits a weak rough eye phenotype indicative of disturbed eye development. Expression of various peptides containing an expanded polyQ tract in all cell types of the retina with the *GMR*-GAL4 driver results in progressive pigment cell degeneration, but the level of neuronal degeneration as distinct from support cell degeneration is more difficult to assess, although the overall severity is evident (e.g., ref. 7–9).

A rhodopsin driver (*Rh1*-GAL4) induces transgene expression during the last half of pupal development in the outer six (R1–R6) photoreceptor neurons of each ommatidium in adult flies [43, 44];

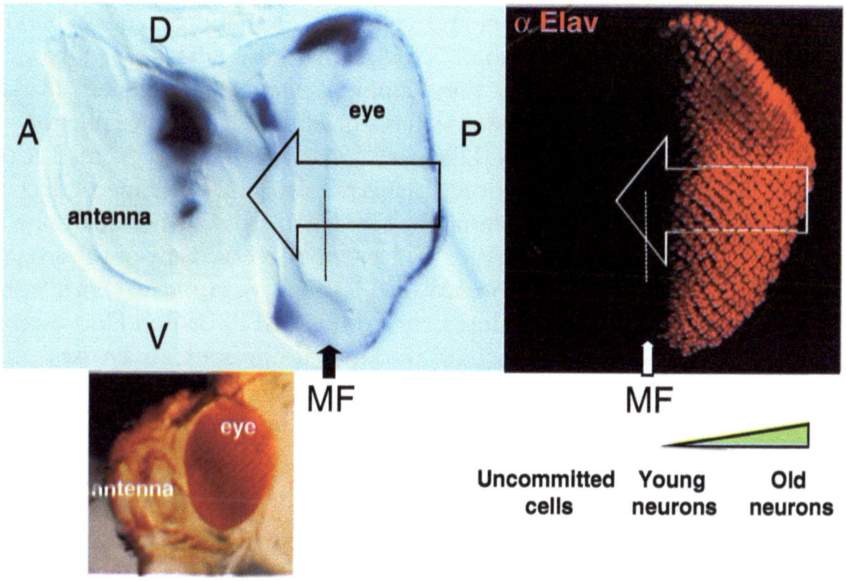

Fig. 6 The fly eye imaginal disc provides a temporal gradient of Htt challenge. In eye imaginal discs, a wave of differentiation (morphogenetic furrow, MF) passes over a field of ~30,000 uncommitted cells from posterior (P) towards anterior (A) (indicated by *open arrow*). The wave (*solid arrow* and *vertical dashed line* indicate the leading edge of the wave) consists of a line of morphogen expression. As this wave of differentiation passes, cells are "born" as neurons and expression of *elav* is initiated (*red stain*), as is expression of the *elav*-GAL4 and with it, the *Htt* transgene. Each row of cells counting from leading edge of the furrow toward posterior has been expressing the *Htt* transgene for 2 h longer than the one anterior to it

thus, not affecting neuronal development but only mature neuron survival. Expression of *Htt* ex1p-Q93 using this driver results in later onset of progressive degeneration with intact rhabdomeres present until approximately day 8 but mild degeneration with the number of rhabdomeres per ommatidium decreasing by day 12 [45].

OK107 drives transgene expression in the mushroom body and in other well-defined subsets of CNS neurons [17, 46]. Flies expressing *Htt* ex1p-93Q under the OK107 driver exhibit a 25 % loss of total mushroom body volume and show specific changes to the structure of the various mushroom body lobes and Kenyon cell bodies [17].

Interestingly, based on our observations, there is no pseudopupil phenotype with tissue-general drivers, such as *Actin* (*Act*), *Armadillo* (*Arm*), and *Daughterless* (*Da*). Typically, when driving the expression of an expanded polyQ Htt peptide, these drivers kill the animals well before eclosion and thus must be reared at lowered temperature to allow survival to adulthood. However, animals reared at low temperatures and shifted to higher temperatures after eclosion do not typically exhibit photoreceptor degeneration, although they do exhibit reduced lifespan. It is worth noting that the relative levels of expression in photoreceptor neurons by these drivers remain to be determined.

3.3.6 Marker and Balancer Effects

Genetic markers and balancers used for genetic stabilization and screening in some crosses may also affect phenotypes of Htt-challenged flies. For example, *Scutoid* (*Sco*), a genetic marker on the second chromosome, has a negative effect on eye phenotype in Htt-challenged flies. Likewise, *Ultrabithorax* (*Ubx*), a marker on several third chromosome balancers, also has a negative effect and thus these markers should be avoided in crosses when analyzing the pseudopupil phenotype. We have not observed any untoward pseudopupil effects with *Stubble* (*Sb*), *Curly* (*Cy*), *Humeral* (*Hu*), *Sternopleural* (*Sp*), Green Fluorescent Protein (GFP) or Red Fluorescent Protein (RFP) when used for tracking chromosomes in crosses.

3.4 Pseudopupil Method of Monitoring Photoreceptor Degeneration

The steps involved in performing a pseudopupil assay and the details for analysis of pseudopupil data are described.

1. Prepare a microscope slide
   Clear fingernail polish is used to mount the fly head to a microscope slide. If the fingernail polish is not very tacky, place a dot of nail polish on the slide before decapitation to allow time to dry. If the nail polish is already tacky, then it can be placed on the slide after decapitation (*see* **Note 3**).
2. Decapitate the fly
   The fly should be alive before decapitation and it is necessary to work quickly through this assay to minimize the amount of cellular degeneration that will occur before the rhabdomeres can be counted. We put the fly to sleep using CO_2 and decapitate the fly with a hypodermic needle attached to a syringe (*see* **Note 4**).
3. Mount the head
   Using forceps to grip the proboscis, place the head onto the tacky drop of nail polish in a position so that it is resting at an angle with the tangent of the eye as close as possible to parallel with the surface of the slide (Fig. 7a).
4. Visualization of rhabdomeres
   The general strategy for visualization is to create a situation where transmitted light is shining on one side of the head such that the majority of light that one sees on the other side must come through the rhabdomeres (Fig. 7a). For example, an upright Nikon Optiphot-2, with a DPlan 50 0.09 oil lens can be used to visualize the rhabdomeres in each ommatidium. The lighting must be adjusted correctly to see the pattern clearly. Make sure that all filters are removed to maximize the amount of light getting through. Focus the condenser. With a low power objective, find and focus on the head, then move the objective aside, add a drop of oil, and swing the oil objective into place (*see* **Note 5**). Close the iris diaphragm to a diameter approximating the size of the head so that all light must go through the object (*see* **Note 6**).

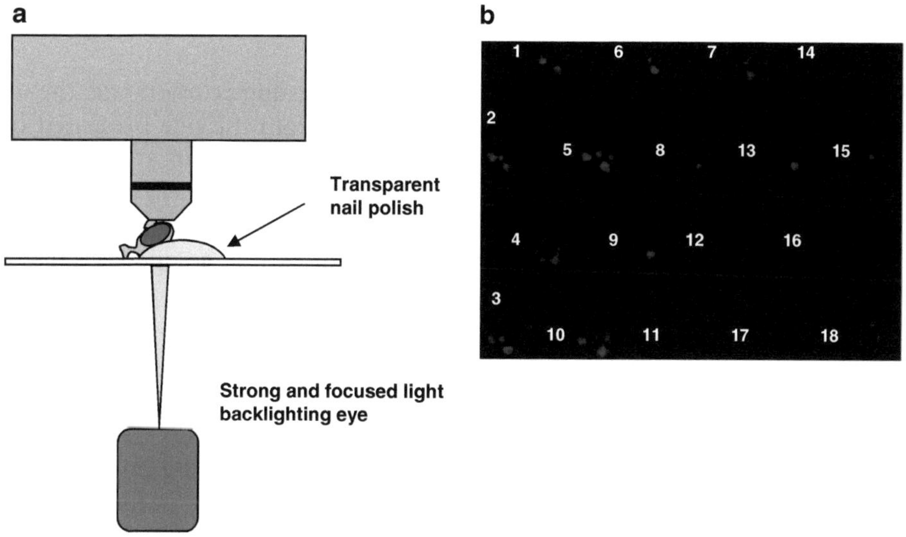

Fig. 7 Schematic of the microscope setup for pseudopupil analysis. (**a**) The fly head is immobilized on a slide with transparent nail polish and the eye is positioned for optimal visualization with a narrow beam of light illuminating the eye from below. (**b**) Example of a counting pattern for the pseudopupil assay. Start at the top of the eye and count downward until ommatidia are no longer clearly visible, then count up the next row in an S-like pattern. Continually adjust the focus of the microscope for optimal visibility while progressing from one ommatidium to another. Avoid the two rows of ommatidia on the edges of the eye where there is a greater amount of degeneration. Any rhabdomere that is not an intact circle is regarded as degenerated. For example, with the caveat that final scoring involves focusing up and down, one might score the second column (ommatidia #3–6) as, 3, 5, 4, 3

5. Count the rhabdomeres

   Start at the top of the eye and count down until you are no longer able to see clearly defined ommatidia, and then count back up the next row (as shown in Fig. 7b, also *see* **Note 7**). Any rhabdomere that is not an intact circle is regarded as degenerated. Be consistent in counting each eye. In an HD fly, the number of visible intact rhabdomeres varies from 0 to 7, and the structure of rhabdomeres in the ommatidium becomes disorganized (Fig. 7b). A given experiment should typically be scored by one individual, and it is best to score the eyes blind. Record at least 30–40 ommatidia per head and score at least five heads for each genotype/treatment.

6. Data analysis

   Sum the total number of ommatidia with different numbers of rhabdomeres (0–7) and obtain the average. Two sets of pseudopupil data from different genotypes or drug groups can be compared for statistical significance using the Student's *t*-test (*see* **Note 8**). It is sometimes useful to graph the distribution of the number of photoreceptors per ommatidium for comparison.

4 Notes

1. For analyzing viability and assessing eclosion rate, the set up of the crosses will be different from the one presented here for pseudopupil analysis. We typically set up at least ten vials of crosses with five pairs of virgins and males in each vial and keep the parents for 4–5 days before emptying the vials. We then count the number of flies of each genotype each day as they eclose and express the eclosion rate as the ratio of *Htt*-expressing flies *vs.* non-expressing controls.
2. One's ability to see the rhabdomeres depends on the level of pigment in the eye. In flies with very little pigment, the rhabdomeres can be impossible to see by this technique; thus, it is better to use a driver that has a strong mini *w+* or cross a wild-type *w+* gene into your *w−* transgenic flies. If pigmented eyes are not possible, another measure of retinal degeneration is to fix and section the eye and measure the retinal thickness [21].
3. We have successfully used Sally Hansen's "Advanced Hard as Nails" clear nail polish. It is best to use an older bottle of fingernail polish in which the polish is slightly tacky. Older nail polish can be mixed with newer nail polish to produce an ideal moderately tacky consistency. This helps keep the eye in its upright position. As an alternative, a glob of Vaseline (petroleum jelly) can be used but is less stable than fingernail polish.
4. Within 15 min after decapitation, the pseudopupil image will begin to fade. To decapitate flies, we use forceps to hold the fly body with its ventral side down or on its side on a microscope slide under a dissecting scope and use a 26 G × 5/8″ needle attached to a 1 cm^3 syringe to cut through the neck, taking care not to touch the eyes at any point during the dissection. Some methods of manipulation cause the proboscis to extend more than others. The proboscis can then be used as a handle to move the head into position without damaging the eyes.
5. It is better to lower the stage a bit by turning the coarse adjustment knob about a quarter turn before changing from a low magnification objective to the oil objective to avoid the objective touching or damaging the head.
6. By closing or opening the diaphragm you can increase or decrease the contrast in the ommatida.
7. It is usually necessary to refocus and re-center the eye in the field of view for optimal visualization. Focus just below the highest point of the curved surface of the eye. Choose a plane of focus where you can clearly see and count the rhabdomeres in ≥30 ommatidia. Continually adjust the focus of the microscope for

optimal visibility as you progress from one ommatidium to another, avoiding the two rows of ommatidia on the edges of the eye where there is a greater amount of variation.

8. A two-tailed unpaired *t*-test is appropriate if one is comparing the effect of a drug or genetic manipulation on flies that already have some level of degeneration (e.g., flies expressing the mutant *Huntingtin* exon 1 peptide where the average number of rhabdomeres per ommatidium is about 4.0). A paired *t*-test would be used in a case where the same fly is tested before and after a specific treatment.

References

1. Paulson HL, Bonini NM, Roth KA (2000) Polyglutamine disease and neuronal cell death. Proc Natl Acad Sci USA 97(24):12957–12958. doi:10.1073/pnas.210395797 210395797 [pii]
2. Gatchel JR, Zoghbi HY (2005) Diseases of unstable repeat expansion: mechanisms and common principles. Nat Rev Genet 6(10):743–755. doi:nrg1691 [pii] 10.1038/nrg1691
3. Macdonald ME, Ambrose CM, Duyao MP, Myers RH, Lin C, Srinidhi L, Barnes G, Taylor SA, James M, Groot N, Macfarlane H, Jenkins B, Anderson MA, Wexler NS, Gusella JF, Bates GP, Baxendale S, Hummerich H, Kirby S, North M, Youngman S, Mott R, Zehetner G, Sedlacek Z, Poustka A, Frischauf AM, Lehrach H, Buckler AJ, Church D, Doucettestamm L, Odonovan MC, Ribaramirez L, Shah M, Stanton VP, Strobel SA, Draths KM, Wales JL, Dervan P, Housman DE, Altherr M, Shiang R, Thompson L, Fielder T, Wasmuth JJ, Tagle D, Valdes J, Elmer L, Allard M, Castilla L, Swaroop M, Blanchard K, Collins FS, Snell R, Holloway T, Gillespie K, Datson N, Shaw D, Harper PS (1993) A novel gene containing a trinucleotide repeat that is expanded and unstable on Huntingtons-disease chromosomes. Cell 72(6):971–983
4. Gusella JF, Macdonald ME (1995) Huntingtons-disease. Semin Cell Biol 6(1):21–28
5. Kremer B, Goldberg P, Andrew SE, Theilmann J, Telenius H, Zeisler J, Squitieri F, Lin BY, Bassett A, Almqvist E, Bird TD, Hayden MR (1994) A worldwide study of the Huntingtons-disease mutation—the sensitivity and specificity of measuring Cag repeats. N Engl J Med 330(20):1401–1406
6. Nagai Y, Fujikake N, Ohno K, Higashiyama H, Popiel HA, Rahadian J, Yamaguchi M, Strittmatter WJ, Burke JR, Toda T (2003) Prevention of polyglutamine oligomerization and neurodegeneration by the peptide inhibitor QBP1 in Drosophila. Hum Mol Genet 12(11):1253–1259
7. Jackson GR, Salecker I, Dong XZ, Yao X, Arnheim N, Faber PW, MacDonald ME, Zipursky SL (1998) Polyglutamine-expanded human huntingtin transgenes induce degeneration of Drosophila photoreceptor neurons. Neuron 21(3):633–642
8. Marsh JL, Walker H, Theisen H, Zhu YZ, Fielder T, Purcell J, Thompson LM (2000) Expanded polyglutamine peptides alone are intrinsically cytotoxic and cause neurodegeneration in Drosophila. Hum Mol Genet 9(1):13–25
9. Warrick JM, Paulson HL, Gray-Board GL, Bui QT, Fischbeck KH, Pittman RN, Bonini NM (1998) Expanded polyglutamine protein forms nuclear inclusions and causes neural degeneration in Drosophila. Cell 93(6):939–949
10. Kazemi-Esfarjani P, Benzer S (2000) Genetic suppression of polyglutamine toxicity in Drosophila. Science 287(5459):1837–1840. doi:8327 [pii]
11. Steffan JS, Bodai L, Pallos J, Poelman M, McCampbell A, Apostol BL, Kazantsev A, Schmidt E, Zhu YZ, Greenwald M, Kurokawa R, Housman DE, Jackson GR, Marsh JL, Thompson LM (2001) Histone deacetylase inhibitors arrest polyglutamine-dependent neurodegeneration in Drosophila. Nature 413(6857):739–743
12. Lee WCM, Yoshihara M, Littleton JT (2004) Cytoplasmic aggregates trap polyglutamine-containing proteins and block axonal transport in a Drosophila model of Huntington's disease. Proc Natl Acad Sci USA 101(9):3224–3229
13. Zhang S, Binari R, Zhou R, Perrimon N (2010) A genomewide RNA interference screen for modifiers of aggregates formation by mutant Huntingtin in Drosophila. Genetics 184(4):1165–1179. doi:genetics.109.112516 [pii] 10.1534/genetics.109.112516

14. Romero E, Cha GH, Verstreken P, Ly CV, Hughes RE, Bellen HJ, Botas J (2008) Suppression of neurodegeneration and increased neurotransmission caused by expanded full-length huntingtin accumulating in the cytoplasm. Neuron 57(1):27–40. doi:S0896-6273(07)00985-3 [pii] 10.1016/j.neuron.2007.11.025
15. Mugat B, Parmentier ML, Bonneaud N, Chan HY, Maschat F (2008) Protective role of Engrailed in a Drosophila model of Huntington's disease. Hum Mol Genet 17(22):3601–3616. doi:ddn255 [pii] 10.1093/hmg/ddn255
16. Franceschini N (1972) Information processing in the visual systems of arthropods. Springer, Berlin
17. Agrawal N, Pallos J, Slepko N, Apostol BL, Bodai L, Chang LW, Chiang AS, Thompson LM, Marsh JL (2005) Identification of combinatorial drug regimens for treatment of Huntington's disease using Drosophila. Proc Natl Acad Sci USA 102(10):3777–3781. doi:10.1073/Pnas.0500055102
18. Brand AH, Perrimon N (1993) Targeted gene-expression as a means of altering cell fates and generating dominant phenotypes. Development 118(2):401–415
19. Marsh JL, Thompson LM (2004) Can flies help humans treat neurodegenerative diseases? Bioessays 26(5):485–496. doi:10.1002/Bies.20029
20. Lin DM, Goodman CS (1994) Ectopic and increased expression of Fasciclin II alters motoneuron growth cone guidance. Neuron 13(3):507–523. doi:0896-6273(94)90022-1 [pii]
21. Fernandez-Funez P, Nino-Rosales ML, de Gouyon B, She WC, Luchak JM, Martinez P, Turiegano E, Benito J, Capovilla M, Skinner PJ, McCall A, Canal I, Orr HT, Zoghbi HY, Botas J (2000) Identification of genes that modify ataxin-1-induced neurodegeneration. Nature 408(6808):101–106
22. Karpilow JM, Pimentel AC, Shamloula HK, Venkatesh TR (1996) Neuronal development in the Drosophila compound eye: photoreceptor cells R1, R6, and R7 fail to differentiate in the Retina aberrant in pattern (rap) mutant. J Neurobiol 31(2):149–165
23. Spradling AC, Rubin GM (1983) The effect of chromosomal position on the expression of the Drosophila xanthine dehydrogenase gene. Cell 34(1):47–57
24. Groth AC, Fish M, Nusse R, Calos MP (2004) Construction of transgenic Drosophila by using the site-specific integrase from phage phi C31. Genetics 166(4):1775–1782
25. Duffy JB (2002) GAL4 system in Drosophila: a fly geneticist's Swiss army knife. Genesis 34(1–2):1–15. doi:10.1002/Gene.1015
26. Matsumoto K, Tohe A, Oshima Y (1978) Genetic-control of galactokinase synthesis in Saccharomyces cerevisiae—evidence for constitutive expression of positive regulatory gene Gal4. J Bacteriol 134(2):446–457
27. Elliott DA, Brand AH (2008) The GAL4 system: a versatile system for the expression of genes. Methods Mol Biol 420:79–95. doi:10.1007/978-1-59745-583-1_5
28. McGuire SE, Le PT, Osborn AJ, Matsumoto K, Davis RL (2003) Spatiotemporal rescue of memory dysfunction in Drosophila. Science 302(5651):1765–1768
29. Ito K, Awano W, Suzuki K, Hiromi Y, Yamamoto D (1997) The Drosophila mushroom body is a quadruple structure of clonal units each of which contains a virtually identical set of neurones and glial cells. Development 124(4):761–771
30. Pignoni F, Zipursky SL (1997) Induction of Drosophila eye development by Decapentaplegic. Development 124(2): 271–278
31. Osterwalder T, Yoon KS, White BH, Keshishian H (2001) A conditional tissue-specific transgene expression system using inducible GAL4. Proc Natl Acad Sci USA 98(22): 12596–12601
32. Roman G, Endo K, Zong L, Davis RL (2001) P{Switch}, a system for spatial and temporal control of gene expression in *Drosophila melanogaster*. Proc Natl Acad Sci USA 98(22): 12602–12607
33. Rorth P, Szabo K, Bailey A, Laverty T, Rehm J, Rubin GM, Weigmann K, Milan M, Benes V, Ansorge W, Cohen SM (1998) Systematic gain-of-function genetics in Drosophila. Development 125(6):1049–1057
34. Rorth P (1996) A modular misexpression screen in Drosophila detecting tissue-specific phenotypes. Proc Natl Acad Sci USA 93(22):12418–12422
35. Ando R, Hama H, Yamamoto-Hino M, Mizuno H, Miyawaki A (2002) An optical marker based on the UV-induced green-to-red photoconversion of a fluorescent protein. Proc Natl Acad Sci USA 99(20):12651–12656. doi:10.1073/Pnas.202320599
36. Ni JQ, Liu LP, Binari R, Hardy R, Shim HS, Cavallaro A, Booker M, Pfeiffer BD, Markstein M, Wang H, Villalta C, Laverty TR, Perkins LA, Perrimon N (2009) A Drosophila resource of transgenic RNAi lines for neurogenetics. Genetics 182(4):1089–1100. doi:10.1534/Genetics.109.103630

37. Pfeiffer BD, Ngo TTB, Hibbard KL, Murphy C, Jenett A, Truman JW, Rubin GM (2010) Refinement of tools for targeted gene expression in Drosophila. Genetics 186(2):735–755. doi:10.1534/Genetics.110.119917
38. Robinow S, White K (1988) The locus elav of *Drosophila melanogaster* is expressed in neurons at all developmental stages. Dev Biol 126(2):294–303
39. Lee WC, Yoshihara M, Littleton JT (2004) Cytoplasmic aggregates trap polyglutamine-containing proteins and block axonal transport in a Drosophila model of Huntington's disease. Proc Natl Acad Sci USA 101(9):3224–3229. doi:10.1534/Genetics.110.119917
40. Bonini NM, Fortini ME (2003) Human neurodegenerative disease modeling using Drosophila. Annu Rev Neurosci 26:627–656. doi:10.1146/annurev.neuro.26.041002.131425 041002.131425 [pii]
41. Ellis MC, Oneill EM, Rubin GM (1993) Expression of Drosophila glass protein and evidence for negative regulation of its activity in nonneuronal cells by another DNA-binding protein. Development 119(3):855–865
42. Sang TK, Li C, Liu W, Rodriguez A, Abrams JM, Zipursky SL, Jackson GR (2005) Inactivation of Drosophila Apaf-1 related killer suppresses formation of polyglutamine aggregates and blocks polyglutamine pathogenesis. Hum Mol Genet 14(3):357–372. doi:ddi032 [pii] 10.1093/hmg/ddi032
43. Kirschfeld K, Feiler R, Franceschini N (1978) Photo-stable pigment within rhabdomere of fly photoreceptors No 7. J Comp Physiol 125(3):275–284
44. Chyb S, Hevers W, Forte M, Wolfgang WJ, Selinger Z, Hardie RC (1999) Modulation of the light response by cAMP in Drosophila photoreceptors. J Neurosci 19(20):8799–8807
45. Slepko N, Bhattacharyya AM, Jackson GR, Steffan JS, Marsh JL, Thompson LM, Wetzel R (2006) Normal-repeat-length polyglutamine peptides accelerate aggregation nucleation and cytotoxicity of expanded polyglutamine proteins. Proc Natl Acad Sci USA 103(39):14367–14372. doi:10.1073/Pnas.0602348103
46. Branco J, Al-Ramahi I, Ukani L, Perez AM, Fernandez-Funez P, Rincon-Limas D, Botas J (2008) Comparative analysis of genetic modifiers in Drosophila points to common and distinct mechanisms of pathogenesis among polyglutamine diseases. Hum Mol Genet 17(3):376–390. doi:ddm315 [pii] 10.1093/hmg/ddm315

Chapter 4

Size Analysis of Polyglutamine Protein Aggregates Using Fluorescence Detection in an Analytical Ultracentrifuge

Saskia Polling, Danny M. Hatters, and Yee-Foong Mok

Abstract

Defining the aggregation process of proteins formed by poly-amino acid repeats in cells remains a challenging task due to a lack of robust techniques for their isolation and quantitation. Sedimentation velocity methodology using fluorescence detected analytical ultracentrifugation is one approach that can offer significant insight into aggregation formation and kinetics. While this technique has traditionally been used with purified proteins, it is now possible for substantial information to be collected with studies using cell lysates expressing a GFP-tagged protein of interest. In this chapter, we describe protocols for sample preparation and setting up the fluorescence detection system in an analytical ultracentrifuge to perform sedimentation velocity experiments on cell lysates containing aggregates formed by poly-amino acid repeat proteins.

Key words Analytical ultracentrifugation, Sedimentation velocity analysis, SEDFIT, Sucrose, Fluorescence detection system (FDS), Green fluorescent protein (GFP), Cell lysate

1 Introduction

The biochemical characterization of protein aggregates formed by tandem amino acid repeats is challenging due to the size and often insoluble nature of the aggregates, as well as the lack of robust techniques for their separation and quantitation. Recent developments in analytical ultracentrifugation methods, however, have opened the way for more precise measurements of such aggregates. The analytical ultracentrifuge has had a long history in studies of protein self-assembly. Sedimentation velocity and equilibrium experiments on the analytical ultracentrifuge have been widely used for characterizing protein complexes, and protein–protein interactions [1–3], and more specifically for the analysis of amyloid oligomers and fibrils [4]. The development of a commercial fluorescence detection system (FDS) for the analytical ultracentri-

Danny M. Hatters and Anthony J. Hannan (eds.), *Tandem Repeats in Genes, Proteins, and Disease: Methods and Protocols*, Methods in Molecular Biology, vol. 1017, DOI 10.1007/978-1-62703-438-8_4, © Springer Science+Business Media New York 2013

fuge has further extended the sensitivity and specificity of these techniques [5, 6]. The FDS is well suited for exploring the sedimentation behavior of fluorescently labeled proteins over a broad range of concentrations. Of particular interest is the development of green fluorescence protein (GFP) constructs as in vivo fluorescence tags, permitting the use of sedimentation velocity studies and fluorescence detection for the analysis of protein aggregation and fibril formation in cell lysates and *ex vivo* [7].

A particularly useful feature of the FDS is the capacity to study how proteins aggregate in the context of their natural environment such as in cells or in extracellular fluids provided the solutions are not (overly) turbid. Early published work with the FDS showed that GFP could be detected specifically in solutions containing high concentrations of a "background" protein (bovine serum albumin), which demonstrated the capacity for its application in complex solutions [5]. FDS was also shown to be useful for the study of fluorescent "tracers" in serum and *E. coli* lysate and for monitoring antibody binding to antigens [6]. More recently, sedimentation velocity experiments using the FDS have been used to monitor the aggregation kinetics of huntingtin protein in mammalian cell lysates [7], successfully resolving three differently sized populations that exist in cells.

This chapter presents protocols for sample preparation and setting up the FDS unit in the analytical ultracentrifuge for analysis. The concept and theory behind sedimentation velocity experiments on the analytical ultracentrifuge have been detailed in reviews elsewhere [4, 8, 9] and are only perfunctorily reviewed here. Briefly, the sedimentation velocity technique monitors the radial concentration distributions (boundaries) of solutes during movement through a centrifugation field in a sample compartment over a period of time. Figure 1 illustrates a typical sedimentation velocity experiment showing a sample compartment containing the protein sample. At the start of the experiment (t0), the protein fluorescence intensity along the distance of the cell is uniform. At various times (i.e., t1, t2) of sedimentation, at constant speed and temperature, the boundary moves down the cell as the solute is depleted from the sample meniscus. Macromolecules of different size and shape will yield a different rate of sedimentation and pattern of sedimentation. The overall sedimentation profile can thus be fitted to known sedimentation models to yield a size distribution plot that describes the size and proportion of all the species in a sample solution. Use of SEDFIT software to perform experimental data analysis and model fitting has been detailed elsewhere and the reader is again directed to them [9–11].

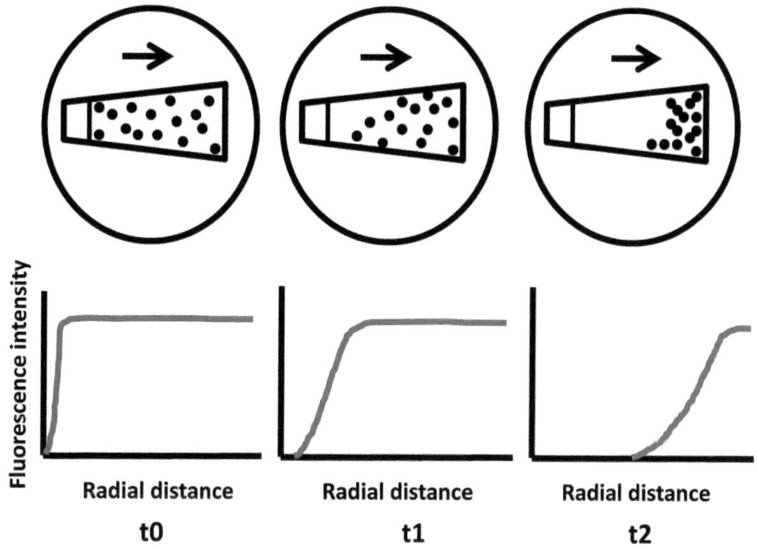

Fig. 1 Sedimentation velocity experiment in an analytical ultracentrifuge. Fluorescently labeled proteins in a sample container are spun at an appropriate speed. *Arrows* indicate the direction of the centrifugal force. At initial time (t0), all the molecules are homogeneously distributed at each point along the sample container ("cells") and the fluorescence intensity at each point after the meniscus is equal. After certain times of centrifugation (t1, t2), the protein molecules sediment towards the bottom of the cell, manifesting in a moving fluorescence concentration boundary along the length of the cell compartment

2 Materials

1. Mammalian cells expressing the protein of interest tagged to EGFP. (More details for the design of the system are explained in Subheading 3.1.)
2. Phosphate-buffered saline.
3. Liquid nitrogen or a dry/ice ethanol bath.
4. TX buffer: 20 mM Tris–HCl pH 8.0, 2 mM $MgCl_2$, 150 mM NaCl, 1 % w/v Triton X-100, benzonase nuclease (Merck, Germany), and a complete EDTA-free protease inhibitor cocktail tablet (Roche Applied Science). For bead lysis, the buffer can omit Triton X-100.
5. 27 Gauge needle and syringe.
6. 0.5 mm Zirconia/Silica beads (DainTree Scientific, Australia) in vials with an O-ring and a screw cap.
7. Fluorescence plate reader.

8. Bicinchoninic acid kits (e.g., from Thermo Scientific).
9. Sucrose.
10. Digital refractometer.
11. Beckman XL-I or XL-A analytical ultracentrifuge equipped with an AVIV Biomedical (Lakewood, NJ) FDS (*see* **Note 1**).
12. Heavy mineral oil (FC43 oil, Fluorinert).

3 Methods

The presently available commercial FDS is equipped with a 488 nm excitation laser, with emission from the sample directed through a pair of 505 nm cutoff filters and into a photomultiplier tube for signal digitization. Thus, the FDS is suitable for detection of dyes such as fluorescein and GFP, as well as other fluorophores with compatible excitation and emission properties.

3.1 Tag the Protein of Interest with GFP or an Appropriate Fluorophore

The fluorescence properties of GFP are well suited for tracking proteins expressed in mammalian cells using the FDS. Other fluorescent proteins—notably cyan fluorescent protein (CFP)—are not sufficiently bright when used in mammalian cell lysates with the standard FDS optics. However, CFP is adequate for detection of purified proteins using the FDS, albeit with far less sensitivity than GFP. It remains to be tested how well other variants such as yellow fluorescent protein (YFP) work. An important consideration is that a monomeric form of GFP is used, such as EGFP containing the A206K mutation, since many early GFP derivatives are known to oligomerize [12]. We have found that the Emerald form of GFP, which contains the mutations S72A, N149K, M153T, I167T, and A206K on the original EGFP [13], is a bright and particularly useful construct for sedimentation analysis using the FDS.

3.2 Protein Expression

1. As a starting point for relatively abundantly expressed proteins (e.g., using a cytomegalovirus promoter to drive high yielding expression), about 2.5×10^6 transiently transfected mammalian cells, or a confluent 25 cm^2 plate, should be sufficient to prepare 2–4 samples for sedimentation analysis using total cellular protein concentrations of 0.2–0.5 mg/mL. We generally use Neuro2A cells for expressing polyglutamine proteins, but have found that higher levels of expression can be obtained with HEK AD293 cells.
2. Harvest cells expressing the fluorescent protein of interest by scraping culture flasks in phosphate-buffered saline.
3. Pellet harvested cells by centrifuging at approximately $250 \times g$ for 5 min.
4. Snap-freeze the pellet in liquid nitrogen or a dry/ice ethanol bath at −80 °C until lysis to avoid protein degradation.

3.3 Triton X-100-Mediated Cell Lysis

1. To lyse cells, thaw the cell pellets on ice. The following steps should be performed on ice with ice-cold buffers. Lyse cells by extrusion through a 27 gauge needle in TX buffer.
2. Snap-freeze using liquid N_2 and store at −80 °C until further use.

3.4 Mechanical Cell Lysis

Some weaker self-associations in protein aggregates may be perturbed by the presence of Triton X-100. In this case, we have alternatively used 0.5 mm Zirconia/Silica beads to lyse cells mechanically. This alternative protocol is detailed below.

1. Resuspend cell pellet in TX buffer lacking Triton X-100.
2. Once resuspended, divide cells into two 0.5 mL aliquots. Add each to separate 2 mL polystyrene screw cap vials containing an approximate 0.5 mL volume of beads. For efficient lysis, a similar ratio of beads to sample and a similar ratio of bead/sample to air are essential.
3. Vortex the samples six times for 45 s each, with intervals of 30 s on ice. Samples should be kept on ice to avoid heating of the sample.
4. Transfer lysate into 1.5 mL microcentrifuge tubes using a 27 gauge needle to avoid transfer of the beads.
5. Snap-freeze the samples and store at −80 °C until further use.

3.5 Standardization of Experimental Parameters Including Lysate Total Protein Concentrations

1. If the protein (or oligomer) of interest is known to be entirely soluble, then spin the samples at 5,000 × *g* for 30 min to reduce baseline turbidity. If the protein of interest is suspected to form large aggregates or complexes, or it is not known whether it does or not, do not spin the crude lysate as this may remove larger complexes from the sample and bias the analysis.
2. Measure the fluorescence of the supernatant (or whole-cell lysate) using a fluorescence plate reader.
3. After matching the samples based on fluorescence, as an optional second control, expression levels can be matched/validated for degradation via Western blot quantitation.
4. A useful internal control for influences of lysate components on sedimentation behavior when comparing two or more proteins/samples is to analyze different experimental samples in exactly the same lysate concentrations. This can be established simply by determination and standardization of total cellular protein concentration (e.g., bicinchoninic acid protein assays) and by adjustment of the transfections to obtain standardized levels of protein-GFP expression or by dilution of lysates with lysates from untransfected cells (*see* **Note 2**).

3.6 Viscosity Adjustment with Sucrose

A key step in the resolution of the very large inclusions formed by huntingtin and other poly-amino acid repeat proteins is the addition of sucrose in the sample to increase its viscosity. This enables slower movement of inclusions through the sample compartment in the ultracentrifuge and allows their sedimentation to be monitored. For inclusions formed by huntingtin and other poly-amino acid repeat proteins, we have found that a final concentration of 2 M sucrose (≈55 % w/w) works well (*see* **Notes 3** and **4**).

1. Prepare an approximately 3 M sucrose stock by weighing out 51.35 g of sucrose and topping up to 50 mL with purified water in a small glass beaker (preferably 100 mL capacity).
2. Place on a heating block and heat to 200 °C for 10 min with constant stirring.
3. Transfer the dissolved solution to a 50 mL tube and leave to cool to room temperature.
4. Measure the refractive index of the sucrose solution using a digital refractometer (*see* **Note 5**). The refractive index (RI) can be used to precisely calculate the molarity (*M*) of the solution and the subsequent weight and volume of the amount of sucrose solution required to create a final concentration of 2 M sucrose using the following formulae [14]:

$$\text{Molarity } (M) = 21.019\text{RI} - 28.036 \tag{1}$$

$$\text{Density of sucrose } (\rho) = 0.1257M + 1.00 \tag{2}$$

$$\textit{Volume of sucrose to get final } 2M = \left(\frac{2}{M}\right) \times \textit{total volume} \tag{3}$$

$$\textit{Weight of sucrose to get final } 2M = \textit{Volume of sucrose} \times \textit{Density of sucrose} \tag{4}$$

5. Use the volume to calculate how much TX buffer is required to get to a final volume of 500 μL. Weigh out the sucrose solution (the solution at this stage is too viscous to pipet accurately) into 1.5 mL microcentrifuge tubes.
6. Aliquot the correct amounts of TX buffer into each microcentrifuge tube and add the corresponding amount of untransfected and fluorescent sample. In the case of 2 M sucrose samples, layer the sample carefully on top of the sucrose.
7. When all components have been added, mix the samples gently using a 1,000 μL pipette. Mix by stirring the sample with the pipette tip. Do not suck up any sample as it is very viscous when unmixed and will stick in the pipette tip. Subsequently, thoroughly resuspend the sample by turning it upside down and flicking it repeatedly. Samples are now ready for analytical ultracentrifugation with the FDS.

3.7 Sedimentation Analysis Using the FDS

Depending on the nature of the aggregates, it may be necessary to perform a number of complementary sedimentation velocity experiments at different rotor speeds to capture the sedimentation properties of all the species of interest present in cells (*see* **Note 4**). Monomers (e.g., masses of ~40 kDa) and low-mass oligomers in cell lysate are resolved by standard high-speed experiments (50,000 rpm) and without using sucrose. Low-speed experiments (3,000 rpm) capture the sedimentation of material up to about 5,000 S in cell lysate, which is equivalent to masses on the scale of millions Dalton. A feature of polyglutamine aggregates is that substantially larger material may also be present. In this case, addition of 2 M sucrose significantly slows sedimentation rates and has been useful to monitor material up to about 320,000 S using a rotor speed of 3,000 rpm [7]. Such material is sufficiently large (e.g., 1.5–8 μm in size) to move in and out of the laser focal point (~10 μm) and this behavior appears as apparent noise in the data [7]. While this profoundly lowers the capacity for quantitative analysis of such large material, approximate sizes can still be estimated. Comparison of the combinatorial sedimentation velocity experiments enables proportions of the different molecular forms to be calculated from the experimental data, effectively enabling aggregation kinetics to be tracked in cells.

1. Load 50 μL of FC43 heavy mineral oil into each compartment of the cell centerpiece using gel-loading tips. Heavy oil is required because scans to the base of the cell (>7.1 cm for a double-sector cell) suffer from significant attenuation in fluorescence intensity as the base is approached. This detection artifact is caused by the cell window holder and screw ring blocking an increasing proportion of the cone-shaped laser beam as it moves towards the cell bottom. Without the use of heavy oil, large aggregates that have sedimented to the bottom of the cell at early scan times may not be observed.
2. Layer 350 μL of sample on top of the heavy mineral oil in each sector of the cell centerpiece using gel-loading tips.
3. Seal the filling holes with provided plastic spacers and screws.
4. Insert and align the cells into the rotor with the filling holes facing the center of the rotor.
5. Install the Aviv FDS unit into the ultracentrifuge following Aviv-provided instructions.
6. Allow the vacuum in the rotor chamber to fall below 100 μm before turning on the laser.
7. Switch on the laser, initiate the provided AOS software, and start the rotor spinning at 3,000 rpm (*see* **Note 6**).
8. When the laser is locked and the magnet angle of the rotor has been established, check the fluorescence signal of the samples,

and adjust the gain setting for the photomultiplier tube accordingly. In our experience, data with the highest signal-to-noise ratio and most useful for sedimentation analysis is obtained when the gain is set to provide intensities above 400 counts (arbitrary units) and below 3,000 counts. Detection of signal intensities is set to cut off at ~4,000 counts to avoid saturation of the detector system.

9. Set the experimental parameters to the desired rotor speeds, temperature, and running time. For monitoring large aggregates and inclusions in sucrose, a rotor speed of 3,000 rpm is typically used. For oligomeric species, buffer that does not contain sucrose and a rotor speed of 3,000 rpm is employed.

   Figure 2a illustrates an example of the sedimentation profile of a GFP-tagged polyglutamine peptide (72Q) in 2 M sucrose. This sample was spun at 3,000 rpm where only very large aggregates and inclusions sediment. Approximately 50 % of the total fluorescence signal is sedimenting under these conditions, indicating that about half of the sample consists of very large aggregates. In contrast, at a centrifugal speed of 50,000 rpm and with no sucrose in solution (Fig. 2b), very large aggregates and inclusions sediment to the bottom almost immediately and do not yield a detectable sedimentation profile. The remaining monomers and small oligomers sediment detectably at this speed.

10. Analysis of experimental data is performed using SEDFIT software (by Peter Schuck), which can be downloaded for free from http://www.analyticalultracentrifugation.com (*see* **Notes 7–11**). To account for differences in solvent density and viscosity, sedimentation coefficients (S) can be converted in terms of a standard solvent (water at 20 °C) ($S_{20,w}$) using the following formula:

$$S_{20,w} = S_{\text{observed}} \left(\frac{\eta_{T,w}}{\eta_{20,w}} \right) \left(\frac{\eta_s}{\eta_w} \right) \left(\frac{1 - \bar{\upsilon} \rho_{20,w}}{1 - \bar{\upsilon} \rho_{T,s}} \right) \tag{5}$$

where $\eta T_{,w}$ and $\eta_{20,w}$ are the viscosities of water at the experimental temperature and at 20 °C, respectively; η_s and η_w are the viscosities of the solvent and water, respectively, at a common temperature; $\bar{\upsilon}$ is partial specific volume; $\rho_{20,w}$ is the density of water at 20 °C; and $\rho_{T,s}$ is the density of solvent at experimental temperature. Further details on experimental analysis can be obtained from the reviews stated in the Introduction. Estimations of sucrose viscosities to calculate η_s at high concentrations and different temperatures can be obtained in [15], and the densities $\rho_{T,s}$ from Eq. 2. The other values for Eq. 5 can be calculated in the program Sednterp

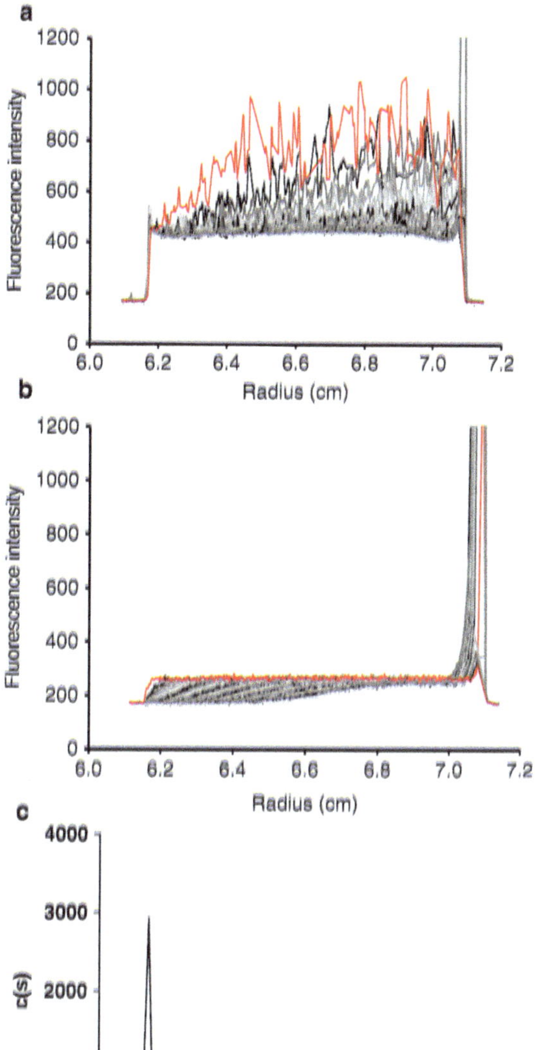

Fig. 2 Sedimentation analysis of GFP-tagged polyQ72 peptide. (**a**) The sample was in TX buffer containing 2 M sucrose, and the centrifugation speed used was 3,000 rpm. Under such conditions, monomers and oligomers do not sediment (manifest in the nonmoving bottom 50 % of the profile). Very large aggregates (inclusions) sediment (manifest in the upper 50 % of the profile). The first scan during the centrifugation experiment is shown in *red*, with subsequent scans shown in *black*, and then *grayscales*. (**b**) The sample is now in TX buffer without sucrose and the centrifugation speed used was 50,000 rpm. Under these conditions, very large aggregates sediment immediately to the bottom of the cell and are not detected by the instrument. The remaining smaller monomers sediment detectably and yield a sedimentation profile. (**c**) Fitting of the experimental sedimentation profile (**b**) into a sedimentation model with no assumptions of heterogeneity (c(s)) yields a size distribution plot indicating a predominant single species at ~2.3S

(by John Philo), which can be downloaded for free from http://www.jphilo.mailway.com/download.htm. From our example in Fig. 2b, fits to this data yielded a size distribution plot of a predominant single species at 2.3S, consistent with a GFP-poly 72Q protein (Fig. 2c).

4 Notes

1. The rotor to be used for the XLI is an An-Ti60 rotor, which holds sample containers ("cells"). Cells should be of a velocity type such as double-sector charcoal-epon centerpieces and either quartz or sapphire windows. A Delrin 5-sector calibration cell manually filled with fluorescein is used for automated calibration of the radial position and angular location of the cell compartments. Operator control of the FDS (including control of the ultracentrifuge itself) is achieved using Advanced Operating Software (Aviv Biomedical).
2. If samples with similar total protein concentrations but vastly different levels of fluorescent protein are being examined, the samples can be matched to allow accurate comparison between samples by measuring the fluorescence associated with 100 μL of 1 g/L of total protein in cell lysate in a 96-well plate in a plate-reader, followed by dilution of these samples with untransfected cell lysate to yield a similar fluorescence level. This step is necessary if protein expression varies significantly between samples. The final sample volume for experiments with the FDS experiments should be 500 μL with a minimum loading volume of 350 μL.
3. It is important to ensure that all sedimentation experiments are run under exactly the same conditions of temperature, buffer density, and viscosity. Experimentally, this often requires using prechilled rotors and cells assembled ahead of time, and up to 2 h of temperature equilibration in the ultracentrifuge prior to analysis. We typically perform our experiments at 11 °C, which is a compromise between cooler temperatures to minimize sample degradation and the practical operation of the instrument under defined conditions. It may also be necessary to standardize the density and viscosity of the buffer/lysate using densitometer and viscometer measurements—sucrose densities and viscosities at high concentrations can be challenging to determine using standard sources, such as Sednterp. Reference 15 provides a handy source of viscosity versus sucrose concentration and temperatures. There are also multichannel cells available that would allow multiple sample conditions to be performed in one run.

4. Lysates of certain cell types may contain components that exhibit intrinsic fluorescence. Control experiments should, therefore, be conducted to determine whether fluorescence and/or sedimentation of lysate components are observed using the FDS. Lysates of the cells used in our experiments (mouse Neuro2A cells and HEK293 cells) overexpressing a non-fluorescently tagged control protein yielded negligible fluorescence and no observable moving boundaries in control experiments [7].
5. When preparing sucrose solutions, we have observed that the larger the beaker, the faster the sucrose will recrystallize, resulting in a nonhomogeneous solution. A convenient way to check for this is by dipping a pipette tip in the solution and letting a drop of the solution fall on a paper tissue. Then smooth out the drop using a gloved finger and feel for any inconsistencies or granular material.
6. Initiation of the FDS instrument entails firing of the laser and subsequent laser-power lock, after which the angular location of the calibration strip is acquired. Once this is achieved, emission gain settings and experiment conditions can be adjusted by the operator. These processes require a rotor speed of 3,000 rpm and can take from ~10 to 30 min to complete. Large aggregates, such as inclusions and amyloid fibrils, may sediment significantly at 3,000 rpm; thus, minimizing this period is particularly important. While laser-power lock and angular position settings are automated by the AOS software, time can be saved in setting of the fluorescence gains using prior knowledge of the approximate fluorescence yields of samples and which samples are matched. In addition, previously saved experimental method conditions may be used. It is also worth noting that firing of the laser requires a centrifuge chamber vacuum of below 150 μm. Thus, it is recommended that the vacuum be allowed to equilibrate to below this value with the rotor stationary.
7. Sedimentation profiles obtained by fluorescence detection do not yield a visible sample meniscus. This is in contrast to data obtained with absorbance detection optics, which yield a spike at the sample meniscus position due to light deflection at the air–water interface. An accurate determination of the sample meniscus position is important for calculating the rate of movement of experimental sedimentation boundaries. To define the meniscus position, a small amount of light mineral oil (~5 μL) containing 0.1 % (w/v) 4,4-difluoro-1,3,5,7,8-pentamethyl-4-bora-3a,4a-diaza-*s*-indacene (BODIPY493/503, Invitrogen) may be layered on top of the sample [16]. One potential problem, however, is that BODIPY may bind the protein of interest and co-sediment with the protein. Alternatively, the meniscus position can be empirically fitted in the SEDFIT

analysis software without the use of extraneous dye. We have found that experimental fitting of the meniscus in the software yields good fits to experimental data in most cases.

8. The baseline of all FDS fluorescence scans is offset from zero due to the presence of a small "dark count" signal. The magnitude of this offset scales with gain settings and can be manually estimated for each experiment from the fluorescence signal obtained within the air-gap above the solution column of each sector.

9. Due to the sensitivity of the FDS and the very low concentrations of fluorescent sample needed, one potential problem is the detection of fluorescent material adhering to the cell windows, thereby contributing to noise in the sedimentation data. This only occurs when using purified protein (i.e., non-cell lysates). This problem can be resolved by including carrier proteins such as bovine serum albumin or hen egg white lysozyme in the sample. The effect of carrier proteins on the labeled protein of interest can be checked by performing sedimentation experiments using varying concentrations of carrier protein. Conversely, using high concentrations of fluorescent sample may result in an inner filtering effect where signal intensity no longer scales linearly with sample concentration. The extent of inner filtering in a sample can be established by a titration curve of the sample over the range of concentrations of interest.

10. An upward-sloping plateau is sometimes observed in the sedimentation profile. This has been attributed to small errors in the tracking of the laser beam such that the focal point of the beam shifts during the scan or is too close the sector walls, resulting in systematic intensity variation [6]. The effects of this on accurate sedimentation analysis can be mitigated by correcting for time-independent (TI) noise in the data fitting software (SEDFIT). Similarly, drifts in the baseline over time (radial-independent (RI) noise) can also be corrected in SEDFIT.

11. One caveat for sedimentation analysis of biologically complex solutions—especially very concentrated solutions typical of biological fluids—is the inherent large effects on the sedimentation of the protein of interest from non-ideality or protein–protein interactions. Nevertheless, with careful controls much information can still be gleaned from experiments performed in biological fluids, and our previous studies have shown the system to be robust for examination of proteins expressed in mammalian cells when compared to the purified recombinant counterpart [7]. A useful control is to perform an experiment on the purified recombinant protein and compare the results to those in cell lysate. Nevertheless, deviations in *s* values from non-ideality should not change broader conclusions about

heterogeneity between vastly different sized aggregate species commonly seen in amyloid-like aggregation kinetics [7]. A further note is that at high centrifugal speeds, sucrose will co-sediment with the protein of interest and form gradients in the sample container, potentially resulting in nonideal sedimentation behavior. We have only employed sucrose when using centrifugal speeds of 3,000 rpm.

References

1. Howlett GJ, Minton AP, Rivas G (2006) Analytical ultracentrifugation for the study of protein association and assembly. Curr Opin Chem Biol 10(5):430–436. doi:S1367-5931(06)00124-4 [pii] 10.1016/j.cbpa.2006.08.017
2. Balbo A, Brown PH, Braswell EH, Schuck P (2007) Measuring protein-protein interactions by equilibrium sedimentation. Curr Protoc Immunol Chapter 18:Unit 18.18. doi:10.1002/0471142735.im1808s79
3. Brown PH, Balbo A, Schuck P (2008) Characterizing protein-protein interactions by sedimentation velocity analytical ultracentrifugation. Curr Protoc Immunol Chapter 18:Unit 18.15. doi:10.1002/0471142735.im1815s81
4. Mok YF, Howlett GJ (2006) Sedimentation velocity analysis of amyloid oligomers and fibrils. Methods Enzymol 413:199–217. doi:S0076-6879(06)13011-6 [pii] 10.1016/S0076-6879(06)13011-6
5. MacGregor IK, Anderson AL, Laue TM (2004) Fluorescence detection for the XLI analytical ultracentrifuge. Biophys Chem 108(1–3):165–185. doi:10.1016/j.bpc.2003 [pii] 10.018 S0301462203003041
6. Kroe RR, Laue TM (2009) NUTS and BOLTS: applications of fluorescence-detected sedimentation. Anal Biochem 390(1):1–13. doi:S0003-2697(08)00797-5 [pii] 10.1016/j.ab.2008.11.033
7. Olshina MA, Angley LM, Ramdzan YM, Tang J, Bailey MF, Hill AF, Hatters DM (2010) Tracking mutant huntingtin aggregation kinetics in cells reveals three major populations that include an invariant oligomer pool. J Biol Chem 285(28):21807–21816. doi:10.1074/jbc.M109.084434
8. Schuck P, Perugini MA, Gonzales NR, Howlett GJ, Schubert D (2002) Size-distribution analysis of proteins by analytical ultracentrifugation: strategies and application to model systems. Biophys J 82(2):1096–1111
9. Pham Cle L, Mok YF, Howlett GJ (2011) Sedimentation velocity analysis of amyloid fibrils. Methods Mol Biol 752:179–196. doi:10.1007/978-1-60327-223-0_12
10. Schuck P (2000) Size-distribution analysis of macromolecules by sedimentation velocity ultracentrifugation and lamm equation modeling. Biophys J 78(3):1606–1619. doi:10.1016/S0006-3495(00)76713-0
11. Schuck P (2004) A model for sedimentation in inhomogeneous media. I. Dynamic density gradients from sedimenting co-solutes. Biophys Chem 108(1–3):187–200. doi:10.1016/j.bpc.2003.10.016
12. Rizzo MA, Springer GH, Granada B, Piston DW (2004) An improved cyan fluorescent protein variant useful for FRET. Nat Biotechnol 22(4):445–449. doi:10.1038/nbt945 [pii] nbt945
13. Tsien RY (1998) The green fluorescent protein. Annu Rev Biochem 67:509–544. doi:10.1146/annurev.biochem.67.1.509
14. Lide DR (ed) (1996) CRC handbook of chemistry and physics, 77th edn. CRC, Boca Raton, FL
15. Longinotti MP, Horacio RC (2008) Viscosity of concentrated sucrose and trehalose aqueous solutions including the supercooled regime. J Phys Chem Ref Data 37(3):1503–1515
16. Bailey MF, Angley LM, Perugini MA (2009) Methods for sample labeling and meniscus determination in the fluorescence-detected analytical ultracentrifuge. Anal Biochem 390(2): 218–220. doi:S0003-2697(09)00221-8 [pii] 10.1016/j.ab.2009.03.045

Chapter 5

A Method for the Incremental Expansion of Polyglutamine Repeats in Recombinant Proteins

Amy L. Robertson and Stephen P. Bottomley

Abstract

The polyglutamine diseases are caused by the expansion of CAG repeats. A key step in understanding the disease mechanisms, at the DNA and protein level, is the ability to produce recombinant proteins with specific length glutamine tracts which is a time-consuming first step in setting up in vitro systems to study the effects of polyglutamine expansion. Described here is a PCR-based method for the amplification of CAG repeats, which we used to incrementally extend CAG length by 3–5 repeats per cycle. This method could be translated into various contexts where amplification of repeating elements is necessary.

Key words Polyglutamine, Triplet repeat expansion, CAG repeat, Cloning, Poly-amino acid

1 Introduction

Nine hereditary diseases have been found to correlate with the expansion of a CAG trinucleotide repeat, encoding glutamine, at a specific site within the open reading frame of a disease gene (reviewed in refs. 1, 2). These disorders are collectively termed the polyglutamine diseases, and include Huntington's disease and a number of spinocerebellar ataxias [1, 2]. Individuals with a CAG repeat number above 35–40 display a neurodegenerative phenotype, with deposition of protein aggregates as neuronal inclusions [3]. Investigating various aspects of CAG and glutamine repeats at the DNA and protein levels, respectively, is important in clarifying the molecular mechanisms leading to pathogenesis in these diseases.

Technical challenges in the cloning and expression of proteins with variable length polyglutamine repeats have rate-limited research progress [4]. Further, the availability of, and potential to produce, recombinant proteins of specific repeat lengths has limited the breadth of experimental studies. Most research groups studying the disease-causing proteins have amplified constructs from disease genes. More recently a number of studies have utilized

Danny M. Hatters and Anthony J. Hannan (eds.), *Tandem Repeats in Genes, Proteins, and Disease: Methods and Protocols*, Methods in Molecular Biology, vol. 1017, DOI 10.1007/978-1-62703-438-8_5,

polyglutamine-containing model proteins to analyze disease-associated pathways [5–8]. In many of these model systems, the polyglutamine segment was also PCR amplified from a disease gene along with adjacent non-polyglutamine-coding regions [9, 10], necessary because of the non-specificity of primer binding to the repeating CAG motif. It is important to note however that recent experimental evidence indicates that amino acid residues surrounding a polyglutamine repeat tract can modulate its behavior [11–13]. As such, the presence of extraneous amino acids has affected the interpretability of model protein studies.

Cassette insertion techniques have been used for the insertion of synthetically prepared polyglutamine segments [14–16]. When using these methods there are issues with annealing specificity, which decreases as the number of repeats in a homogeneous CAG tract becomes larger. An additional difficulty is that commercial companies, which specialize in gene construction, struggle to synthetically produce CAG tracts longer than 40–50 repeats. Homogeneous CAG repeats are very unstable, and commonly contract in repeat number [17, 18]. This phenomenon is particularly prevalent in bacterial cells, and a number of groups have taken advantage of this for the uncontrolled amplification of homogeneous repeats [17]. Control of repeat number is forfeited, and large expansions in repeat lengths have been observed [18]. In vivo amplification methods are therefore not ideal if small, incremental, or specific changes in repeat number are required.

A dual-fragment amplification and ligation method for the extension of polyglutamine repeats in huntingtin exon 1 has been previously described by Peters and Ross [19]. This strategy involves the concurrent amplification of two overlapping fragments, both containing the complete polyglutamine-encoding region. The two fragments are digested with a Type 11S restriction endonuclease, enzymes for which the cleavage position is located outside of the recognition sequence (reviewed in ref. 20), reducing PCR-based problems associated with introducing restriction sites juxtaposed to a repeating DNA segment. The coordinated positioning of the unique Type 11S sites leaves fragments with "sticky ends" located at either the 5′ or 3′ end of the CAG repeat. Subsequent fragment and vector ligation effectively doubles the trinucleotide repeat number. The efficiency of this method depends upon repeat location, vector design, and fragment size.

We extended on the approach described by Peters and Ross [19] to construct a polyglutamine-containing model protein system, comprising the *Staphylococcus aureus* protein A B domain (SpA), located N-terminal to various length polyglutamine repeats (Qn) (Fig. 1) [21, 22]. The repeat amplification method described here was efficient in constructing a polyglutamine model protein system. This method has benefits over other approaches for cloning polyglutamine repeats. Compared with insertion techniques,

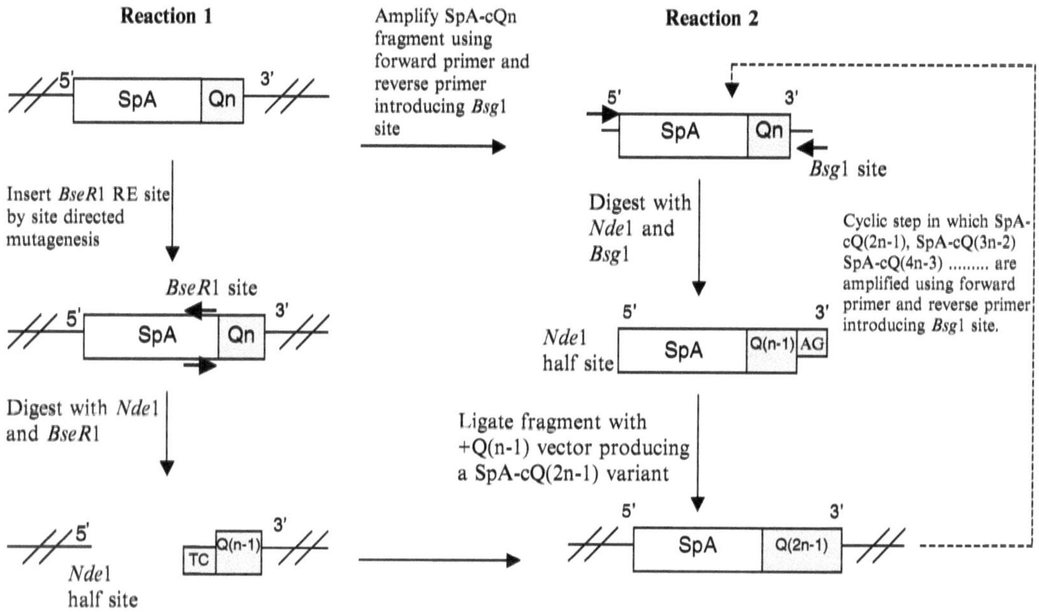

Fig. 1 There are two reactions involved in the amplification of consecutive CAG repeats. Reaction 1 involves the insertion of a Type IIS restriction site (*BseR*1) at a specific location 7 bp upstream from the beginning of the CAG_n (Qn) repeat. The vector is then digested with *BseR*1 and *Nde*1, for which the cleavage site is located 5′ to the beginning of the SpA coding sequence. Digestion leaves a linearized vector with Q(n − 1) repeats, and an antisense TC overhang. Reaction 2 involves the amplification of the SpA-cQn coding region. The reverse amplification primer introduces another Type IIS recognition site (*Bsg*1) 15 bp downstream from the Qn repeat. Subsequent digestion of this fragment with *Bsg*1 and *Nde*1 leaves a fragment comprising SpA-cQ(n − 1) repeats, with an AG overhang 3′ on the sense strand. The products of Reactions 1 and 2 are ligated, yielding a construct comprising SpA-cQ(2n − 1). This product can then be fed back into the beginning of Reaction 2. Subsequent products are amplified in a cyclic process, yielding constructs SpA-cQ(2n − 1), SpA-cQ(3n − 2), SpA-cQ(4n − 3), SpA-cQ(5n − 4), etc

this approach does not introduce extraneous amino acids in the regions flanking the repeat region. The method described here improves upon that previously described in reference 19, by reducing the dual-fragment amplification and ligation strategy to a single-fragment method. This method offers an efficient alternative for the controlled amplification of polyglutamine repeats and has potential for use in variable cloning applications, particularly in cases where repeating DNA motifs hinder the introduction of restriction sites directly adjacent to or within a repeat.

2 Materials

1. Template plasmid DNA.
2. Complementary primers introducing *B*seR1 seven base pairs upstream of the first CAG codon in the template DNA.
3. 100 mM dNTP stock.

4. *Pfu* DNA polymerase (Promega).
5. 10× *Pfu* DNA polymerase buffer (Promega).
6. Thermal Cycler (PCR machine).
7. EzyVision stain (Amresco).
8. Agarose to prepare 1 and 2 % agarose gels. Dissolve 1 or 2 % (w/v) agarose in TAE buffer (40 mM Tris–acetate, 1 mM EDTA, pH 8.5).
9. UV detector.
10. *Dpn*1 enzyme.
11. Novablue GigaSingles competent cells (Novagen).
12. Ampicillin.
13. LB/agar/ampicillin plates.
14. DNA extraction kit.
15. *BseR*1 restriction enzyme (New England Biolabs).
16. *Nde*1 restriction enzyme (Promega).
17. *BseR*1/*Nde*1 digest buffer: 10 mM Tris–HCl, 10 mM $MgCl_2$, 50 mM NaCl, 1 mM DTT, pH 7.9, supplemented with 100 μg/ml bovine serum albumin.
18. DNA-gel purification kit.
19. *Bsg*1 restriction enzyme (New England Biolabs).
20. Primers for fragment amplification. The reverse amplification primer must include a *B*sg1 recognition sequence 14 base pairs downstream from the final CAG repeat.
21. *Bsg*1/*Nde*1 digest buffer: 20 mM Tris–acetate, 10 mM magnesium acetate, 50 mm potassium acetate, 1 mM DTT, pH 7.9, supplemented with 100 μg/ml bovine serum albumin and S-adenosylmethionine.
22. T4 DNA ligase (Promega).

3 Methods

3.1 Overview of Polyglutamine Repeat Amplification Methodology

In an example of the amplification process, the original template for Reactions 1 and 2 was a synthetically prepared construct with 5 CAG repeats (SpA-cQ5). Upon introduction of the *BseR*1 site and restriction enzyme digestion the product of Reaction 1 was a linearized vector containing 4 CAG repeats and an antisense TC overhang. This vector was used as a reactant for a number of ligation reactions with products from Reaction 2, yielding the SpA-cQ9 and SpA-cQ13 constructs. Concurrently, to speed up the amplification process, a *BseR*1 site was introduced into the SpA-cQ13 construct, undergoing Reaction 1, yielding a linearized Q12 vector. This was then combined with the digested

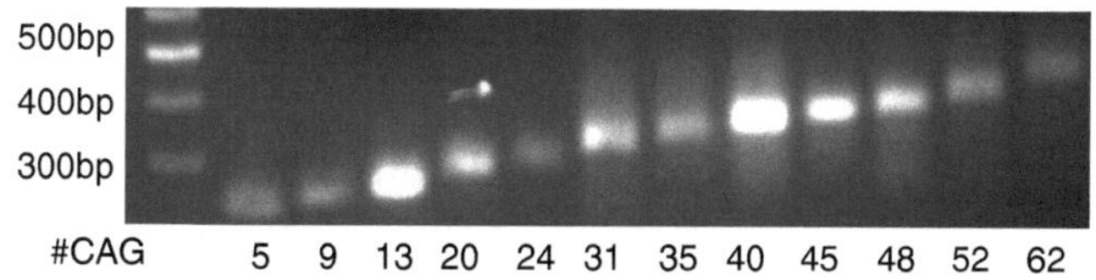

Fig. 2 2 % agarose gel of PCR-amplified SpA-cQn fragments

SpA-cQ9 and SpA-cQ13 fragments to produce SpA-cQ20 and SpA-cQ24. A cyclic process was then used to generate genes containing repeat lengths of up to 62 glutamines (Fig. 2).

We found that upon ligation, some repeat lengths deviated from the anticipated size. This is probably a result of the intrinsically unstable behavior of CAG repeats [23]. The genetic instability implicated in the polyglutamine diseases is reflected in vitro, as the repeating nature of the DNA lends to the formation of irregular conformations, with base pair misalignment causing strand slippage and the formation of hairpin structures, resulting in fluctuations in repeat length [23, 24] (*see* **Note 1**). The recombinant proteins comprising polyglutamine repeat lengths of between 5 and 52 glutamines were successfully expressed in a soluble form and readily purified [21]. SpA-cQ62 was not expressed in *E. coli* cells, probably due to the rapid aggregation propensity and toxicity of this protein.

3.2 Amplification of Homogeneous CAG Repeats

The main focus of this chapter is to demonstrate the controlled incremental amplification of homogeneous CAG repeats. There are several Type 11S enzymes that could be utilized, including *Bsa*1, *BsmB*1, *BseR*1, and *Bsg*1. The specific enzymes selected depend on the plasmid and fragment DNA sequences, as it is required that these sites are novel upon their insertion. The remainder of this section describes the methodology using a system involving the *BseR*1 and *Bsg*1 restriction enzymes. We describe the construction of a polyglutamine-containing model protein system, in which different length polyQ repeats are located at the C-terminus of a soluble fusion partner, the *Staphylococcus aureus* protein A B domain (SpA). Using this method, we have relative control of the repeat length, and have incrementally expanded the repeat number by 3–5 glutamines per cycle. The amplification process can be described in two reactions (Fig. 1).

3.2.1 Cyclic Amplification Reaction 1

This method involves an initial once-off site-directed mutagenesis step for the introduction of a *BseR*1 recognition sequence 7 bp upstream of the first CAG codon (Reaction 1, Fig. 1). Subsequent digestion with *BseR*1 and *Nde*1, for which the unique restriction site is located upstream of the SpA coding sequence, cleaves out the SpA gene and leaves a Q(n – 1) repeat, and a TC overhang on the antisense strand (Fig. 1). The end product of Reaction 1 is thus a linearized vector comprising Q(n − 1) repeats.

Site-Directed Mutagenesis for *BseR*1 Insertion

1. Set up PCR in a 25 μl reaction mixture containing 50 ng of template SpA-cQ5 pET21a, 500 nM of each primer, 200 μM of each dNTP, and 1.45 U of Pfu DNA polymerase in 1× Pfu DNA polymerase buffer. Thermal cycling was performed as follows: initial denaturation at 95 °C for 1 min; 18 cycles at 95 °C for 1 min; annealing at 5 °C below the melting temperature of the primers for 1 min and 72 °C for 12 min; and one final extension cycle at 72 °C for 12 min.
2. To check for the presence of PCR product, 5 μl from each reaction is mixed with 1 μl of EzyVision stain and loaded onto a 1 % (w/v) agarose gel. Run samples on the agarose gel at a voltage of 100 V for 25 min (*see* **Note 2**).
3. Methylated template strand is digested with 10 U of Dpn1 and incubated at 37 °C for 2 h.
4. 5 μl of product can then be transformed into 50 μl NovaBlue GigaSingles competent cells. Transformations are performed according to the manufacturer's instructions. Transformants are then plated onto selective agar plates, for example ampicillin; however the antibiotic used depends on the resistance markers of the expression vector used. Colonies are then grown overnight at 37 °C.
5. Select five to ten colonies to screen. Pick the colonies from the plate with a wire loop and inoculate selective medium. Grow overnight at 37 °C.
6. Purify the plasmid DNA using a commercial DNA purification kit.
7. Screen colonies initially by performing a small-scale digest with *BseR*1, which will be a novel site in positive colonies. To set up the small-scale digest, incubate 5 μl of purified DNA with 1 μl of *BseR*1, in 10 mM Tris–HCl (pH 7.9), 10 mM $MgCl_2$, 50 mM NaCl, 1 mM DTT, and supplemented with 100 μg/ml bovine serum albumin for 1 h at 37 °C. Run 5 μl of digest product on a 1 % agarose gel (*see* **Note 3**).
8. Clones that are positive for the *BseR*1 site can be further screened by DNA sequencing to verify the CAG repeat number, as long repeat segments are prone to fluctuate during PCR.

*BseR*1/*Nde*I Restriction Enzyme Digest

1. The positive clone with *BseR*1 selected in the previous step can be digested with *BseR*1 and *Nde*I (or an alternative site that digests near the 5′ end of the coding sequence of interest). The digest is performed in 10 mM Tris–HCl, 10 mM $MgCl_2$,

50 mM NaCl, and 1 mM dithiothreitol, pH 7.9, supplemented with 100 μg/ml bovine serum albumin, for 2 h at 37 °C.

2. Digested plasmid is purified using a commercially available gel purification kit (according to the manufacturer's instructions), and DNA quantified using $\varepsilon_{260} = 50$ ng/μl. In our example, this digestion cuts 5′ to SpA, leaving an *Nde*1 half-site, and leaves only 4 CAG repeats, plus a TC overhang on the antisense strand, corresponding to the *BseR*1 cleavage site (+Q(n – 1) vector).

3.2.2 Cyclic Amplification Reaction 2

In Reaction 2 (Fig. 1) the entire SpA-cQn coding region is PCR-amplified using a 5′ sequence-specific primer, and a template-specific primer introducing a *Bsg*1 recognition sequence 14 bp downstream from the final CAG repeat. Cleavage with *Bsg*1 leaves Q(n – 1) repeats, and an AG overhang at the 3′ end of the sense strand. The digested fragment can then be subcloned into the Q(n – 1) linearized vector from Reaction 1, creating a construct comprising Q(2n – 1) repeats. This product is then used as input for Reaction 2, with subsequent rounds proceeding in a cyclic process, whereby the length of the polyglutamine repeat is extended by Q(n – 1) repeats in each cycle (*see* **Note 4**).

PCR Amplification of Fragment

1. Fragments comprising the polyQ repeat are PCR-amplified from plasmid DNA. Of key importance in this step is the introduction of a *Bsg*1 recognition sequence 14 bp downstream from the final CAG repeat in the reverse amplification primer. Perform fragment amplification in a 25 μl reaction volume, containing 100 ng of template SpA-cQn pET21a, 500 nM of each primer, 200 μM of each dNTP, and 1.45 U of *Pfu* DNA polymerase (Promega) in *Pfu* DNA polymerase buffer (Promega). The thermal cycler was programmed as follows: initial denaturation at 72 °C for 1 min; 35 cycles at 95 °C for 1 min; annealing at 5 °C below the primer melting temperature for 1 min and 72 °C for 1 min, 30 s; and final extension at 72 °C for 1 min, 30 s.
2. Confirm the presence of PCR product by running a small aliquot on a 2 % agarose gel.
3. PCR-amplified fragments are then digested with *Nde*1 and *Bsg*1, in 20 mM Tris–acetate, 10 mM magnesium acetate, 50 mM potassium acetate, and 1 mM dithiothreitol, pH 7.9, supplemented with 100 μg/ml bovine serum albumin and S-adenosylmethionine, for 2 h at 37 °C. Digested fragments can then be purified using a gel purification kit (according to the manufacturer's instructions), and DNA quantitated using $\varepsilon_{260} = 50$ ng/μl. This digestion leaves a fragment with a 5′ *Nde*1 half-site and an AG overhang at the 3′ end corresponding to the *Bsg*1 cleavage site.

3.2.3 Ligations and Colony Screening

Digested SpA-cQn fragments are ligated with the + Q(n − 1) vector using T4 DNA ligase.

1. Set up 10 μl ligation reactions containing 100 ng digested vector, 500 ng fragment, and 1 μl of T4 DNA ligase in ligase buffer (30 mM Tris–HCl (pH 7.8), 10 mM $MgCl_2$, 10 mM DTT, 1 mM ATP) (*see* **Note 5**).
2. Incubate at room temperature for 2 h.
3. Directly transform ligation products into competent JM107 cells, plate onto selective agar plates, and incubate at 37 °C overnight.
4. An initial screen in this case is a *BseR*1 digestion, as positive clones that have successfully ligated with fragment DNA are *BseR*1 negative. The full sequence of positive clones can then be verified by DNA sequencing.

3.3 Construction of Imperfect Repeats with Interrupting Codons/Residues

This incremental amplification method can also be utilized for the introduction of alternative glutamine- and non-glutamine-encoding codons at coordinated positions within the CAG repeat. A technical challenge in inserting alternative nucleotides and/or codons within a repeating CAG tract is the non-specificity of the primer for the repetitive sequence. These problems can be overcome utilizing a variation of this repeat amplification method in which CAA codons can be inserted into various positions of the repeat tract. Heterogeneous CAG/CAA repeat tracts are more stable compared to homogeneous CAG repeats [25]. We have utilized this method to insert cysteine residues at various positions within the polyglutamine repeat. Also, prolines have been inserted at various positions within the polyglutamine stretch employing a method involving Type 11S restriction endonucleases [26].

3.3.1 Insertion of Cysteine Residues Directly 5′ to CAG Repeat and Digestion

1. To insert a TGC (encoding cysteine) codon directly 5′ to the first CAG repeat, design complementary primers. The primers should be complementary to the approximately 17 base pairs preceding the first CAG, followed by the TGC insertion, and the beginning 12 base pairs of the CAG repeat.
2. Perform the site-directed mutagenesis reaction and sequencing following the basic instructions in Subheading 3.2.1.
3. After screening and identifying positive clones, perform a reaction to introduce a strategically positioned *BseR*1 restriction site such that *BseR*1 cleaves within the cysteine-encoding codon, to leave a CG overhang on the antisense strand.
4. As in Subheading 3.2.1, perform the *BseR*1/*Nde*l digest.

3.3.2 Introduction of Cysteine Residue Directly 3′ to CAG Repeat and Fragment Preparation

1. Similarly to the above reaction to introduce a cysteine 5′ to the CAG repeat, a complementary primer pair to introduce a TGC codon directly 3′ to the last CAG repeat must be designed. The primers should again be complementary to approximately

12 base pairs of the CAG repeat, followed by the TGC codon and approximately 17 base pairs after this.

2. The remainder of the steps in fragment preparation and subsequent ligation with the *BseR*1/*Nde*1-digested vector are described in Subheadings 3.1.2 and 3.1.3.

3.4 Frameshifting of Homogeneous Repeats to Prepare Other Homopolymeric Amino Acids

When a single nucleotide is deleted or inserted in the region preceding a homogeneous CAG repeat, a frameshift occurs such that the repeat tract can encode a polyalanine or polyserine repeat. As long as the frameshift is corrected for after the repeat tract such that the downstream sequence and termination codon are in frame, recombinant proteins with polyalanine or polyserine proteins are translated. We have utilized this method to rapidly generate a library of polyalanine-encoding constructs from our existing polyglutamine-encoding library. Maintenance of trinucleotide homogeneity therefore allows for the possibility of frameshift modulation of the repeat tract to form polyalanine or polyserine segments.

4 Notes

1. The instability noted highlights the need for careful monitoring of the tract length, in our case the small repeat length fluctuations were manageable, with DNA sequencing validating the CAG repeat number. Therefore, verify the repeat number after each round of polyglutamine addition.
2. If at this stage you do not see any PCR product, optimization and/or troubleshooting of the PCR reaction is necessary. This may involve adjustment of the annealing temperature (it is often useful to run a temperature gradient if you have a PCR cycler equipped with this function), the primer, and/or dNTP concentration.
3. To check for *BseR*1-digested products in screen, always run uncleaved plasmid DNA as a control.
4. An alternative approach for generating a fragment for insertion utilizes an oligonucleotide-derived cassette [27]. This approach may be preferable for some whose system is amenable to this, for example in cases where the insertion fragment incorporating the CAG repeat is small.
5. For ligations we find that a 1:5 ratio of vector:fragment is optimal. In initial polyglutamine elongation cycles it may be of benefit to test a range of vector:fragment ratios to determine what is optimal for your system.

Acknowledgement

The authors would like to thank Dr. Lisa Cabrita for advice regarding the methodology and manuscript.

References

1. Margolis RL, Ross CA (2001) Expansion explosion: new clues to the pathogenesis of repeat expansion neurodegenerative diseases. Trends Mol Med 7(11):479–482
2. Zoghbi H, Orr H (2000) Glutamine repeats and neurodegeneration. Annu Rev Neurosci 23:217–247
3. Gusella JF, MacDonald ME (2000) Molecular genetics: unmasking polyglutmaine triggers in neurodegenerative disease. Nat Rev Neurosci 1:109–115
4. Chow MK, Ellisdon AM, Cabrita LD, Bottomley SP (2006) Purification of polyglutamine proteins. Methods Enzymol 413:1–19
5. Tanaka M, Morishima I, Akagi T, Hashikawa T, Nukina N (2001) Intra- and intermolecular beta-pleated sheet formation in glutamine-repeat inserted myoglobin as a model for polyglutamine diseases. J Biol Chem 276(48): 45470–45475
6. Ladurner AG, Fersht AR (1997) Glutamine, alanine or glycine repeats inserted into the loop of a protein have minimal effects on stability and folding rates. J Mol Biol 273:330–337
7. Ignatova Z, Gierasch LM (2006) Extended polyglutamine tracts cause aggregation and structural perturbation of an adjacent beta barrel protein. J Biol Chem 281(18):12959–12967
8. Nagai Y, Inui T, Popiel HA, Fujikake N, Hasegawa K, Urade Y, Goto Y, Naiki H, Toda T (2007) A toxic monomeric conformer of the polyglutamine protein. Nat Struct Mol Biol 14:332–340
9. Masino L, Kelly G, Leonard K, Trottier Y, Pastore A (2002) Solution structure of polyglutamine tracts in GST-polyglutamine fusion proteins. FEBS Lett 513:267–272
10. Nagai Y, Tucker T, Ren H, Kenan DJ, Henderson BS, Keene JD, Strittmatter WJ, Burke JR (2000) Inhibition of polyglutamine protein aggregation and cell death by novel peptides identified by phage display screening. J Biol Chem 275(14):10437–10442
11. Nozaki K, Onodera O, Takano H, Tsuji S (2001) Amino acid sequences flanking polyglutamine stretches influence their potential for aggregate formation. Neuroreport 12(15): 3357–3364
12. Bhattacharyya A, Thakur AK, Chellgren VM, Thiagarajan G, Williams AD, Chellgren BW, Creamer TP, Wetzel R (2006) Oligoproline effects on polyglutamine conformation and aggregation. J Mol Biol 355(3):524–535
13. Burke MG, Woscholski R, Yaliraki SN (2003) Differential hydrophobicity drives self-assembly in Huntington's disease. Proc Natl Acad Sci USA 100(24):13928–13933
14. Chen YW (2003) Site-specific mutagenesis in a homogeneous polyglutamine tract: application to spinocerebellar ataxin-3. Protein Eng 16(1): 1–4
15. Chow MKM, Ellisdon AM, Cabrita LD, Bottomley SP (2004) Polyglutamine expansion in Ataxin-3 does not affect protein stability. J Biol Chem 279(46):47643–47651
16. Tobelmann MD, Kerby RL, Murphy RM (2008) A strategy for generating polyglutamine "length libraries" in model host proteins. Protein Eng Des Sel 21(3):161–164. doi:gzm078[pii]10.1093/protein/gzm078
17. Ohshima K, Kang S, Wells RD (1996) CTG triplet repeats from human hereditary diseases are dominant genetic expansion products in *Escherichia coli*. J Biol Chem 271(4): 1853–1856
18. Sarkar PS, Chang HC, Boudi FB, Reddy S (1998) CTG repeats show bimodal amplification in *E. coli*. Cell 95(4):531–540
19. Peters MF, Ross CA (1999) Preparation of human cDNas encoding expanded polyglutamine repeats. Neurosci Lett 275(2): 129–132
20. Roberts RJ, Belfort M, Bestor T, Bhagwat AS, Bickle TA, Bitinaite J, Blumenthal RM, Degtyarev S, Dryden DT, Dybvig K, Firman K, Gromova ES, Gumport RI, Halford SE, Hattman S, Heitman J, Hornby DP, Janulaitis A, Jeltsch A, Josephsen J, Kiss A, Klaenhammer TR, Kobayashi I, Kong H, Kruger DH, Lacks S, Marinus MG, Miyahara M, Morgan RD, Murray NE, Nagaraja V, Piekarowicz A, Pingoud A, Raleigh E, Rao DN, Reich N, Repin VE, Selker EU, Shaw PC, Stein DC, Stoddard BL, Szybalski W, Trautner TA, Van Etten JL, Vitor JM, Wilson GG, Xu SY (2003) A nomenclature for restriction enzymes, DNA methyltransferases, homing endonucleases

and their genes. Nucleic Acids Res 31(7): 1805–1812

21. Robertson AL, Horne J, Ellisdon AM, Thomas B, Scanlon MJ, Bottomley SP (2008) The structural impact of a polyglutamine tract is location-dependent. Biophys J 95(12):5922–5930. doi:S0006-3495(08)82007-3[pii] 10.1529/biophysj.108.138487
22. Robertson AL, Bate MA, Buckle AM, Bottomley SP (2011) The rate of polyQ-mediated aggregation is dramatically affected by the number and location of surrounding domains. J Mol Biol 413(4):879–887. doi:S0022-2836(11)01020-5 [pii]10.1016/j.jmb.2011.09.014
23. Bacolla A, Wells RD (2004) Non-B DNA conformations, genomic rearrangements, and human disease. J Biol Chem 279(46): 47411–47414
24. Pearson CE, Tam M, Wang YH, Montgomery SE, Dar AC, Cleary JD, Nichol K (2002) Slipped-strand DNAs formed by long (CAG)*(CTG) repeats: slipped-out repeats and slip-out junctions. Nucleic Acids Res 30(20):4534–4547
25. Dorsman JC, Bremmer-Bout M, Pepers B, van Ommen GJ, Den Dunnen JT (2002) Interruption of perfect CAG repeats by CAA triplets improves the stability of glutamine-encoding repeat sequences. Biotechniques 33(5):976–978
26. Popiel HA, Nagai Y, Onodera O, Inui T, Fujikake N, Urade Y, Strittmatter WJ, Burke JR, Ichikawa A, Toda T (2004) Disruption of the toxic conformation of the expanded polyglutamine stretch leads to suppression of aggregate formation and cytotoxicity. Biochem Biophys Res Commun 317(4):1200–1206. doi:10.1016/j.bbrc.2004.03.161[pii] S0006291X0400662X
27. Scior A, Preissler S, Koch M, Deuerling E (2011) Directed PCR-free engineering of highly repetitive DNA sequences. BMC Biotechnol 11:87. doi:1472-6750-11-87[pii] 10.1186/1472-6750-11-87

Chapter 6

Pulse Shape Analysis (PulSA) to Track Protein Translocalization in Cells by Flow Cytometry: Applications for Polyglutamine Aggregation

Yasmin M. Ramdzan, Rebecca Wood, and Danny M. Hatters

Abstract

Pulse shape analysis (PulSA) is a flow cytometry-based method that can be used to study protein localization patterns in cells. Examples for its use include tracking the formation of inclusion bodies of polyglutamine-expanded proteins and other aggregating proteins. The method can also be used for phenomena relating to protein movements in cells such as translocation from the cytoplasm to the nucleus, trafficking from the plasma membrane to the Golgi, and stress granule formation. An attractive feature is its capacity to quantify these parameters in whole-cell populations very quickly and in high throughput. We describe the basic experimental details for performing PulSA using expression of GFP-tagged proteins, endogenous proteins labelled immunofluorescently, and organelle dyes.

Key words Flow cytometry, Fluorescence-activated cell sorting (FACS), FlAsH, ReAsH, Immunostaining

1 Introduction

Proteins move between different subcellular locations as part of normal cellular activities or as a consequence of abnormal circumstances such as aggregation [1]. Changes in protein localization are traditionally studied using microscopy-based imaging, which, while information-rich, has limited throughput due to the complexity of data collection and analysis. Recent developments in high content imaging have partly addressed the throughput limitation, but specialized instrumentation and relatively complex methods of analysis are still required [2].

The flow cytometry-based pulse shape analysis (PulSA) method provides a simple alternative to imaging for analysis of protein localization changes. PulSA additionally offers new capabilities including the recovery of cell populations with distinct protein localizations or aggregation states [3]. The PulSA method is compatible with standard flow cytometry instrumentation making

Danny M. Hatters and Anthony J. Hannan (eds.), *Tandem Repeats in Genes, Proteins, and Disease: Methods and Protocols*, Methods in Molecular Biology, vol. 1017, DOI 10.1007/978-1-62703-438-8_6,

the technique amenable in a standard laboratory setting. PulSA relies on standard pulse height and width parameters of a fluorescence channel, which are combined to provide the area parameter that defines particle brightness [4, 5]. The width measures the duration of a fluorescent signal as a particle traverses past the laser beam whereas the height is the peak intensity across that width. Hence a particle contained within a subcellular compartment will have different width and height parameters to, for example, a whole cell. Fluorescence pulse and height widths can be collected simultaneously with different fluorescent channels, enabling multidimensional analysis or internal calibrations for cell or organelle size. We have utilized PulSA to provide quantitative information on shifts in various proteins from a diffuse cytosolic distribution into punctate or dispersed aggregates, and protein movements from the plasma membrane to the Golgi, and from the cytoplasm to the nucleus [3].

A particularly useful application relevant to this book theme is for tracking the formation of inclusion bodies of aggregating proteins, such as polyglutamine(polyQ)-expanded Htt. In practice, the profound condensation of Htt molecules in a cell leads to a large reduction in pulse width and an increase in pulse height, which enables cells to be readily separated into two populations by gates on a plot of pulse height versus pulse width [3].

Here we describe simple PulSA data acquisition, interpretation, and tips to experimental design.

2 Materials

1. Cells ready for flow cytometry. For this, the protein of interest must be fluorescently labelled. For details *see* Subheading 3.1.
2. Control samples for setting photomultiplier tube (PMT) voltage:
   (a) Negative control: Cells not exhibiting the fluorescence of choice; for example, cell not expressing the protein of interest.
   (b) Positive control: Cells exhibiting the highest level of fluorescence likely to be obtained.
3. Windows XP-based computer.
4. Flow cytometry analysis software (e.g., FACSDiva software (BD Biosciences) or FlowJo (Tree Star, Inc)); *see* **Note 1**.
5. FCS Extract 1.02 software; *see* **Note 2**.
6. Microsoft Office Excel.
7. Flow cytometer capable of collecting pulse width and height (e.g., LSRFortessa flow cytometer (BD Biosciences)).
8. Sorting flow cytometer capable of collecting pulse width and height (e.g., FACSAria cell sorter (BD Biosciences)).

3 Methods

Before beginning an experiment, it is important to understand the behavior of the localization pattern of the protein in a cell. Hence, microscopy data is useful to have at hand to validate PulSA with the expected protein localization of the protein and how it changes upon a treatment of interest. Also of importance is consideration of controls to set up gates in the flow cytometer. This may be done using controls for different states (such as aggregated versus non-aggregated) or by standardizing the protein localization pattern against another cellular feature (for example, using a nuclear stain).

3.1 Sample Preparation

1. The protein of interest must be fluorescently labelled either by fusion to a fluorescent protein (e.g., GFP, mCherry, CFP) or through immunofluorescence (*see* **Note 3**). In addition or alternatively, organelles and other structures can be labelled with dyes (e.g., Hoechst 33342). Multiple colors can be used simultaneously provided there is no major bleedthrough of the fluorescence channels. At least 1×10^6 cells per sample are required. If the cells are to be sorted and recovered, we recommend increasing this amount to at least 6×10^6 cells (*see* **Note 3**).
2. Cells can be analyzed live or fixed (*see* **Note 4**).

3.2 Parameters for PulSA

1. Ensure that the flow cytometer is equipped with the appropriate laser and filter sets to analyze the fluorescent markers of choice. This can be easily determined under the default cytometer configuration on the flow cytometer.
2. The flow cytometer must have the option of collecting pulse height, area, and width parameters for each channel of interest (*see* **Note 5**).
3. The flow rate of choice can vary. We have obtained similar results using any of the flow rates on the BD LSR flow cytometer.
4. We recommend collecting data for 50,000–150,000 cells of interest (*see* Subheading 3.3).
5. The signal detection parameters on the flow cytometer should be carefully adjusted to determine background fluorescence and minimum sample fluorescence [6]. The dynamic range should be modified by adjusting the voltage of the PMTs such that the sample signal spans the full detector range. To set the PMTs a positive control exhibiting the fluorescence of choice and a nonfluorescent negative control are required. By examining histograms of the area parameter (i.e., total cell brightness) for positive and negative controls, an appropriate threshold can be chosen (Fig. 1a).

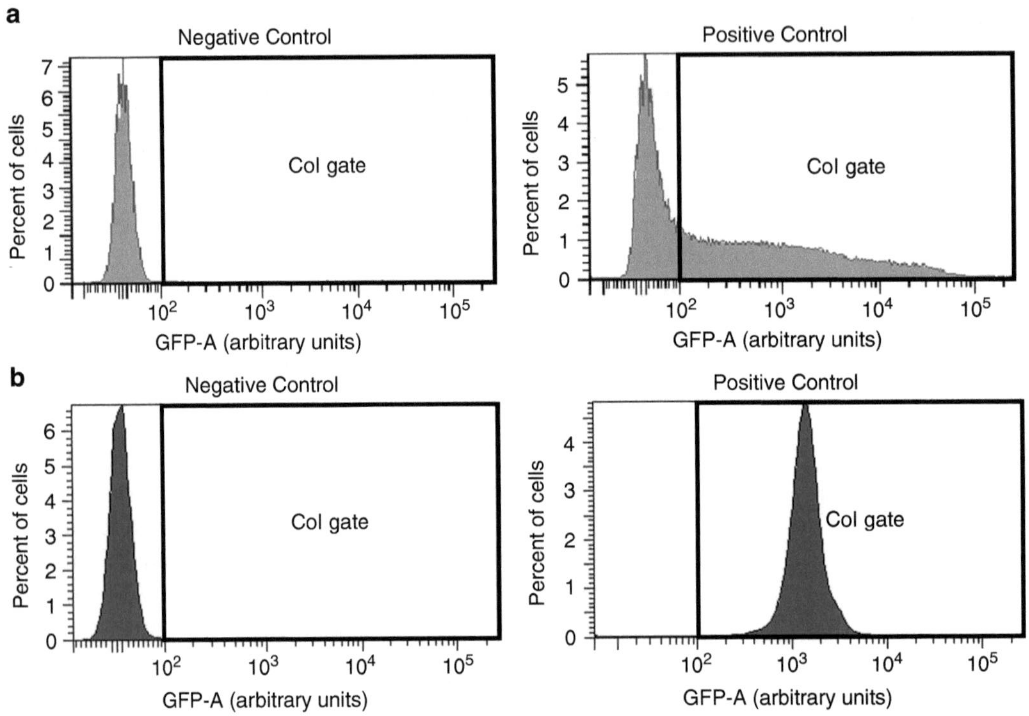

Fig. 1 Selecting the cells with the target protein. The histogram of the GFP-A of the negative and positive controls enables the threshold and dynamic range to be established for the target protein. Here, the cells of interest (CoI) gate is indicated. (**a**) Typical histograms for a transfected target protein. The signal of the positive control was set to span the full detector range by adjusting the PMT voltage. (**b**) Typical histograms for endogenous markers; that is, proteins detected by immunofluorescence (or for the use of organelle stains)

3.3 Initial Gating Considerations

We typically establish our gates to track the following:

1. *Single and uniform cells*: To ensure selection of representative cells and to remove doublet cells, clumps or sheared cells, and debris, we first gate the "main" cell population based on the forward scatter (FSC-A) and side scatter (SSC-A) plots. Forward scatter is proportional to the size of the particles and side scatter is proportional to inner complexity or granularity of the particle. The data can be best viewed on a contour plot of FSC-A versus SSC-A plot.
2. *Cells containing the target protein [cells of interest (CoI) gate]*: The second gate should select fluorescently labelled cells. Typically, for transient transfections of say a GFP-tagged protein, there will be a population of cells with background fluorescence whereas the transfected cells will form a substantially brighter population of cells. These populations can be observed by viewing a histogram of the area (A) fluorescence parameter of the target (e.g., GFP). Using a negative control (cells that do not express the fluorescent protein) and a positive control (cells that express the fluorescent protein) enables a

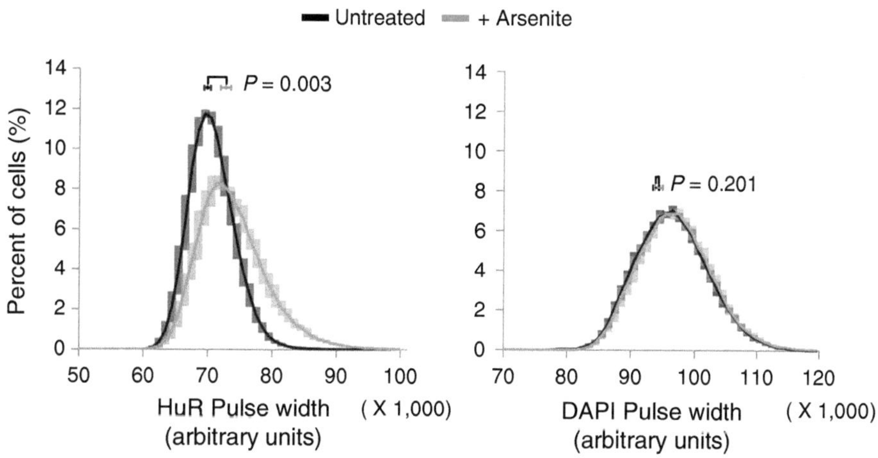

Fig. 2 Histogram of pulse widths of nucleus and HuR. When comparing an untreated sample (no stress granules; protein is in the nucleus) versus cells treated with arsenite (stress granules; protein forms several dispersed puncta in cytoplasm), there is a noticeable shift in the pulse width of HuR detected by immunofluorescence. There is no shift in the nucleus, stained with DAPI. Data are adapted from [3]

boundary to be set for a gate (Fig. 1a). When studying endogenous proteins detected by immunofluorescence or using an organelle stain, all cells are likely to shift to higher fluorescence relative to the unlabelled cells (Fig. 1b).

3.4 Histogram Pulse Width Analysis

There are two variations of PulSA we typically use, which can emphasize different features of the data. The first strategy relies on the population pulse width values, which can provide a clear view of small or subtle changes in the localization patterns. In this analysis, pulse widths are examined as histograms, which we typically perform directly on exported flow cytometry data (*see* **Note** 7). As an example, we illustrate how pulse width analysis can detect a change in cellular distribution of the human antigen R protein (HuR) but not DAPI upon stress granule formation (Fig. 2).

3.4.1 Import Pulse Width Values into Excel

1. Export FCS files of the CoI gate, selecting only the pulse width readouts.
2. Convert FCS files into Microsoft Office Excel comma-delimited text files using FCS Extract 1.02 software.
3. Open files in Microsoft Office Excel. We recommend copying replicates into adjacent columns of a single worksheet and different treatment groups into different worksheets of a single Excel file.

3.4.2 Generate a Frequency Table

As an example, Fig. 3 shows the construction of the frequency table for the first several bins of the untreated cell DAPI pulse width. Where required, formulae are entered in the first row of the appropriate column. For simplicity, only two of the three replicates are displayed.

	A	B	C	D	E	F	G	H	I	J	K
1	RAW DATA		BINS	CUMULATIVE FREQUENCY		ABSOLUTE FREQUENCY		PERCENT OF CELLS		AVERAGE	SD
2	Replicate1	Replicate 2		Replicate1	Replicate 2	Replicate1	Replicate 2	Replicate1	Replicate 2		
3	90376	98118	70000	=FREQUENCY(A$3:A$40003, $C3)			.	.	.		
4	94977	92686	71000	0	0	=D4-D3	0	=F4/COUNT(A$3:A$40003)*100		=AVERAGE(H4,I4)	=STDEV(H4,I4)
5	93634	88347	72000	0	0	0	0	0.00	0.00	0.00	0.00
6	92710	94232	73000	0	0	0	0	0.00	0.00	0.00	0.00
7	100609	89811	74000	0	0	0	0	0.00	0.00	0.00	0.00
8	95316	104645	75000	0	0	0	0	0.00	0.00	0.00	0.00
9	108344	85266	76000	1	0	1	0	0.00	0.00	0.00	0.00
	100106	102236	77000	3	0	2	0	0.01	0.00	0.00	0.00
	92460	92484	78000	8	1	5	1	0.01	0.00	0.01	0.01
	93640	102108	170000	38876	38873	0	0	0.00	0.00	0.00	0.00
	99740	89830	171000	38876	38873	0	0	0.00	0.00	0.00	0.00

Fig. 3 Example of histogram worksheet from Excel. The raw flow cytometry data has in this case 40,000 values (not all rows are shown; columns A and B). Data are assigned to bins, defined by their upper limits as listed in column C. The cumulative frequency (columns D and E) is calculated from the raw data and then converted to absolute frequency (columns F and G) and finally percent of cells (columns H and I). Replicate percent values are averaged (column J) and the standard deviations are calculated (column K). Example formulae are given for the first bin for replicate A

1. Choose a bin width (i.e., the magnitude of each discrete interval) for the histogram. The bin width will determine how smooth the distribution will appear and will affect the ability to meaningfully compare treatment groups. As a starting point we recommend the Rice rule of thumb which advises using $2 \times N^{1/3}$ bins (where N is the number of events; here, cells) spanning the data range. However, trial of different bin widths such that the histogram meaningfully conveys that the distribution of pulse widths may be required. Most importantly, it is essential that all treatment groups to be compared are plotted according to the same bin widths. In the example provided, bin width was set at 1,000.
2. Enter the upper limit of each bin into the column adjacent to the raw data values. This can be done easily by entering the first two values and then using the auto fill function to enter the remaining values.
3. Calculate the cumulative frequency in each bin. Cumulative frequency is defined as the number of events whose values lie below a given bin upper limit; it can be calculated using the formula "=*FREQUENCY*(*data range, bin upper limit*)". In our example, up to 40,000 cells were analyzed for each replicate, with values entered in cells from A3 downwards. The first bin upper limit was entered in C3. Therefore, cumulative frequency for the first bin was calculated as "= *FREQUENCY*(*A$3:A$40003, $C3*)" (*see* D3, Fig. 3); *see* **Note 6**.
4. Convert each bin's cumulative frequency to absolute frequency by subtracting the cumulative frequency of the preceding bin (*see* F4 and G4, Fig. 3).

5. Convert absolute frequency to percent of cells as shown in cell H4 of Fig. 3. Note that using the "*COUNT*(*data range*)" function overcomes the need to change the formula if the number of cells in each replicate varies.
6. Calculate the average percent of cells belonging to each bin using the formula "=*AVERAGE*(*data range*)," where the data range contains the percent of cells for each of the replicates. Calculate the standard deviation from this average using the formula "=*STDEV*(*data range*)."

3.4.3 Plot the Histogram Outputs as a Scatter Plot

1. Plot bin value versus average percentage of cells.
2. Add error bars showing ± standard deviation. In our example (Fig. 3), *x*-values are in column C, series *y*-values are in column J, and custom error bars are defined as column K.

3.4.4 Statistical Analysis of Differences in Pulse Width

1. Calculate the median pulse width for all cells in each dataset. This is done using the formula "=*MEDIAN*(*data range*)" where the data range includes all raw pulse width values. In our example, the median pulse width for replicate one of the untreated DAPI pulse width would be calculated as "=*MEDIAN*(*A3*:*A40003*)."
2. For each treatment group, calculate the average and standard deviation of pulse width medians. The average median pulse width is calculated using the formula "=*AVERAGE*(*Md1*, ..., *Mdn*)" where Md_1 to Md*n* are the median pulse widths of *n* replicates.
3. Compare treatment groups using a student's *t*-test (two treatment groups) or ANOVA (greater than two treatment groups). The Excel Analysis ToolPak add-in has tools for either analysis. For most applications, we recommend using the *t-test: two sample assuming equal variance* tool or the *ANOVA: single factor* tool as appropriate.

3.5 Two-Dimensional Analysis of Pulse Shapes

To more completely view the relationship between the pulse height values and the pulse width values of each cell, a dot or a contour plot of height versus width can be performed. For very large changes in protein localization—such as protein condensation arising from inclusion formation, the large shifts in height and width values enable a clear demarcation between the cells with aggregates and the cells without. This analysis is particularly useful in that it enables 2-dimensional gating strategies to clearly separate the two populations of cells, which can thus enable recovery by cell sorting for further analysis if desired.

1. Create a dot plot or a contour plot of pulse heights versus pulse widths of the fluorescence of the protein of interest.

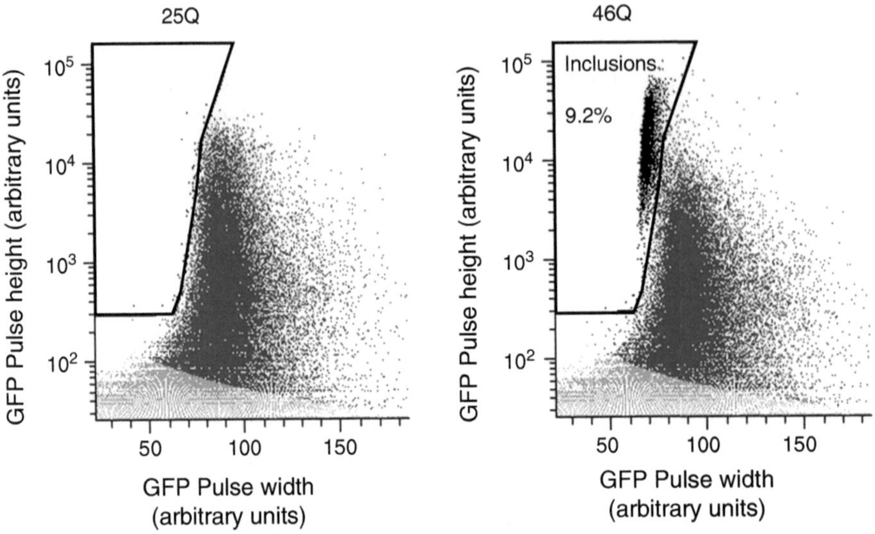

Fig. 4 Two-dimensional gating strategy for pulse shape analysis. Data shows Neuro2a cells transfected with huntingtin exon 1 fusions to GFP in wild-type (25Q) and the aggregation-prone (46Q) form. The 46Q form shows a unique population characteristic with inclusions

2. By comparing control samples of cells without aggregates versus cells with aggregates, it should be possible to identify the unique population characteristic of aggregates (low widths and high heights; Fig. 4—*see* **Note 8**).
3. This population can be gated for quantitative analysis.

4 Notes

1. We use FACSDiva software for our applications and routine analysis, and in some cases where FACSDiva is not able to analyze the data, we export the FCS data for analysis in Excel (Microsoft Office) or Sigmaplot (Systat).
2. FCS Extract 1.02 software is from Earl F Glynn, Stowers Institute for Medical Research, and can be downloaded free of charge at research.stowers-institute.org/efg/ScientificSoftware/Utility/FCSExtract.
3. Detailed methods for immunostaining cells are not covered in this chapter and can be found elsewhere (e.g., [7]). When performing immunofluorescence for PulSA, we recommend starting with at least 6×10^6 cells, since the extensive washing steps tend to cause cells to be lost during preparation. We also recommend optimizing the primary antibody concentration using microscopy. It is important that this is done on cells prepared in suspension (rather than relying on adherent cells) as labelling may differ.

4. We highly recommend fixing cells with 2–4 % (w/v) paraformaldehyde rather than with organic solvents, which can cause shrinkage of the cells and potential difficulties in interpreting the pulse width values.
5. Some cytometers may not be able to provide pulse height and pulse width values on multiple lasers. For the LSRFortessa, the pulse area (A), width (W), and height (H) can be collected under the Inspector—Cytometer Settings menu item.
6. Use of the "$" signs enables the formula to be filled down and copied to the column describing replicate 2 while maintaining a correct data range and bin upper limit.
7. It is useful to do replicates of each treatment, to test for reproducibility of the readout from the flow cytometer. But note that FACSDiva software does not allow direct assessment of the triplicates. Therefore it is useful to reassess the data manually and perform a histogram analysis.
8. Our example is of cells expressing wild-type (H25Q) and mutant polyglutamine (polyQ)-expanded (H46Q) huntingtin which causes Huntington's disease. The wild-type huntingtin is diffuse through the cytosol whereas the mutant forms bright puncta in a subset of cells. The method shown also can be adapted for immunofluorescence of untagged Htt as described in our recent work [3].

Acknowledgements

This work was funded by grants to D.M.H. from the Australian Research Council (DP120102763) and the Hereditary Disease Foundation. D.M.H. is a Grimwade Research fellow, funded by the Miegunyah Trust.

References

1. Tyedmers J, Mogk A, Bukau B (2010) Cellular strategies for controlling protein aggregation. Nat Rev Mol Cell Biol 11(11):777–788
2. Abraham VC, Taylor DL, Haskins JR (2004) High content screening applied to large-scale cell biology. Trends Biotechnol 22(1):15–22. doi:10.1016/j.tibtech.2003.10.012
3. Ramdzan YM, Polling S, Chia CP, Ng IH, Ormsby AR, Croft NP, Purcell AW, Bogoyevitch MA, Ng DC, Gleeson PA, Hatters DM (2012) Tracking protein aggregation and mislocalization in cells with flow cytometry. Nat Methods 9(5):467–470. doi:10.1038/nmeth.1930
4. Hoffman RA (2009) Pulse width for particle sizing. Curr Protoc Cytom Chapter 1:1.23.21–21.23.17. doi:10.1002/0471142956.cy0123s50
5. Shapiro HM (2003) Practical Flow Cytometry, 4th ed., Wiley, Hoboken, New Jersey. ISBN: 0-471-41125-6
6. Baumgarth N, Roederer M (2000) A practical approach to multicolor flow cytometry for immunophenotyping. J Immunol Methods 243(1–2):77–97
7. Jung T, Schauer U, Heusser C, Neumann C, Rieger C (1993) Detection of intracellular cytokines by flow cytometry. J Immunol Methods 159(1–2):197–207. doi:10.1016/0022-1759(93)90158-4

Chapter 7

Characterizing Social Behavior in Genetically Targeted Mouse Models of Brain Disorders

Emma L. Burrows and Anthony J. Hannan

Abstract

Fragile X syndrome, the leading inherited cause of mental retardation and autism spectrum disorders worldwide, is caused by a tandem repeat expansion in the FMR1 (fragile X mental retardation 1) gene. It presents with a distinct behavioral phenotype which overlaps significantly with that of autism. Emerging evidence suggests that tandem repeat polymorphisms (TRPs) might also play a key role in modulating disease susceptibility for a range of common polygenic disorders, including the broader autism spectrum of disorders (ASD) and other forms of psychiatric illness such as schizophrenia, depression, and bipolar disorder [1]. In order to understand how TRPs and associated gene mutations mediate pathogenesis, various mouse models have been generated. A crucial step in such functional genomics is high-quality behavioral and cognitive phenotyping. This chapter presents a basic behavioral battery for standardized tests for assaying social phenotypes in mouse models of brain disorders, with a focus on aggression.

Key words Autism, Fragile X syndrome, Mouse model, Aggression

1 Introduction

Autism spectrum disorders (ASDs) are childhood-onset developmental disorders characterized by impairment of social and communication skills, and repetitive behavior. In addition to these core symptom domains, patients with ASDs frequently exhibit a variety of comorbid behavioral symptoms, including irritability marked by aggression and self-injurious behavior which adversely impact on patient quality of life. Fragile X syndrome, caused by a trinucleotide repeat expansion in the *Fmr1* gene, is the most common inherited cause of developmental disability and the most common known single gene cause of ASDs [2]. Similar to idiopathic ASDs, irritable behavior and aggression is often exhibited by patients with fragile X syndrome; however, research to date in this disorder has not focused on this target symptom cluster. Mouse models with strong face validity to the diagnostic symptoms in

Danny M. Hatters and Anthony J. Hannan (eds.), *Tandem Repeats in Genes, Proteins, and Disease: Methods and Protocols*, Methods in Molecular Biology, vol. 1017, DOI 10.1007/978-1-62703-438-8_7,

particular aggression provide robust tools for understanding the biological basis of behavioral traits within ASD and facilitate development of treatment strategies.

Aggression in mice is a robust, innate, social behavior and serves to assist the acquisition of social ranking and resources from the environment. For example, in mice dominance hierarchies are established and maintained through confrontations between rival males. Aggressive behaviors in wild-type mice exist in several different forms. Territorial aggression between males can be measured utilizing the resident–intruder test whereby a juvenile intruder mouse is introduced to the home cage of a test mouse [3, 4]. This paradigm mimics many elements of the behavior displayed by resident males to exclude other breeding males from their home territory and their mates [5, 6]. Escalated, pathological, and abnormal forms of aggression can emerge in rodent models following genetic or pharmacologic manipulation and are characterized by prolonged and frequent attacks (including biting on sensitive regions such as the flanks and/or face) and brief latencies to attack following the introduction of an intruder mouse [7, 8]. The atypical antipsychotics risperidone [9] and aripiprazole [10] are currently prescribed for treatment of aggressive and irritable behaviors in children with ASD; however, side effects significantly limit the use of these drugs. Pretreatment of rodent ASD models with clinically effective antipsychotics prior to the assessment of aggression provides predictive validity, increasing the robustness of the model as a tool for understanding the neurobiology of aggression and the development of novel therapeutics.

The resident–intruder test has been widely employed as a standard test for assaying territorial aggression in rodents [11]. The test is based on species-typical adaptive aggression that results from the tendency of mice to mark and patrol their territory and vigorously confront all intruders. In order to reliably observe this confrontation in the resident–intruder test, younger but sexually mature mice from submissive strains are selected for intruder mice. Resident mice are isolation housed for a period of time in order to allow marking and the development of territory.

2 Materials

1. Resident male mice between 10 weeks and 6 months of age (*see* **Note 1**).
2. Intruder male mice (novel mice), e.g., C57BL/6J 129/SvImJ adult mice (The Jackson Laboratory) between 8 weeks and 6 months of age, preferably 2–4 weeks younger than resident mice and approximately body weight matched, as the subjects (*see* **Note 2**).

3. Home cage of resident male mouse.
4. Test room, with minimal cues visible to the subject.
5. Holding area: dedicated room or quiet area near the testing room.
6. Cameras (preferably CCTV security cameras, e.g., Panasonic WV-CP280) with mounting bracket.
7. DVD recorder and blank DVDs or capacity to digitize with a digital video converter and saved to a high-capacity computer drive, thus eliminating the need for a DVR and DVDs.
8. Marking pen (dark) and (if preferred) paw tattoos, ear punches, ear tags.
9. Stopwatches without beepers or with beepers silenced.
10. Drug and vehicle. 30 gauge needles and syringes (*see* **Note 3**).

3 Methods

3.1 Pretest Preparations

1. To allow home cages to be marked with the scent from the male resident only, remove male resident mice from group housing and place them in a clean home cage with fresh bedding and house individually for 1 week. Home cages are not changed during this period of isolation.
2. Set up the testing room environment to minimize stress and bias due to visual, olfactory, and auditory cues (*see* **Note 4**).
3. Set up video recording device and place camera directly above the cage.

Due to the complicated nature of interactions between resident and intruder, video recording of the experimental session is essential for later analysis. In order to accurately assess all aggression parameters, each test session may need to be viewed more than once.

3.2 Test

1. On the test day, all mice are habituated to the holding room for an hour prior to testing.
2. If resident mice are drug-treated, inject a non-sedative dose of drug or vehicle and place resident back into home cage for period of time until drug has reached its peak activity.
3. A male intruder mouse is placed into the home cage of a resident mouse, and behavior is video-recorded for 5 min (Fig. 1). Each session is closely monitored and trials are aborted if the experimenter observes tufts of hair being removed from either animal. Further details on planning are detailed in the notes section (*see* **Note 5**).

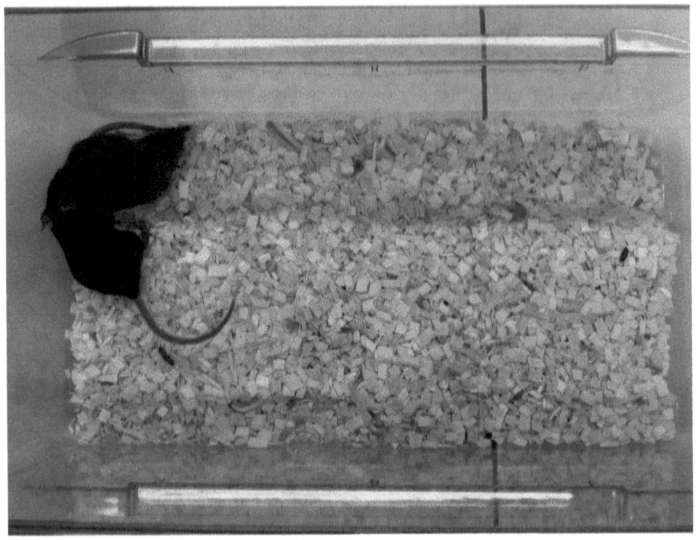

Fig. 1 Resident–intruder set up

4. Due to high variability in aggressive behavior, the test should be repeated on 4 consecutive days until a stable level of aggression is achieved in each mouse. Stability has been previously defined as no more than 15 % variation in the frequency of attacks from the previous test [12].

3.3 Anticipated Results

Latency to first attack, attack incidence and nonaggressive social interactions initiated by the resident male are scored from videotapes of each test session by an experimenter blind to mouse genotype and group. An attack is defined as a bite directed at the back or flanks of the intruder, and nonaggressive social interactions are defined as contact initiated by the resident mouse, specifically sniffing, climbing on, and grooming the intruder. Trials in which the intruder confronts the resident mouse must be excluded. Not all C57BL/6 mice will attack their intruder pair, allowing heightened and reduced levels of aggression in strains/genetic mutants to be clearly delineated (Fig. 2). To observe the evolution of behavior over time score each parameter in time blocks of 1 min.

3.4 Statistical Analysis Considerations

1. *Sample size must be preplanned.* As part of your experimental design, it is critical to run a power analysis in order to calculate the required sample size for statistically power. Often, if this is the first time a study has been conducted, no prior estimate of effect size is available and thus no power analysis can be conducted. A sample size of 10–15 in each group is a reasonable start.
2. *The same animal will be sampled over 4 days.* Due to the fact that animals are repeatedly tested over 4 days, each observation is correlated within a given animal. Statistical tests to appropriately account for this correlation (e.g., ANOVA with repeated

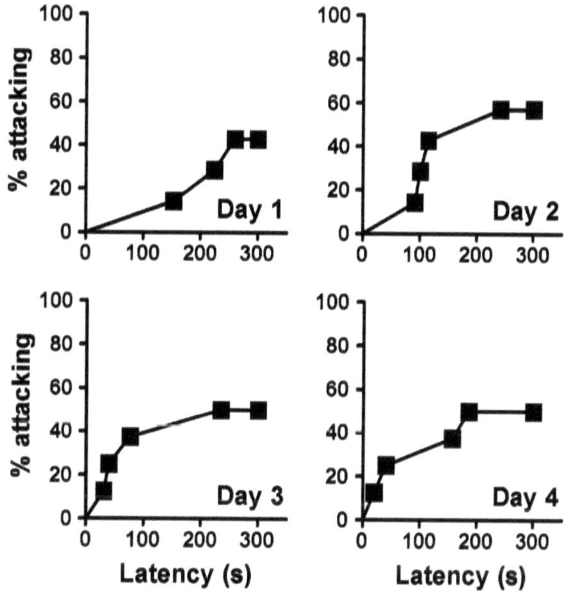

Fig. 2 Percentage of C57BL/6 mice attacking

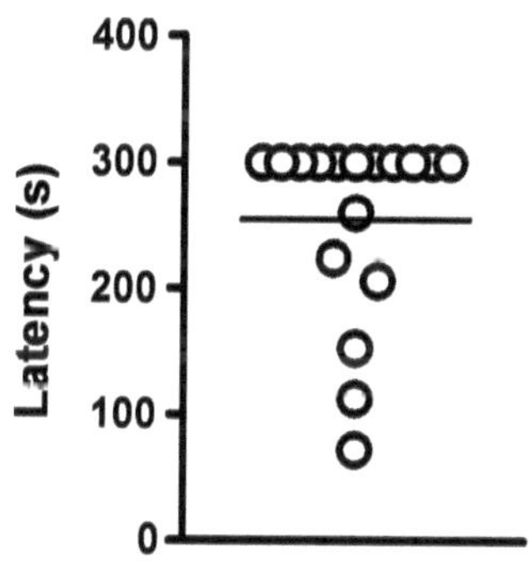

Fig. 3 Attack latencies of C57Bl/6 mice

measures if normally distributed or hierarchical regression models if normality cannot be assumed) can be used to estimate the effect of treatment on aggression over time.

3. *Is your population normally distributed?* The standard statistical analysis reported for the resident–intruder test is a repeated-measures ANOVA. While aggression is a normal and innate behavior in wild-type mice, not every mouse will reliably attack an intruder. This may result in a ceiling effect, with a large proportion of those animals not attacking possessing latencies of 300 s (Fig. 3). This population is not normally distributed and thus an ANOVA is not appropriate. If data is unable to be transformed to a normal distribution using standard transformations (logarithmic, square root, reciprocal), an alternative

is to run a survival analysis (i.e., Cox regression model) that compares the probability that an event will occur at any given time between two or more groups. This model will take into account the fact that animals with latencies of 300 s may have actually attacked if the test was conducted over a longer time period [13].

3.5 Considerations of Whether the Behavior Is True Aggression or an Artifact

Changes in aggressive behavior may be reflective of a more general impairment in social function, and it is recommended that social interaction is quantified [14]. In rodents, social interaction and aggression depends largely on olfactory cues. Time spent investigating nonsocial and social olfactory cues is commonly observed to indicate that olfactory function in subjects is intact [15]. Reduced investigation of social odors compared with controls may be consistent with disruptions in social or aggressive behavior. Furthermore, abnormal stress responses may precipitate aggressive behavior and can be assessed using the elevated plus maze [16]. Ultimately, when characterizing a genetically targeted mouse model, it is informative to use a broad battery of social, cognitive, affective, and motor behavior tests, to avoid potential confounds and facilitate interpretation of the phenotype.

4 Notes

1. At 4 weeks of age, mice are weaned and transferred to group housing in standard cages. Group housing can consist of two, three, or four mice per cage and if possible the number of gene-mutant, drug-treated mice per cage should be kept balanced in each cage. Late weaning and then subsequent mixing of adult males may result in excess aggression in home cages. This situation mimics that of social-defeat stress, in which a mouse is repeatedly exposed to an aggressive male and develops higher levels of anxiety and should be averted [17].

   Resident mice do not have to be experimentally naive and can be tested in low-stress tests (e.g., developmental milestones, juvenile social interaction, elevated plus maze, light/dark exploration, general health battery, olfactory sensitivity) without confounding their response in the resident–intruder test. Testing animals in highly stressful tests (e.g., prepulse inhibition, forced-swim test, fear conditioning, Morris water maze) prior to the resident–intruder test is not recommended. Sexual experiences and social isolation have strong influences on social behaviors in adult mice [18–21]. To avoid these confounding factors, sexually naive and previously group-housed mice are preferred for this task.
2. If using mice purchased from a commercial supplier, when animals arrive in the facility, transfer them to standard cages

immediately. Do not mix adult males from different shipping compartments since unfamiliar adult males could be very aggressive toward one another. Allow the animals to acclimate to the facility for at least 1 week before they are used as intruders.

Selecting a younger and more submissive intruder mouse will facilitate the expression of territorial aggression. Intruder mice should be 2–4 weeks younger than resident mice and approximately matched for body weight. The expression of intermale aggression differs with inbred strains [22]. Mice from the 129/SvImJ inbred strain are relatively nonaggressive and are commonly used as intruders in resident–intruder aggression tests [14]. Other inactive strains (e.g., AJ, LP/J, 129/SvJac) could be used instead of 129/SvImJ. Conversely, the BALB/cJ inbred mouse strain exhibits relatively high levels of intermale aggression and should not be selected as an intruder mouse [23]. It is advised to rotate intruders to fight with a different resident each day to spread the variability that may arise from different resident–intruder pairs and to remove the potential for habituation. Testing each resident–intruder pair at the same time of the day over each of the 4 consecutive days controls for fluctuations in circadian rhythm.

3. If animals are drug-treated prior to resident–intruder testing, ensure that the dose selected is not sedative in an independent cohort of mice. Assess distance travelled by drug-treated animals compared to vehicle-treated controls in locomotor activity arenas (Fig. 4). Some antipsychotic drugs are highly lipophilic and require a solvent to dissolve into solution, and it is advisable to avoid all solvents that cause irritation or alter locomotion.
4. It is recommended that each resident–intruder test be conducted in isolation from the remainder of the cohort to be

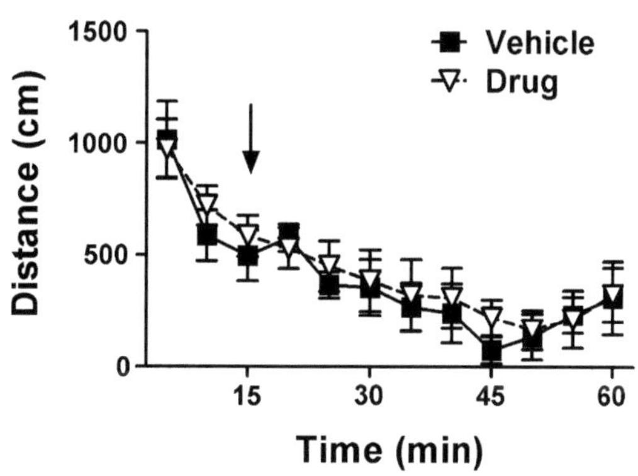

Fig. 4 Locomotor activity following drug administration. Distance travelled was measured in Truscan (Coulbourn Instruments)

tested. In addition to audible squeaks, ultrasonic vocalizations are emitted during male–male encounters [24, 25] and have the potential to influence behavior of other animals if housed in the same room.

It is critical to conduct the resident–intruder test in a room that is free from bias or distraction. The goal is to remove any salient room cues, uneven illumination, and odors that could direct the attention of the resident mouse from the intruder. This is particularly important for the resident–intruder test as mice are unable to be habituated to the testing room.

Minimize room cues that are visible to the subject. When the cage is placed on a bench, the ceiling and plain walls should only be in the subject's view. Store all supplies under the bench. If for practical reasons there is no way to make the room plain and symmetrical, use an enclosure or curtain to shield the cage. Ideally, the experimenter will be able to remove themselves from the test room and view the experiment live on a screen outside the room. If this is not possible, the experimenter must sit inside the room to observe each fight and intervene when specified by the Institutional Ethics Committee guidelines.

It is advisable to conduct the experiment in a sound-attenuated quiet room to avoid auditory distractions. Do not conduct the test in a room that has been previously used for housing or experimenting on rats or culling as the residual odor may cause distress to resident or intruder mice.

5. The resident–intruder test takes only 5 min to complete. Allow a further 15 min to return the intruder mouse to its cage, set up the next resident's cage, inject resident mouse, take notes, and start/stop DVD recorder. Altogether, this method takes one person 15 ± 5 min to finish for each mouse. It is reasonable to test a cohort of up to 25 mice a day beginning testing 1 h after lights on and finishing 1 h prior to lights off. It is not recommended to conduct any behavioral test (unless testing circadian rhythms) 1 h prior to lights on or off. If resident mice are drug-treated, the time between injection and when the drug has reached its peak activity will have to be incorporated into the time schedule and the number of mice that can be tested per day adjusted accordingly. As each resident–intruder pair requires close supervision, and potential intervention if the intruder mouse is injured, a high-throughput setup is not possible for the resident–intruder test unless more experimenters are present. If testing a small number of mice, it is recommended to test them all as close together as possible, to remove any confounds due to differences in activity during the first half of the day vs. latter half.

Acknowledgements

This work was funded by grants to A.J.H. from the National Health and Medical Research Council (APP1034785 and APP1047674). A.J.H. is an Australian Research Council Future Fellow (FT3).

References

1. Hannan AJ (2010) Tandem repeat polymorphisms: modulators of disease susceptibility and candidates for 'missing heritability'. Trends Genet 26(2):59–65. doi:10.1016/j.tig.2009.11.008
2. Hannan AJ (2012) Tandem repeat polymorphisms: genetic plasticity, neural diversity and disease. Landes Bioscience and Springer Science+Business Media, New York. Springer series: Advances in Experimental Medicine and Biology 769:1–208.
3. Crawley JN, Schleidt WM, Contrera JF (1975) Does social environment decrease propensity to fight in male mice? Behav Biol 15:73–83
4. Miczek KA, O'Donnell JM (1978) Intruder-evoked aggression in isolated and nonisolated mice: effects of psychomotor stimulants and L-dopa. Psychopharmacology (Berl) 57(1):47–55
5. Brain P, Benton D (1979) The interpretation of physiological correlates of differential housing in laboratory rats. Life Sci 24(2):99–115
6. Miczek KA, Maxson SC, Fish EW, Faccidomo S (2001) Aggressive behavioral phenotypes in mice. Behav Brain Res 125(1–2):167–181
7. Miczek KA, Fish EW, De Bold JF (2003) Neurosteroids, GABAA receptors, and escalated aggressive behavior. Horm Behav 44(3):242–257
8. Miczek KA, Fish EW, De Bold JF, De Almeida RM (2002) Social and neural determinants of aggressive behavior: pharmacotherapeutic targets at serotonin, dopamine and gamma-aminobutyric acid systems. Psychopharmacology 163(3–4):434–458. doi:10.1007/s00213-002-1139-6
9. Chavez B, Chavez-Brown M, Rey JA (2006) Role of risperidone in children with autism spectrum disorder. Ann Pharmacother 40(5):909–916. doi:10.1345/aph.1G389
10. Erickson C, Stigler K, Wink L, Mullett J, Kohn A, Posey D, McDougle C (2011) A prospective open-label study of aripiprazole in fragile X syndrome. Psychopharmacology 216(1):85–90. doi:10.1007/s00213-011-2194-7
11. Miczek KA, Faccidomo S, De Almeida RMM, Bannai M, Fish EW, Debold JF (2004) Escalated aggressive behavior: new pharmacotherapeutic approaches and opportunities. Ann N Y Acad Sci 1036(1):336–355. doi:10.1196/annals.1330.021
12. Velez L, Sokoloff G, Miczek KA, Palmer AA, Dulawa SC (2010) Differences in aggressive behavior and DNA copy number variants between BALB/cJ and BALB/cByJ substrains. Behav Genet 40(2):201–210. doi:10.1007/s10519-009-9325-5
13. Cleves M, Gould W, Gutierrez R, Marchenko Y (2008) An introduction to survival analysis using STATA, 2nd edn. Stata Press, College Station, TX
14. Yang M, Silverman JL, Crawley JN (2011) Automated three-chambered social approach task for mice. Current protocols in neuroscience, Chapter 8:Unit 8.26. doi:10.1002/0471142301.ns0826s56
15. Yang M, Crawley JN (2009) Simple behavioral assessment of mouse olfaction. Current protocols in neuroscience, Chapter 8:Unit 8.24. doi:10.1002/0471142301.ns0824s48
16. Komada M, Takao K, Miyakawa T (2008) Elevated plus maze for mice. J Vis Exp (22):e1088. doi:10.3791/1088
17. Kudryavtseva NN, Bondar NP, Avgustinovich DF (2004) Effects of repeated experience of aggression on the aggressive motivation and development of anxiety in male mice. Neurosci Behav Physiol 34(7):721–730
18. Bouet V, Lecrux B, Tran G, Freret T (2011) Effect of pre- versus post-weaning environmental disturbances on social behaviour in mice. Neurosci Lett 488(2):221–224. doi:10.1016/j.neulet.2010.11.033
19. Kercmar J, Budefeld T, Grgurevic N, Tobet SA, Majdic G (2011) Adolescent social isolation changes social recognition in adult mice. Behav Brain Res 216(2):647–651. doi:10.1016/j.bbr.2010.09.007
20. Koike H, Ibi D, Mizoguchi H, Nagai T, Nitta A, Takuma K, Nabeshima T, Yoneda Y, Yamada K (2009) Behavioral abnormality and pharmacologic response in social isolation-reared mice. Behav Brain Res 202(1):114–121. doi:10.1016/j.bbr.2009.03.028

21. Rawleigh JM, Kemble ED, Ostrem J (1993) Differential effects of prior dominance or subordination experience on conspecific odor preferences in mice. Physiol Behav 54:35–39
22. Guillot PV, Chapouthier G (1996) Intermale aggression and dark/light preference in ten inbred mouse strains. Behav Brain Res 77(1–2):211–213
23. Dow HC, Kreibich AS, Kaercher KA, Sankoorikal GM, Pauley ED, Lohoff FW, Ferraro TN, Li H, Brodkin ES (2011) Genetic dissection of intermale aggressive behavior in BALB/cJ and A/J mice. Genes Brain Behav 10(1):57–68. doi:10.1111/j.1601-183X.2010.00640.x
24. Chabout J, Serreau P, Ey E, Bellier L, Aubin T, Bourgeron T, Granon S (2012) Adult male mice emit context-specific ultrasonic vocalizations that are modulated by prior isolation or group rearing environment. PLoS One 7(1):e29401. doi:10.1371/journal.pone.0029401
25. Lumley LA, Sipos ML, Charles RC, Charles RF, Meyerhoff JL (1999) Social stress effects on territorial marking and ultrasonic vocalizations in mice. Physiol Behav 67(5):769–775

Chapter 8

PCR Amplification and Sequence Analysis of GC-Rich Sequences: *Aristaless*-Related Homeobox Example

May H. Tan, Jozef Gécz, and Cheryl Shoubridge

Abstract

PCR amplification (followed by mutation scanning or direct sequencing) is a technique widely used in mutation detection and molecular studies of disease-causing genes, such as *ARX*. PCR amplification of high GC-rich regions encounters difficulties using conventional PCR procedures. Here, we present the strategies to amplify and sequence these GC-rich regions for the purposes of mutation screening and other molecular analyses.

Key words Polyalanine tracts, GC-rich, PCR, Sequencing, *ARX*

1 Introduction

Despite GC-rich regions accounting for only ~3 % of the human genome, important regulatory domains including promoters, enhancers, and control elements all have high GC content [1]. GC-rich sequences can be found in promoter regions of most housekeeping genes, tumor suppressor genes, and ~40 % of tissue-specific genes [2]. Repeat-associated disorders caused by expansions of trinucleotide repeats can occur in either noncoding sequences, transcribed but not translated, or within translated sequences for homomeric stretches of either glutamine or alanine amino acids. On the human X chromosome, expansion of GC-repeats in folate-sensitive fragile sites is associated with fragile X syndrome, which is the most common form of familial intellectual disability [3]. A range of inherited human diseases, associated with neurocognitive or neurodegenerative phenotypes, are caused by expansion of GC-rich trinucleotide repeats encoding polyalanine (polyA) or polyglutamine (polyQ) domains (Table 1). Polymerase chain reaction (PCR) amplification of GC-rich sequences can be difficult [4]. Ineffective or failing PCR

Danny M. Hatters and Anthony J. Hannan (eds.), *Tandem Repeats in Genes, Proteins, and Disease: Methods and Protocols*, Methods in Molecular Biology, vol. 1017, DOI 10.1007/978-1-62703-438-8_8, © Springer Science+Business Media New York 2013

Table 1
PolyA- or polyQ-codon-containing genes with repetitive GC-rich sequences

Gene (MIM)	Gene name[a]	GC content (%) of ORF	Trinucleotides encoded domains (PolyA or PolyQ)	Disease	References
ARX (300382)	*Aristaless*-related homeobox	72.5	4 polyA domains	NS-XLID; PRTS; ISSX/WS; IEDE; OS	[23]
PHOX2B (603851)	Paired-like homeobox 2b	66.1	2 polyA domains	CCHS	[26]
SOX3 (313430)	SRY (sex determining region Y)-box 3	70	4 polyA domains	XH	[27]
HTT (613004)	Huntingtin	52.6	1 polyQ	HD	[28]
ATXN1 (601556)	Ataxin 1	62.3	1 polyQ	SCA1	[29]

CCHS congenital central hypoventilation syndrome (MIM 209880), *HD* Huntington disease (MIM 143100), *IEDE* infantile epileptic–dyskinetic encephalopathy (MIM 308305), *ISSX*(*WS*) X-linked infantile spasms (West syndrome) (MIM 308350), *NS-XLID* non-syndromic X-linked intellectual disability (MIM 300419), *OS* Ohtahara syndrome—early infantile epileptic encephalopathy (MIM 308350), *PRTS* Partington syndrome—intellectual disability with dystonic movements, ataxia, and seizures (MIM 309510), *SCA1* spinocerebellar ataxia 1 (MIM 164400), *XH* X-linked hypopituitarism (MIM 300123)

[a]Gene names are official full names provided by HGNC

amplification of GC-rich regions complicates molecular analyses of these regions. The difficulties to PCR amplify these regions in vitro together with a relatively high frequency of errors may simply reflect the in vivo situation when the cells replicate these types of sequences.

The major hindrance to PCR amplification of GC-rich templates is the formation of secondary structures such as hairpin loops of single-stranded GC-rich sequences [4]. Many approaches have been developed to overcome such problems. Addition of organic molecules such as dimethyl sulfoxide (DMSO), glycerol, polyethylene glycerol, betaine, formamide, nonionic detergents, 7-deaza-dGTP, and dUTP into the PCR reaction mixture helps to resolve the complex secondary structure formation, thereby reducing the melting temperature of the primers and the templates [5–17]. Different DNA polymerases [2], template denaturation with NaOH, hot start PCR, stepdown PCR, slow down PCR, and primer modification have also shown to improve PCR amplification of high GC-rich DNA templates [18–22]. In addition, adjustment of magnesium concentration, pH of reaction buffer, PCR cycle temperature, and cycling numbers could have an effect. Incorporating the above-mentioned factors besides being labor and time intensive is only effective in some applications. In this chapter we outline a step-by-step approach to robustly amplify regions that are GC-rich. These optimized PCR conditions are a starting point for mutation screening or to generate or modify such templates for molecular cloning.

An example of an important disease-causing gene with mutations expanding GC-rich sequences is the *Aristaless*-related homeobox (*ARX*) gene (GenBank: NM_139058.2) (MIM 300382). *ARX* is one of the most frequently mutated genes in X-linked intellectual disability. The gene is predominantly expressed in fetal and adult brain, testis, skeletal muscle, and pancreas. This paired-type homeodomain transcription factor is crucial for early embryonic development. Mutations in *ARX* include nonsense and missense mutations and recurrent expansions of the GC-rich regions encoding the first two of four polyalanine tracts [23]. Mutations in these N-terminal polyalanine tracts contribute to ~60 % of all mutations reported in *ARX*. A frequent mutation in the first polyalanine tract (PA1), c.304ins$(GCG)_7$, expands this tract by seven alanine residues, resulting in a 23 alanine tract. The most common mutation in *ARX* occurs in the second polyalanine tract (PA2), c.429_452dup (24 bp) (~40 % of all reported *ARX* mutation), and involves addition of eight alanine residues to the 12 alanine tract. Both polyalanine expansions result in a range of clinical phenotypes (Table 1). Coding sequences of genes with high GC content, such as *ARX*, are not routinely covered by sequence capture followed by massively parallel sequencing, most likely due to the difficulty to sufficiently enrich the GC-rich regions (J.G., unpublished data). We present our data on the *ARX* gene as an

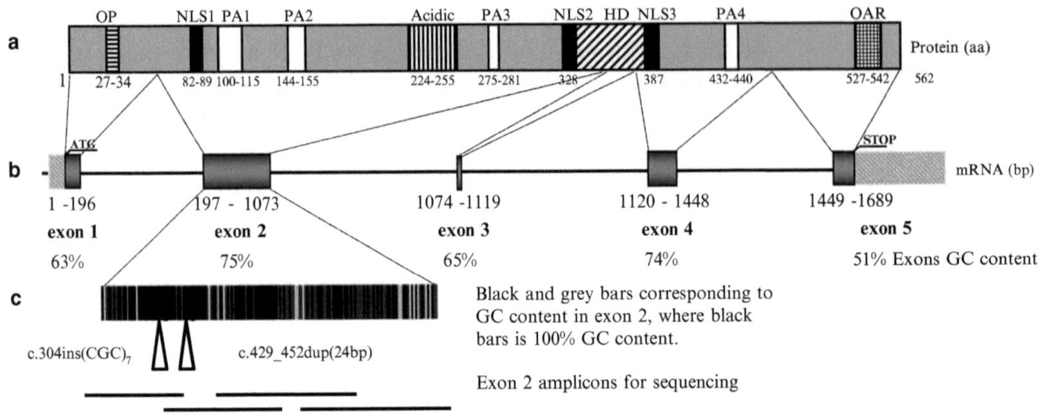

Fig. 1 Schematic diagram of the ARX homeobox transcription factor. (**a**) Known functional domains of ARX protein are highlighted; octapeptide (OP) in *horizontal stripes*, three nuclear localization sequences (NLS) in *black*, four polyalanine tracts (PA) in *white*, acidic domain in *vertical stripes*, homeodomain (HD) in *diagonal stripes*, and *Aristaless* (OAR) domain in *crosshatched*. Amino acid positions of functional domains are indicated below the protein structure. (**b**) Exon–intron structure of *ARX* gene with its five coding exons in *black boxes*; open reading frame in *dark gray*, ATG, STOP codon, and 5′ and 3′ untranslated regions in *diagonal light gray stripes*; base pair sequence of ORF is listed below each exon. (**c**) The percentage of GC content for exon 1–5 is indicated. The positions of *ARX* polyalanine expansion mutations are shown by *triangles* above the amplicons

example of the strategy used to interrogate the most difficult and mutation-prone GC-rich regions in this important disease-causing gene (Fig. 1).

2 Materials

Use autoclaved ultrapure water and analytical grade reagents.

2.1 GC-Rich Polymerase Chain Reaction

1. 100–500 ng of genomic DNA template (*see* **Note 1**).
2. 50 pmol/μl of forward and reverse primers (*see* Table 2).
3. Expand Long Template Enzyme mix (5 U/μl) (Roche) in enzyme storage buffer: 20 mM Tris–HCl, 100 mM KCl, 1 mM dithiothreitol (DTT), 0.1 mM EDTA, 0.5 % (v/v) Nonidet P-40, 0.5 % (v/v) Tween 20, 50 % (v/v) glycerol, pH 7.5 (*see* **Note 2**).
4. FailSafe™ PCR 2× PreMix J buffer (Epicentre Technologies): 100 mM Tris–HCl, 100 mM KCl, 400 μM of each dNTP, 3 mM $MgCl_2$, and 8× FailSafe PCR Enhancer, pH 8.3.

2.2 Gel Electrophoresis

1. 5× TBE buffer: 54 g Tris base, 27.5 g of boric acid, 20 ml of 0.5 M EDTA, pH 8.0, and ultrapure water up to a liter.
2. Agarose gel cast and comb.
3. Molecular Biology grade agarose.

Table 2
***ARX*-specific primers used for amplification and sequencing of *ARX* coding regions**

ARX exon	Primer name	5′ to 3′ sequence	Size (bp)	PCR product (bp)	PCR annealing Tm (°C)	Amplicon GC content (%)
1	ARXe1-F	GTC CAC TAC ACT TGT TAC CGC	21	520	60	61
	ARXe1-R	AAT TGA CAA TTC CAG GCC ACT G	22			
2	ARXe2-F1	CTG ATA GCT CTC CCT TGC CC	20	262	60	81
	ARXe2-R1	GCG GCC CCT GCG CCG TCC GGC CGT TC	26			
	ARXe2-F2	CCC CTC CGC CGC CAC CGC CAA C	22	313	60	81
	ARXe2-R2	TCC TCC TCG TCG TCC TCG GTG CCG GT	26			
	ARXe2-F3	GCA AGT CGT ACC GCG AGA ACG	21	371	60	74
	ARXe2-R3	CAG CTC CTC CTT GGG TGA CA	20			
	ARXe2-F4	AAC TGC TGG AGG ACG ACG AGG	21	392	60	68
	ARXe2-R4	TGC GCT CTC TGC CGC TGC GA	20			
3	ARXe3-F	GAA ATA GCT GAG AGG GCA TTG C	22	231	60	61
	ARXe3-R	TCT CTT GGT TTT GTG AAG GGG AT	23			
4	ARXe4-F	GAC GCG TCC GAA AAC AAC CTG AG	23	551	60	72
	ARXe4-R	CCC CAG CCT CTG TGT GTA TG	20			
5	ARXe5-F	ACA GCT CCC GAG GCC ATG AC	20	347	60	71
	ARXe5-R	GAG TGG TGC TGA GTG AGG TGA	21			

4. Ethidium bromide: 50 μg/ml in water.
5. PCR products.
6. DNA marker (e.g., 1 kb plus marker (Invitrogen)).
7. 6× loading buffer containing dyes: 25 mg bromophenol blue, 25 mg xylene cyanol, 3.3 ml 100 % glycerol, and ultrapure water up to 10 ml.
8. Electrophoresis tank and DC power source.

2.3 PCR Purification

1. 100 bp to 10 kb double-stranded PCR product.
2. QIAquick PCR Purification Kit (Qiagen), which contains:
   2.1 Buffer PB (DNA binding buffer).
   2.2 Buffer PE (wash buffer).
   2.3 QIAquick spin columns (DNA-capture columns).
   2.4 Collection tubes.
   2.5 Ethanol (absolute) for dilution of buffer PE.

2.4 Sequencing

1. 800–1,500 ng/μl of double-stranded DNA PCR products or plasmid DNA.
2. BigDye® Terminator v3.1 Cycle Sequencing Kit (PerkinElmer), which contains:
   2.1 BigDye Terminator v.3.1 Sequencing Buffer (5×).
   2.2 BigDye Ready Reaction Mix.
3. Dimethyl sulfoxide (DMSO).
4. 0.2 nM $MgSO_4$ ethanol solution: 70 ml of absolute ethanol, 20 μl of 1 M $MgSO_4$, and ultrapure water up to 100 ml (*see* **Note 3**).
5. 70 % (v/v) ethanol solution.
6. Heating block set at 37 °C.
7. Sequencing analysis software (DNASTAR, Lasergene software package) or CLC Sequence Viewer 6 program (freeware available from http://www.clcbio.com/index.php?id=28).

3 Methods

3.1 PCR of GC-Rich Templates

Purified genomic DNA is used as template for the amplification of coding and flanking noncoding regions of *ARX*.

3.1.1 Oligonucleotides Design

Primer design is a crucial factor for successful PCR amplification. Manual design of each primer pair allowed us to accommodate stretches of difficult GC-rich sequence and achieve approximately

50 % GC content and a melting temperature of at least 60 °C for each primer. The sequences of both forward and reverse primers, size of the amplified gene product, and the GC content of the gene products amplified are listed in Table 2 (*see* **Note 4**).

In order to screen the coding region of the five *ARX* exons, each exon is amplified by a specific set of primers. The exception to this is exon 2, which requires four overlapping amplicons (Fig. 1c) to achieve robust amplification of GC-rich regions coding for polyalanine tract 1–3 (*see* **Note 5**). The overall GC content of exon 2 is 75 %; however, the GC-rich sequences are not evenly distributed along the exon but are instead centered in three regions coding for polyalanine tracts. The GC content of exon 2 is further increased by mutations leading to the expansion of the first two polyalanine tracts. For instance, the c.304ins$(GCG)_7$ expansion mutation in polyalanine tract 1 increases the GC content of amplicon 1 of exon 2 up to 83 % over a 283 bp region. The length of 100 % GC base pairs coding for polyalanine tract 1 increases from 48 to 69 bp.

Previously, only three sets of primer pairs were used to amplify and sequence exon 2 of *ARX* during mutation screening [24, 25]. The amplicon generated by the first primer set of exon 2 consists of polyalanine tract 1 and 2 and has a GC content of 81 % over a 514 bp region. The PCR amplification using this primer set is often inconsistent. Although it is possible to amplify this target from genomic DNA of controls (C1 and C2), the PCR amplification from genomic DNA of a patient with expanded polyalanine tract (P1) has failed (Fig. 2a). When a PCR product is successfully generated from the genomic DNA of patient with polyalanine tract mutation, instead of a strong, clean PCR product, a doublet PCR product is observed (P2) in gel electrophoresis analysis (Fig. 2b). The inconsistency of the PCR amplification, especially from patient's genomic DNA, is most likely due to the presence of complicated secondary structures formed by GC-rich sequences in exon 2.

Sequencing of the PCR products (P1, P2, and C1) in Fig. 2a, b generates a sequencing chromatogram (Fig. 2c). Secondary structures arising from GC-rich sequences may disrupt the sequencing process leading to either a sudden abrupt signal drop-off or attenuation of signal strength after a run of clear sequence signal. This phenomenon is illustrated in PA1 and PA2 regions of the sample from P2, where the sequence signal drops off at the polyalanine tract regions. A reduction in sequencing signal is also seen in PA1 region of C2, even when a good, clean PCR product (C2) was used as a sequencing template. The sequencing results generated from these amplicons are often inconsistent, requiring multiple re-sequencing analyses.

By generating smaller PCR products across this exon, we routinely achieve robust single PCR products of the expected size as observed by gel electrophoresis. Each amplicon designed for exon 2 of *ARX* is less than 400 bp and consists of only one polyalanine tract.

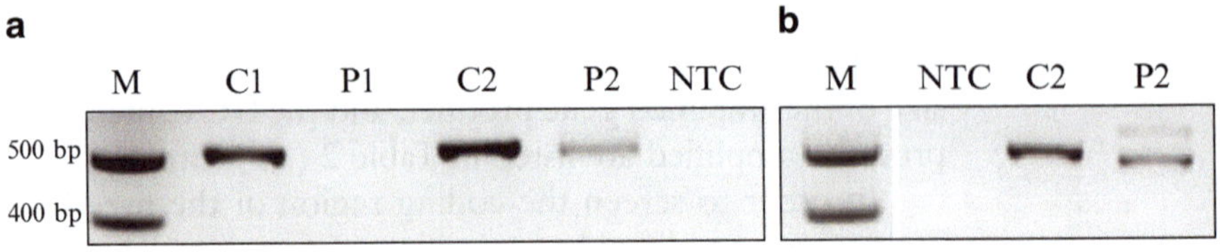

Lanes:		Expected PCR product
M	Marker	-
C1	Control 1	514 bp
C2	Control 2	514 bp
P1	Patient 1 with c.304ins$(GCG)_7$	514 + 21 = 535 bp
P2	Patient 2 with c.429_452dup (24bp)	514 + 24 = 538 bp
NTC	Non-template control	-

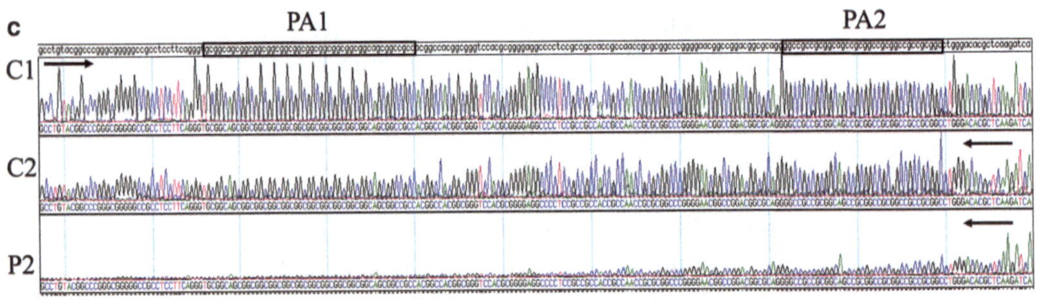

Fig. 2 *ARX* mutation detection using PCR amplification and sequencing. (**a**) *ARX*-specific primers are used to amplify N-terminal region of exon 2 from genomic DNA samples of controls (C1, C2) and patients (P1, P2). The amplicons generated consist of both polyalanine tract 1 and 2 of *ARX*. Strong and clear DNA bands at expected sizes are observed in controls but not in patients. No DNA product is detected in the non-template negative control lane. (**b**) Repeat PCR on sample P2 highlights the inconsistent size of PCR products generated from patients with expanded polyalanine tract mutations in *ARX*. (**c**) DNA sequence chromatogram showing signal drop-off at GC-rich region of *ARX*. An abrupt signal drop-off is observed after initial excellent sequence chromatogram, near the GC-rich region encoding for polyalanine tracts in C2 and P2. Additional sequencing using a forward primer is required to confirm the GC-rich sequence in the signal drop-off region. The *arrows* indicate the direction of sequencing using forward (*left*) or reverse (*right*) primers. Our current protocol uses two sets of *ARX*-specific primers to generate the first (**d**) or second (**e**) amplicon of exon 2 from genomic DNA samples from controls (C1) and patients (P1, P2). DNA bands at expected sizes are observed in both control and patient (listed in *bottom panel*), indicating successful amplification of target sequence. No amplification is present in the non-template negative control lane. The amplicons generated from patients' DNA migrates higher than control, indicating the presence of polyalanine tract expansion. (**f**) Sequence chromatogram of *ARX* exon 2 in control and patient with expanded polyalanine tract mutation c.304ins$(GCG)_7$. Sequence chromatogram from the mutant DNA sample shows 21 additional nucleotides, in the form of seven GCG repeats, in the region coding for the first polyalanine tract. Corresponding amino acids are indicated below the DNA sequences, demonstrating the expansion from 16 alanine residues in the control to 23 residues in the patient. Position of polyalanine tract 1 (PA1) is highlighted in *white box*

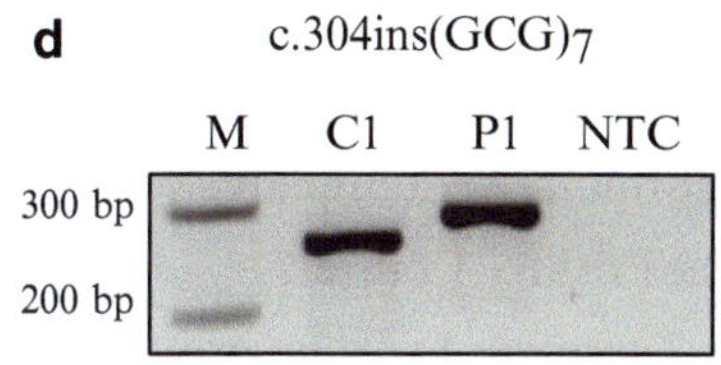

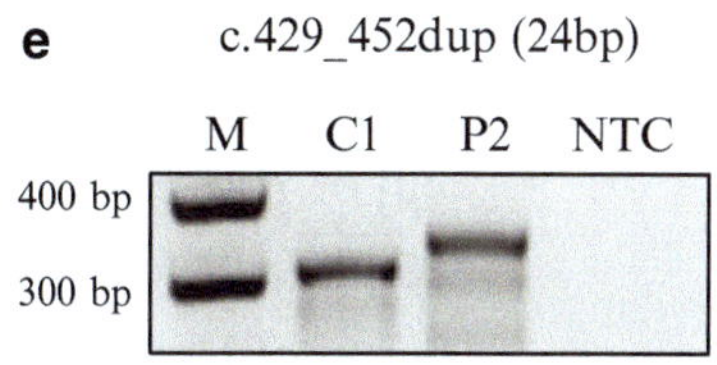

Lanes:		Expected PCR product
M	Marker	-
NTC	Non-template control	-
(A) c.304ins(GCG)$_7$		
C1	Control 1	262 bp
P1	Patient 1 with c.304ins(GCG)$_7$	262 + 21 – 283 bp
(B) c.429_452dup (24bp)		
C1	Control 1	313 bp
P2	Patient 2 with c.429_452dup (24bp)	313 + 24 = 337 bp

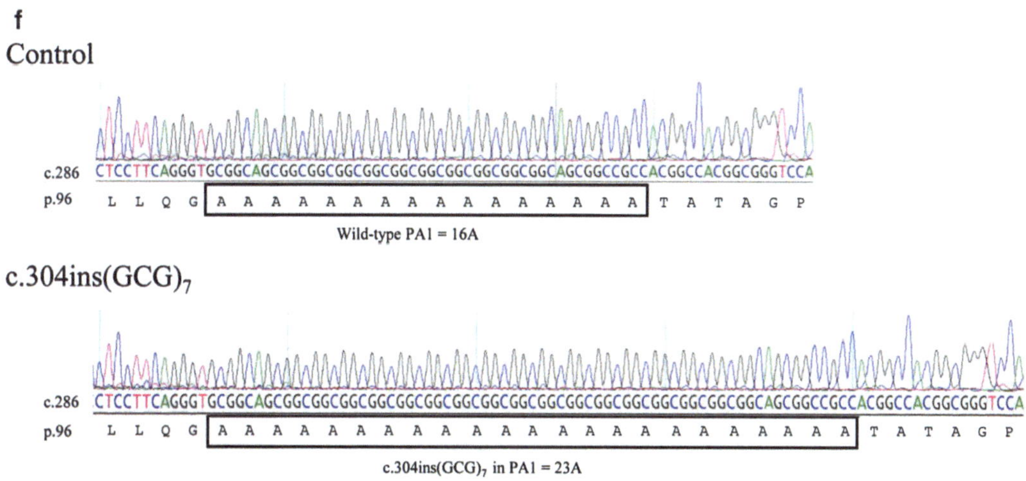

Fig. 2 (continued)

PCR amplification of amplicons 1 and 2 of exon 2 is shown in Fig. 2d, e. Subsequent sequencing outcomes from these products are consistent and reliable, reducing the need for repeat analysis (Fig. 2d–f). The sequencing outcomes from these PCR products (C1 and P1 in Fig. 2d) clearly indicate the difference between a control and a patient with polyalanine expansion mutation (Fig. 2f).

3.1.2 PCR Mixtures

Prepare a 50 μl PCR reaction on ice by adding the following reagents into a 0.2 or 0.5 ml tube:

*x*μl of 100–500 ng genomic DNA template.

25 μl FailSafe™ PCR 2× PreMix J buffer.

0.5 μl of Expand Long Template Enzyme mix.

1 μl of 50 pmol/μl forward primer.

1 μl of 50 pmol/μl reverse primer.

*y*μl of ultrapure water up to 50 μl ($y = 22.5\ \mu l - x$).

Prepare a master mix with all components, but without the template. Add DNA template to each tube before adding an aliquot of master mix. Use of a master mix reduces the error due to differences in pipetting small volumes of multiple components across individual samples. Add DNA template to each tube. Then aliquot the master mix, flick mix, and briefly centrifuge to return sample to the bottom of the tubes before placing in preheated PCR machine. Do not vortex the reaction mixture containing the polymerase/enzyme (*see* **Note 6**).

3.1.3 PCR Conditions

Perform PCR using the following cycling conditions (*see* **Note 7**) in a preheated PCR machine:

Denaturing	94 °C	2 min	
	94 °C	30 s	× 35 cycles
Annealing	60 °C	30 s	× 35 cycles
Elongation	68 °C	2 min	× 35 cycles
	68 °C	10 min	

3.2 Gel Electrophoresis

The presence and size of the expected PCR amplicons is confirmed by gel electrophoresis. Use a 2 % (w/v) agarose gel as the size of expected PCR product is under 1 kb.

1. Mix 6 g of agarose powder with 300 ml of 1× TBE to make a 2 % (w/v) solution.
2. Heat agarose using a microwave oven to boil gently so the agarose dissolves. Let hot agarose cool down to 55–60 °C before addition of ethidium bromide (*see* **Note 8**).
3. Add 0.5 μg/ml of ethidium bromide into 200 ml warm agarose solution, swirl gently to mix, and pour immediately into a gel cast. Ethidium bromide is added to make DNA visible under UV light (*see* **Note 9**).
4. Whilst still hot, place a comb in the gel cast. This forms wells in the gel for PCR samples loading.
5. Prepare 12 μl PCR samples in 1× loading buffer by adding 2 μl of 6x concentrated loading buffer to 10 μl of PCR product.

Flick mix and briefly centrifuge to return sample to the bottom of the tube before loading to gel.

6. When gel solidifies, remove the comb gently from the gel.
7. Place the gel in cast into the electrophoresis tank, and submerge in 1× TBE buffer.
8. Load the DNA marker and PCR samples into individual wells of the gel. The DNA standard is used as a reference to estimate the size of PCR products migrating in the agarose.
9. Close the lid of the electrophoresis tank and connect the tank to electric current. Ensure the negative (black) end and positive (red) end of the current are placed correctly.
10. Turn on the electric current from 90 to 110 V. The DNA strands move away from negative end of the tank. The shorter DNA products move through the gel quicker than the longer DNA strands.
11. Turn off the electric current when the bromophenol blue DNA tracking dye has migrated past halfway down the gel (*see* **Note 10**) (approximately an hour on a 2 % (w/v) agarose gel at 100 V). Remove the gel from the tank.
12. Observe the DNA bands of DNA standard and PCR samples using UV light (*see* **Note 11**).

3.3 PCR Purification

1. Add 5 volumes of DNA binding buffer, PB buffer into per volume of PCR product.
2. Transfer resulting mixture solution into a DNA-capture QIAquick column in a collection tube and centrifuge for a minute.
3. Discard the flow-through and wash the column with 750 μl of wash buffer, PE buffer.
4. Centrifuge for a minute at 17,900 × *g* and discard the wash.
5. Centrifuge for an additional minute to remove residual ethanol from the column (*see* **Note 12**).
6. Elute DNA into a clean 1.5 ml labelled microcentrifuge tube by adding 30 μl of ultrapure water to the center of membrane.
7. Let the column stand at room temperature for a minute.
8. Centrifuge for a minute to obtain purified DNA.
9. Determine the concentration of purified DNA using a spectrophotometer such as NanoDrop spectrophotometer ND-1000 (Thermo Scientific). Absorbance at 260 nm is multiplied by a factor of 50 for double-stranded DNA to get DNA concentration (ng/μl).
10. Store the purified DNA at −20 °C (*see* **Note 13**).

Table 3
***ARX*-specific primers used for sequencing of *ARX* from plasmid**

ARX exon	Primer name	5′ to 3′ sequence	Size (bp)
1–2	ARXex1/2F	TGC AAG GCT CCC CTA AGA GCA	21
2	ARXex2-R2	TCC TCC TCG TCG TCC TCG GTG CCG GT	26
3–4	ARXex3/4F	CCG AGT CCA GGT CTG GTT CCA	21
4	ARXcR1	CAG TCC AAG CGG AGT CGA GCG	21

3.4 Bidirectional Sequencing of GC-Rich Template

3.4.1 Oligonucleotides

The primers used for PCR amplification (Table 2) are also used to sequence the purified PCR products. We routinely use a different set of primers to sequence plasmid DNA containing the open reading frame of *ARX* (no intron and untranslated region). These primers are listed in Table 3.

3.4.2 Sequencing Mixtures

Prepare a 20 μl sequencing reaction on ice by adding the following reagents into a 0.2 or 0.5 ml tube:

x μl of 800–1,500 ng of PCR product or plasmid DNA.

3 μl of BigDye Terminator v.3.1 Sequencing Buffer (5×).

1 μl of BigDye Ready Reaction Mix.

1 μl of DMSO (5 % (v/v) final concentration) (*see* **Note 14**).

1 μl of 3.2 pmol/μl of forward OR reverse primer (*see* **Note 15**).

y μl of ultrapure water up to 20 μl ($y = 14$ μl $- x$).

Make a master mix with all components minus the template. Add DNA template to each tube before adding an aliquot of master mix. Mix (do not vortex) and pulse the samples by brief centrifugation to return samples to the bottom of the tube before placing in preheated PCR machine.

3.4.3 Bidirectional Sequencing Conditions

Perform sequencing amplification under the following cycling conditions:

Denaturing	96 °C	1 min	
	96 °C	10 s	×25 cycles
Annealing	50 °C	5 s	
Elongation	60 °C	4 min	

3.4.4 Precipitation of Sequencing Products

1. Allow tubes to equilibrate at room temperature.
2. Add 75 μl of 0.2 nM $MgSO_4$ ethanol solution to each sample.
3. Mix samples by vortexing and incubate samples for 15 min at room temperature (keep samples in the dark where possible).

4. Centrifuge samples for 15 min at 17,900 × *g*.
5. Remove tubes from microcentrifuge and gently invert the tubes over a paper towel for 1–2 min.
6. Add 100 μl of 70 % ethanol to wash the pellet and centrifuge again for 15 min at 17,900 × *g*. Do not vortex the tubes.
7. Allow tubes to air-dry in a 37 °C heating block for 5 min (*see* **Note 16**). To avoid samples being dislodged, do not aspirate remaining solution.
8. Submit dry samples to DNA sequencing service facility for capillary electrophoresis for size separation, detection, and recording of incorporated dye fluorescence and data output as fluorescent peak trace chromatograms.
9. Analyze the sequencing results using commercial software such as EditSeq and SeqMan programs (DNASTAR, Lasergene software package) or CLC Sequence Viewer 6 program (freeware available from http://www.clcbio.com/index.php?id=28).

4 Notes

1. The quality and purity of template DNA is important and will affect the outcome of PCR amplification.
2. Expand Long Template Enzyme mix is a unique blend of *Taq* DNA polymerase and a thermostable high fidelity *Tgo* DNA polymerase with proofreading activity. Together, these DNA polymerases synergistically generate PCR products with greater yield and three times higher fidelity than *Taq* DNA polymerase alone. GC-rich regions with high melting temperature domains can act as permanent termination sites and have profound effects in the amplification mimicking competitive PCR; when *Taq* DNA polymerase is used in the presence of a proofreading enzyme, this effect may be reduced or eliminated [4].
3. Prepare 0.2 nM $MgSO_4$ stock solution on a weekly basis. Store stock solution at room temperature, do not refrigerate or freeze the solution.
4. To ensure the primer set detects only the desired *ARX* exon (or exon region) when screening genomic DNA, each primer set is submitted for analysis by BLAST (http://www.ncbi.nlm.nih.gov/tools/primer-blast/). BLAST will find sequences in the database, which are similar to the sequences submitted in the query. The results will be displayed in (a) a graph showing the hits found, (b) a table showing sequence identifiers for each hit (scoring related data), (c) alignments for the sequence of interest, and (d) corresponding BLAST scores for each hit

obtained. An ideal primer set will have a 100 % best match to desired target and with no cross homology.

5. Due to high GC base composition at the N-terminal of exon 2, overlapping amplicons with GC content of 69–82 % are designed across exon 2. These overlapping amplicons enable robust PCR efficiency and sequencing coverage. This design is particularly helpful in the case of PCR amplification and sequencing of templates from affected individuals with expanded polyalanine tract mutations. The products generated will have higher GC content than the amplicons generated from controls.
6. Mixing by vortex of PCR reactions containing the enzyme blend will damage the enzyme integrity and cause PCR failure.
7. Optimum initial denaturation temperature for human genomic DNA is 93–95 °C. Expand Long Template Enzyme mix is use for efficient amplification of GC-rich DNA template. Instead of standard PCR elongation temperature at 72 °C, the optimal primer extension temperature is reduced to 68 °C to avoid enzyme loss during prolonged extension times.
8. Direct skin contact with hot agarose can cause severe burns. Always wear protective gloves, goggles, and a lab coat while preparing and casting agarose gels. Swirling of hot agarose leads to boil over. Let hot molten agarose cool down to 55–60 °C before addition of ethidium bromide to prevent boil over.
9. Ethidium bromide is a suspected carcinogen and at high concentration is irritating to the eyes, skin, mucous membranes, and upper respiratory tract. Handle ethidium bromide with extreme care. Always wear protective gloves, goggles, and a lab coat during handling. Gels and solutions containing ethidium bromide should be disposed appropriately as hazardous waste. Refer to Material Safety Data Sheet (MSDS) for more information.
10. On a 1 % agarose gel, xylene cyanol and bromophenol blue typically migrate about the same as a 4,000 bp and 200–400 bp DNA fragment, respectively.
11. Overexposure to ultraviolet (UV) radiation can lead to different stages of erythema (sunburn) and photokeratitis (eye inflammation due to lesions on the cornea and conjunctiva). These symptoms are not immediate and normally appear 4–24 h post exposure. Exposure to UV radiation can be minimized by having adequate skin and eye protection. Always wear a UV-resistant full face shield, goggles with side shields, nitrile gloves, and lab coats when working with UV radiation.

12. Residual ethanol from wash buffer will not be totally removed if flow-through from previous centrifugation is not discarded before this step (as recommended by Qiagen).
13. For long-term storage, store DNA at −20 °C as DNA may degrade in water in the absence of a buffering agent.
14. DMSO can improve the outcome of sequencing reactions from difficult templates such as those with GC-rich regions by relaxing complex secondary structure formations such as hairpins.
15. Primer for the sequencing reaction is freshly prepared at the required dilution each time it is required as DNA primers at low concentration are more prone to degradation through freeze-thaw cycles. Sequencing reactions are conducted with only one primer in each reaction.
16. Do not allow samples to overdry in 37 °C heating block as samples will be difficult to resuspend for subsequent analysis.

Acknowledgments

This work was supported by grants from the National Health and Medical Research Council of Australia (Project Grant 1002732 to C.S.; Principle Research Fellowship 508043 to J.G.) and MS McLeod Foundation Fellowship to C.S. and postgraduate scholarship to M.H.T.

References

1. Wilson AG, Symons JA, McDowell TL, McDevitt HO, Duff GW (1997) Effects of a polymorphism in the human tumor necrosis factor alpha promoter on transcriptional activation. Proc Natl Acad Sci USA 94(7):3195–3199
2. Mamedov TG, Pienaar E, Whitney SE, TerMaat JR, Carvill G, Goliath R, Subramanian A, Viljoen HJ (2008) A fundamental study of the PCR amplification of GC-rich DNA templates. Comput Biol Chem 32(6):452–457. doi:S1476-9271(08)00088-1 [pii] 10.1016/j.compbiolchem.2008.07.021
3. Sutherland GR, Richards RI (1995) The molecular basis of fragile sites in human chromosomes. Curr Opin Genet Dev 5(3):323–327
4. McDowell DG, Burns NA, Parkes HC (1998) Localised sequence regions possessing high melting temperatures prevent the amplification of a DNA mimic in competitive PCR. Nucleic Acids Res 26(14):3340–3347. doi:gkb479 [pii]
5. Musso M, Bocciardi R, Parodi S, Ravazzolo R, Ceccherini I (2006) Betaine, dimethyl sulfoxide, and 7-deaza-dGTP, a powerful mixture for amplification of GC-rich DNA sequences. J Mol Diagn 8(5):544–550. doi:S1525-1578(10)60343-1[pii] 10.2353/jmoldx.2006.060058
6. Ralser M, Querfurth R, Warnatz HJ, Lehrach H, Yaspo ML, Krobitsch S (2006) An efficient and economic enhancer mix for PCR. Biochem Biophys Res Commun 347(3):747–751. doi:S0006-291X(06)01470-7 [pii] 10.1016/j.bbrc.2006.06.151
7. Baskaran N, Kandpal RP, Bhargava AK, Glynn MW, Bale A, Weissman SM (1996) Uniform amplification of a mixture of deoxyribonucleic acids with varying GC content. Genome Res 6(7):633–638
8. Chakrabarti R, Schutt CE (2001) The enhancement of PCR amplification by low molecular-weight sulfones. Gene 274(1–2):293–298. doi:S0378111901006217 [pii]
9. Guerrini R, Moro F, Kato M, Barkovich AJ, Shiihara T, McShane MA, Hurst J, Loi M, Tohyama J, Norci V, Hayasaka K, Kang UJ, Das S, Dobyns WB (2007) Expansion of the first PolyA tract of ARX causes infantile spasms and status dystonicus. Neurology 69(5):427–433.

doi:69/5/427 [pii] 10.1212/01.wnl.0000266594.16202.c1

10. Henke W, Herdel K, Jung K, Schnorr D, Loening SA (1997) Betaine improves the PCR amplification of GC-rich DNA sequences. Nucleic Acids Res 25(19):3957–3958. doi:gka606 [pii]
11. Kang J, Lee MS, Gorenstein DG (2005) The enhancement of PCR amplification of a random sequence DNA library by DMSO and betaine: application to in vitro combinatorial selection of aptamers. J Biochem Biophys Methods 64(2):147–151. doi:S0165-022X(05)00082-5 [pii] 10.1016/j.jbbm.2005.06.003
12. Mutter GL, Boynton KA (1995) PCR bias in amplification of androgen receptor alleles, a trinucleotide repeat marker used in clonality studies. Nucleic Acids Res 23(8):1411–1418. doi:4s0766 [pii]
13. Pomp D, Medrano JF (1991) Organic solvents as facilitators of polymerase chain reaction. Biotechniques 10(1):58–59
14. Sidhu MK, Liao MJ, Rashidbaigi A (1996) Dimethyl sulfoxide improves RNA amplification. Biotechniques 21(1):44–47
15. Sun Y, Hegamyer G, Colburn NH (1993) PCR-direct sequencing of a GC-rich region by inclusion of 10 % DMSO: application to mouse c-jun. Biotechniques 15(3):372–374
16. Turner SL, Jenkins FJ (1995) Use of deoxyinosine in PCR to improve amplification of GC-rich DNA. Biotechniques 19(1):48–52
17. Weissensteiner T, Lanchbury JS (1996) Strategy for controlling preferential amplification and avoiding false negatives in PCR typing. Biotechniques 21(6):1102–1108
18. Agarwal RK, Perl A (1993) PCR amplification of highly GC-rich DNA template after denaturation by NaOH. Nucleic Acids Res 21(22):5283–5284
19. Bachmann HS, Siffert W, Frey UH (2003) Successful amplification of extremely GC-rich promoter regions using a novel 'slowdown PCR' technique. Pharmacogenetics 13(12): 759–766. doi:10.1097/01.fpc.0000054140.14659.61
20. Li LB, Yu Z, Teng X, Bonini NM (2008) RNA toxicity is a component of ataxin-3 degeneration in Drosophila. Nature 453(7198):1107–1111. doi:nature06909 [pii] 10.1038/nature06909
21. Frey UH, Bachmann HS, Peters J, Siffert W (2008) PCR-amplification of GC-rich regions: 'slowdown PCR'. Nat Protoc 3(8):1312–1317. doi:nprot.2008.112 [pii] 10.1038/nprot.2008.112
22. Sahdev S, Saini S, Tiwari P, Saxena S, Singh Saini K (2007) Amplification of GC-rich genes by following a combination strategy of primer design, enhancers and modified PCR cycle conditions. Mol Cell Probes 21(4):303–307. doi:S0890-8508(07)00028-X [pii] 10.1016/j.mcp.2007.03.004
23. Shoubridge C, Fullston T, Gecz J (2010) ARX spectrum disorders: making inroads into the molecular pathology. Hum Mutat 31(8):889–900. doi:10.1002/humu.21288
24. Kato M, Das S, Petras K, Kitamura K, Morohashi K, Abuelo DN, Barr M, Bonneau D, Brady AF, Carpenter NJ, Cipero KL, Frisone F, Fukuda T, Guerrini R, Iida E, Itoh M, Lewanda AF, Nanba Y, Oka A, Proud VK, Saugier-Veber P, Schelley SL, Selicorni A, Shaner R, Silengo M, Stewart F, Sugiyama N, Toyama J, Toutain A, Vargas AL, Yanazawa M, Zackai EH, Dobyns WB (2004) Mutations of ARX are associated with striking pleiotropy and consistent genotype-phenotype correlation. Hum Mutat 23(2):147–159. doi:10.1002/humu.10310
25. Fullston T, Finnis M, Hackett A, Hodgson B, Brueton L, Baynam G, Norman A, Reish O, Shoubridge C, Gecz J (2011) Screening and cell-based assessment of mutations in the Aristaless-related homeobox (ARX) gene. Clin Genet 80(6):510–522. doi:10.1111/j.1399-0004.2011.01685.x
26. Matera I, Bachetti T, Puppo F, Di Duca M, Morandi F, Casiraghi GM, Cilio MR, Hennekam R, Hofstra R, Schober JG, Ravazzolo R, Ottonello G, Ceccherini I (2004) PHOX2B mutations and polyalanine expansions correlate with the severity of the respiratory phenotype and associated symptoms in both congenital and late onset Central Hypoventilation syndrome. J Med Genet 41(5):373–380
27. Laumonnier F, Ronce N, Hamel BC, Thomas P, Lespinasse J, Raynaud M, Paringaux C, Van Bokhoven H, Kalscheuer V, Fryns JP, Chelly J, Moraine C, Briault S (2002) Transcription factor SOX3 is involved in X-linked mental retardation with growth hormone deficiency. Am J Hum Genet 71(6):1450–1455
28. A novel gene containing a trinucleotide repeat that is expanded and unstable on Huntington's disease chromosomes. Huntington's Disease Collaborative Research Group (1993). Cell 72(6):971–983
29. Banfi S, Servadio A, Chung MY, Kwiatkowski TJ Jr, McCall AE, Duvick LA, Shen Y, Roth EJ, Orr HT, Zoghbi HY (1994) Identification and characterization of the gene causing type 1 spinocerebellar ataxia. Nat Genet 7(4):513–520. doi:10.1038/ng0894-513

Chapter 9

Challenges of "Sticky" Co-immunoprecipitation: Polyalanine Tract Protein–Protein Interactions

T.R. Mattiske, May H. Tan, Jozef Gécz, and Cheryl Shoubridge

Abstract

Co-immunoprecipitation (Co-IP) (followed by immunoblotting) is a technique widely used to characterize specific protein–protein interactions. Investigating interactions of proteins containing "sticky" polyalanine (PolyA) tracts encounters difficulties using conventional Co-IP procedures. Here, we present strategies to specifically capture proteins containing these difficult PolyA tracts, enabling subsequent robust detection of interacting proteins by Co-IP.

Key words Polyalanine tract, ARX, Co-immunoprecipitation, Protein–protein interaction

1 Introduction

Expansions of repeating trinucleotide sequences in the genome above a certain length have been linked to a growing number of human diseases. The trinucleotide repeats implicated in these disorders reside in either coding or noncoding sequences, transcribed but not translated sequences or within translated sequences for homomeric tracts of either glutamine or alanine amino acids [1]. In the case of the alanine tract expansion disorders, there are at least nine such hereditary diseases, eight of which involve transcription factors, including the *Aristaless*-related homeobox gene (*ARX*) [1, 2].

ARX [NM_139058.2] (MIM 300382) is a member of the paired-type homeodomain transcription factor family with critical roles in embryonic development [3–5]. Mutations in the *ARX* gene are frequent and with a broad range of clinical expressivity, giving rise to non-syndromic intellectual disability, infantile spasms, or developmental brain malformations [1, 3]. ARX has four polyalanine tracts and expansions of the first two tracts accounts for over half (~60 %) of all patients with *ARX* mutations.

Danny M. Hatters and Anthony J. Hannan (eds.), *Tandem Repeats in Genes, Proteins, and Disease: Methods and Protocols*, Methods in Molecular Biology, vol. 1017, DOI 10.1007/978-1-62703-438-8_9,

Alanine (A) is a hydrophobic, nonpolar amino acid often buried inside the core of proteins. In vitro studies have shown polyalanine peptides containing 7–15 alanines undergo variable levels of conformational transition from a monomeric α-helix to a predominant macromolecular β-sheet [6]. These sticky β-sheets formed from alanine starches are extremely resistant to chemical denaturation and enzymatic degradation. An increase in alanines leads to a conformation rich in beta-sheets, with peptides harboring more than 15 alanines completely converted from monomer to β-sheets [7, 8]. The length of polyalanine tracts is restricted in eukaryotes, usually less than 10 residues and not exceeding 20 residues [9]. Experimental evidence indicates disease-causing expanded polyalanine tracts beyond a certain threshold may induce mis-folding and aberrant protein interactions, degradation, mis-localization and/or aggregation [2, 10–13]. In the case of ARX, when longer polyalanine tract expansion mutations are overexpressed in vitro, these proteins stick together resulting in aggregation within the cell [14–16]. Hence, disease-causing expansions to polyalanine tracts may expose these hydrophobic amino acids to the outside of the protein.

There are several different mutations in *ARX* that lead to a range of expanded polyalanine tract lengths [1]. The 16 and 12 residues of tract 1 and 2, respectively, can be increased to as many as 27 for tract 1 and 21 in tract 2. The severity of the clinical phenotypes associated with these mutations generally increases with the length of the polyalanine tract [17]. Hence, to elucidate the mechanism(s) behind the pathogenesis of these mutations, we need to determine how expanded polyalanine tracts influence the function of the ARX homeodomain transcription factor. One such avenue of investigation is to determine if the interaction of ARX protein with specific protein partners is altered due to expanded polyalanine tracts. The technique of co-immunoprecipitation (Co-IP) is an invaluable tool to study these types of interactions.

Co-IP is a widely used complementary method for the identification and characterization of protein–protein interactions. Co-IP is particularly well suited to identify or examine protein interactions as this technique can be performed in vivo and in vitro. This procedure is unbiased and can lead to the identification of new interacting partners and complexes of proteins. This approach relies on the ability of an antibody (IP antibody) to specifically bind complexes containing a bait protein. A specific antibody against the protein of interest is essential. If this is not available, tagged protein (with, e.g., V5 or Myc) can be used together with a commercial antibody against the tag to capture the protein of interest. The antibody provides a means of immobilizing these complexes to a solid matrix, which in the protocol presented here is accomplished through interaction with protein-A sepharose. Irrelevant proteins are then removed with varying stringency of the wash buffer. Specific proteins interacting with the bait protein are then eluted with the bait protein and subsequently identified by techniques such as immunoblotting or even mass spectrometry.

The high number of proteins in cell lysates means that nonspecific binding to the protein-A sepharose or IP antibody can confound the detection of immunoprecipitated target(s). Additionally, proteins that are normally separated into discrete cellular compartments are mixed together and nonphysiological binding to the target complex may occur. To prevent nonspecific binding to the protein-A sepharose usually a preclearing step is undertaken to remove "sticky" proteins from the lysate. Lysates are incubated with a small volume of protein-A sepharose in the absence of the IP antibody. The protein-A sepharose with bound proteins is then discarded. The desired protein (along with any interacting partners) is then bound to the IP antibody before pulldown with a fresh aliquot of protein-A sepharose. However, proteins that contain PolyA tracts are particularly "sticky" and may cause the protein to bind to the protein-A sepharose in the absence of an antibody. In this case, preclearing the samples would diminish these proteins and any interacting partners in the subsequent IP. Therefore, an alternative procedure needs to be considered when working with sticky proteins such as those containing PolyA tracts. We have found that pre-blocking the protein-A sepharose with whole-cell extract (non-transfected cells) considerably minimizes nonspecific binding of PolyA tract containing ARX protein to the protein-A sepharose allowing antibody-specific pulldown to be achieved [15, 18]. In this chapter, we provide a practical starting point to overcome some of the problems associated with PolyA tract containing proteins when using this technique, highlighting potential modifications to optimize the procedure.

2 Materials

2.1 Co-IP

1. Protein-A sepharose™ CL-4B (GE Healthcare, Cat No# 17-0780-01) (*see* **Note 1**). Prior to starting co-immunoprecipitation, resuspend 500 mg of protein-A sepharose powder in 10 ml of phosphate-buffered saline solution (PBS) (130 mM NaCl and 20 mM sodium phosphate, pH 7.5) with gentle mixing at room temperature for 5 min and store at 4 °C overnight. The next day, spin at 700 × *g* for 5 min and remove clear liquid above settled sepharose by pipetting. Wash in PBS three times by adding 10 ml of PBS and repeat centrifugation as stated above, discarding the PBS supernatant each time. Resuspend in 2 to 2.5 ml of 0.1 % (w/v) sodium azide in PBS in a ratio of 50 % settled gel to 50 % buffer (v/v) and store at 4 °C until you are ready to start (*see* **Note 2**).

2. Lysis buffer: 50 mM Tris–HCl, 120 mM NaCl, 0.5 % (v/v) Nonidet P-40, pH 8.0. Weigh 3.51 g of NaCl and 3.026 g of Tris and transfer to a glass beaker. Add water to a volume of 400 ml. Mix and adjust pH with HCl. Add 2.5 ml Nonidet P-40 and make up to 500 ml with Milli-Q water. Store at 4 °C.
3. 25× Protease Inhibitor Cocktail (Sigma Aldrich, Cat # P8340-5ML) with 25 μg/ml each of aprotinin, leupeptin, and pepstatin.
4. 200 mM of activated sodium vanadate (Na_3VO_4) (stored at −20 °C) (*see* **Note 3**).
5. 200 mM sodium fluoride (NaF) (stored at 4 °C).
6. 200 mM phenylmethylsulfonyl fluoride (PMSF) (stored at 4 °C).
7. Wash Buffer: 250 mM NaCl (*see* **Note 4**), 20 mM Tris–HCI, 1 mM ethylenediaminetetraacetic acid (EDTA), O.5 % (v/v) Nonidet P-40, pH 8.0.
8. 21 G needle.
9. 3 ml syringe.
10. Cell scrapers.
11. Empty spin columns (VWR International, Cat # 27-3565-01) (*see* **Note 5**).
12. Screw cap 1.5 ml tubes.
13. IP antibody (antibody against protein or epitope of interest) (*see* **Note 6**).
14. Plastic paraffin film.
15. Rotating wheel.
16. Heat block at 95 °C.
17. NuPAGE® LDS Sample Buffer (4×): 247 mM Tris–HCl, pH 8.0, 2 % (w/v) lithium dodecyl sulfate (LDS), 0.51 mM EDTA, 0.22 mM SERVA® Blue G250, 0.175 mM phenol red, 10 % (v/v) glycerol.
18. 1× SDS loading buffer: 25 μl of NuPAGE® LDS Sample Buffer, 5 μl of 1 M dithiothreitol (DTT), and ultrapure water up to 100 μl. Preheat to 65 °C before use.

3 Methods

A flowchart illustrating the timing and order of different aspects of the Co-IP method are shown in Fig. 1.

3.1 Protein Extraction Using Lysis Buffer

Before starting the Co-IP, you should be certain that you can detect the protein to be precipitated by immunoblot with the lysis conditions used for Co-IP.

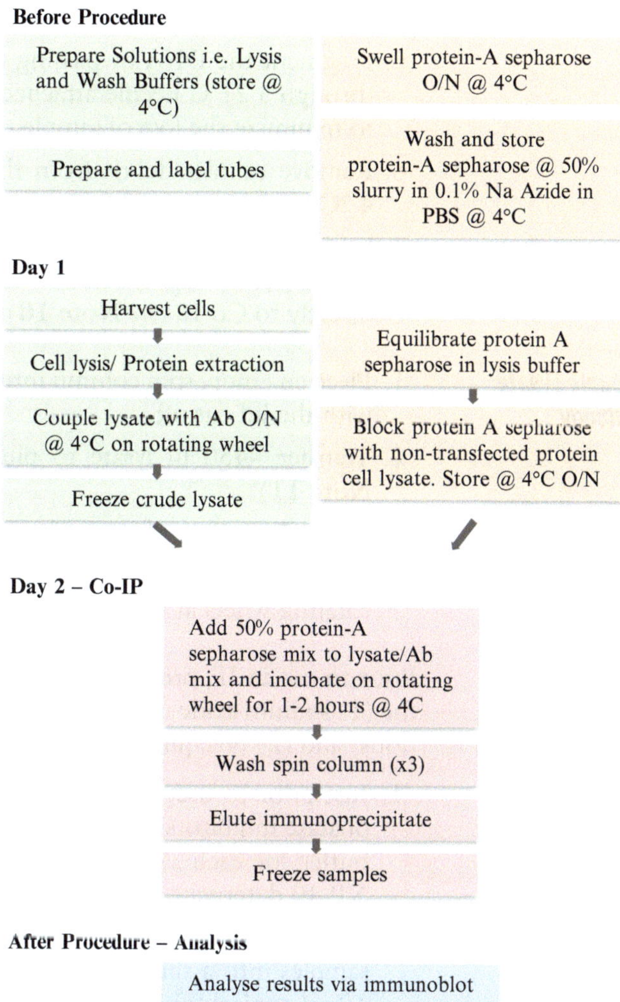

Fig. 1 Flow chart showing experimental flow in real time to indicate how procedures can be run in tandem to optimize time efficiently

After harvest, the cells and lysates should be maintained at 4 °C from that point on.

1. To harvest cells, suction off media and add 200 μl PBS to each well. Scrape cells and transfer to pre-labelled 1.5 ml tubes. Gently pellet cells via centrifugation (700 × *g*) for 3 min and carefully remove fluid from pellet (*see* **Note 7**).
2. Make up 1 ml of activated lysis buffer per sample by adding 40 μl 25× Protease Inhibitor Cocktail + 5 μl 200 mM Na_3VO_4 + 5 μl 200 mM NaF + 5 μl 200 mM PMSF to 1 ml of Lysis Buffer (*see* **Note 8**).
3. Add 1 ml of fresh activated lysis buffer to each sample and pipette up and down gently ten times to ensure complete dispersion and lysis of cells (*see* **Note 9**).

4. Place tubes horizontal in ice and shake cells for 15 min.
5. Fragment the DNA by passing the lysed suspension ten times through a 21 G needle attached to a 3 ml syringe taking care to minimize the loss of sample.
6. Remove cellular debris from the lysates by centrifugation at 4 °C for 30 min at 13,000 × *g*.
7. Without delay, collect supernatant into fresh, labelled, pre-chilled screw cap tubes and leave on ice while proceeding directly to Co-IP (*see* **Note 10**).

3.2 Couple Lysate and Antibody

1. Place an empty spin column into a labelled 1.5 ml flip top tube with the lid cut off.
2. Transfer ~450 μl lysate to plugged micro-spin column (*see* **Note 11**).
3. Add 0.5 μg IP antibody to lysate column (*see* **Note 12**).
4. Firmly secure the lid and plastic paraffin film top and place on rotating wheel at 4 °C overnight.

3.3 Equilibrate Protein-A sepharose in Lysis Buffer

Prior to use in IP, pre-swollen protein-A sepharose needs to have the (1) sodium azide removed (as this can inhibit antibody interactions) and (2) be equilibrated in lysis minus buffer.

1. Just prior to use, prepare lysis minus buffer by adding phosphatase inhibitors (add 5 μl Na_3VO_4 + 5 μl NaF to 1 ml of lysis buffer for each sample), but not the other additives such as NP-40 detergent.
2. Transfer enough 50 % (v/v) protein-A sepharose slurry for all samples into a suitably large tube for equilibration (such as a 10 ml conical tube). We routinely prepare as much as 90 μl of slurry per sample at this stage.
3. Spin at 700 × *g* for 5 min, remove, and discard PBS supernatant via pipetting.
4. Wash protein-A sepharose slurry in 2× volume lysis minus buffer.
5. Repeat centrifugation and wash in **steps 3** and **4**.
6. Resuspend protein-A sepharose pellet in equal volume lysis minus buffer to make 50 % (v/v) slurry.

3.4 Blocking Protein-A sepharose

To avoid binding of "sticky" PolyA tract containing target proteins to the protein-A sepharose, we have developed an alternative blocking strategy. By preincubating the protein-A sepharose with cell lysate minus the protein of interest (non-transfected cell lysate), we have overcome the binding of our PolyA tract containing proteins to the protein-A sepharose. When pre-blocked protein-A sepharose is added to the protein–antibody immune complex sample,

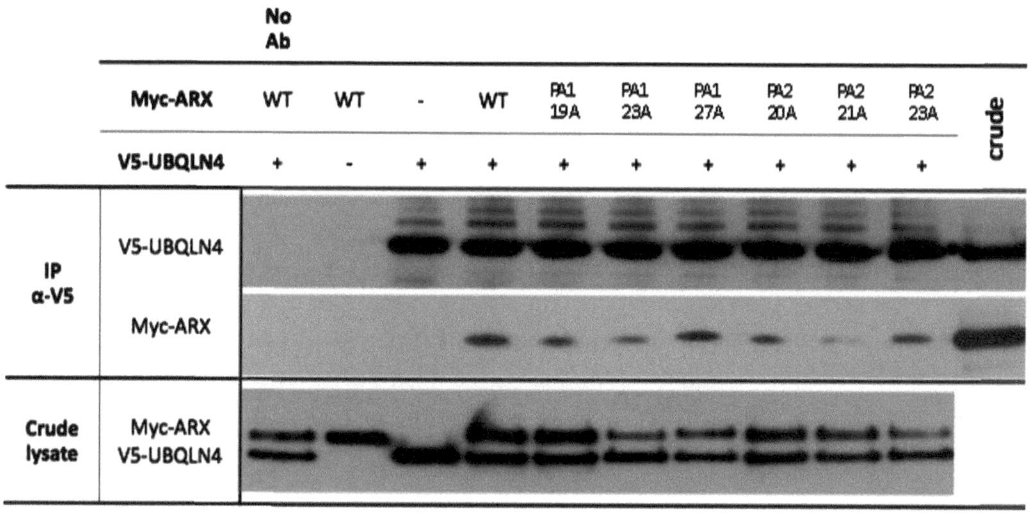

Fig. 2 Polyalanine tract mutant ARX co-immunoprecipitates with UBQLN4 in mammalian cells. HEK293T cells transfected with Myc-ARX-Wt or PA mutant constructs and V5-UBQLN4 were lysed and protein immunoprecipitated (IP) with antibodies against the V5. Samples were loaded onto 4–12 % SDS-PAGE gels and analyzed for the presence of Co-IP proteins. Detection of Myc-ARX proteins bound to V5-UBQLN4 by immunoblotting with mouse anti-Myc horseradish-peroxidase (HRP)-conjugated antibody. All samples transfected with V5-UBQLN4 showed a protein band of the predicted size upon blotting with anti-V5 HRP-conjugated antibody. Specific IP of each overexpressed protein was achieved with no band present in samples from cells transfected with both Myc-ARX and V5-UBQLN4 but no IP antibody added. Cells transfected with Myc-ARX alone or V5-UBQLN4 alone were used as negative controls. V5-UBQLN4 (~64 kDa) and Myc-ARX (~62 kDa) are both present in protein lysates (*bottom panel*)

we get consistent and robust enrichment of our target protein with negligible pulldown in our "no antibody" control (Fig. 2).

1. Blocking of enough protein-A sepharose for all samples can be done in one batch. For each sample, add 50 μg of protein from non-transfected whole-cell lysate to 90 μl of protein-A sepharose in lysis minus slurry in a tube large enough for subsequent washing steps.
2. Firmly secure the lid with plastic paraffin film and incubate for 1 h at 4 °C on a rotating wheel.
3. Centrifuge pre-blocked protein-A sepharose at 700×*g* for 5 min, remove, and discard the supernatant.
4. Wash pellet in 2× volume lysis minus buffer, centrifuge as above and discard supernatant. Repeat wash (*see* **Note 13**).
5. Resuspend pre-blocked protein-A sepharose pellet in equal volume lysis minus buffer.

3.5 Immuno-precipitation

1. 50 μl of 50 % (v/v) pre-blocked protein-A-sepharose slurry in lysis minus buffer was added to lysate–antibody mix in an empty spin column. Incubate on rotating wheel at 4 °C for 2 h (*see* **Note 14**).
2. Loosen lid on screw top and remove plug from spin column. Keep the plugs for **step 8**.
3. Place spin column in a collection tube and remove all fluid from protein-A sepharose by centrifugation (2,000 × *g*) for 10 s. Transfer the spin column to a new collection tube. Keep the flow through (for this and subsequent wash steps) if required for analysis of wash buffer stringency.
4. Add 400 μl of wash buffer, replace lid loosely, and centrifuge (2,000 × *g*) for 10 s. Transfer spin column to a new collection tube (wash 1).
5. Repeat **step 4** (wash 2).
6. Preheat 1× SDS elution buffer to 65 °C.
7. Repeat **step 4**, but incubate in wash buffer for 5 min before centrifugation (wash 3).
8. Transfer spin columns to labelled collection tubes. Re-plug spin columns and add 40 μl of preheated 1× SDS elution buffer to each column and incubate for 2–3 min while gently mixing (*see* **Note 15**).
9. Remove plugs and centrifuge column in labelled collection tube for 10 s to elute bound antibody and bound proteins. Samples can now be stored at −20 °C until analysis by immunoblot (*see* **Note 16**).

4 Notes

1. Protein-A is commonly used; however, protein-G binds a broader range of Ig subtypes at higher efficiency. For antibodies used in this protocol, protein-A or protein-G both work well.
2. Make up at least 25 % more than minimum required accounting for loss during washing steps and pipetting. When stored at 4 °C in the presence of sodium azide, the sepharose is stable for months; however we routinely make fresh and use within 1–2 weeks.
3. To make 200 mM activated Na_3VO_4 dissolve 368 mg of sodium orthovanadate in 9 ml of purified water in a 50 ml conical tube and mix by vortexing. Adjust to pH 10 with HCl, with stirring. At pH 10, solution will be yellow. Boil the solution until it turns colorless (approximately 10 min). All of the crystals should be dissolved. Cool to room temperature. Readjust to

pH 10 and repeat **steps 3** and **4** until solution remains colorless and pH stabilizes at 10. Adjust the final volume to 10 ml with purified water. Store the activated sodium orthovanadate in 500 μl aliquots and freeze at −20 °C.

4. To reduce nonspecific protein binding to the target protein, increasing the ionic strength (stringency) of the wash buffer can be optimized. Higher concentration of salt in the wash buffer will reduce protein–protein interactions. To find the optimal ionic strength of the wash buffer that disrupts nonspecific protein–protein interactions but still allows specific interactions with your target protein, titrate the salt concentration starting from 120 mM increasing up to a strength that gives no signal in the "no antibody" control, which can be as high as 900 mM NaCl.
5. We have found using empty spin column tubes allows us to increase washing efficiency, enables a more effective elution of antigen and associated protein, and eliminates resin loss, yielding more consistent results.
6. When endogenous protein expression is very low, cells can be transiently transfected with a plasmid encoding the protein of interest under control of an appropriate promoter. If the antibody against the protein of interest is not sufficient for immunoprecipitation, the target sequence can be cloned in frame with a short tag such as MYC or nV5 to produce a fusion protein. There are commercially available antibodies against these types of tags which have two significant advantages: (1) proven specific pulldown in IP procedures and (2) commercially available versions that are directly conjugated to HRP secondary antibodies, enabling elimination of contamination by heavy and light chains in subsequent immunoblot detection of the Co-IP eluate.
7. Two types of controls are routinely included to ensure the interaction of the proteins is specific. (1) To ensure that the binding and stringency of the washing buffers are optimized to detect protein binding specifically to the target protein precipitated by the specific antibody and not binding to protein-A sepharose directly, we routinely include a "no antibody" control using a cell lysate known to contain the target protein. Ensure adequate cell lysis is available. (2) An additional control if required would be to control for the potential of nonspecific binding of proteins to the antibody used to pull down the target. For this control, a sample of cell lysate known to be deficient for the protein of interest can be included. (3) Once a Co-IP relationship is established, the relationship can be confirmed using reciprocal immunoprecipitation (i.e., immunoprecipitate the interacting protein directly and look for pulldown of the original target protein).

8. For a successful co-precipitation, you need to generate whole-cell extracts that optimize the yield and activity of the proteins to be analyzed. Global inhibition of proteolysis through the inclusion of multiple classes of protease inhibitors is usually essential. We include both protease (protease cocktail mix and PMSF) and phosphatase (NaF and Na_3VO_4) inhibitors in the activated lysis buffer. If problems arise with extract preparations, it is recommended to vary the amount of salt (NaCl 0–500 mM) and nonionic detergent (0–1 %) or employ additional methods of breaking open cells, such as multiple freeze–thaw cycles.
9. To give an example of sample handling, concentrations, and volumes used, we routinely plate HEK293T cells in a 6-well plate at 8×10^5 cells for transfection of our target 24 h later as previously described [18, 19]. Cells are harvested 24 h post-transfection. Six wells for each construct are harvested together and lysed in 1 ml of lysis buffer. This routinely yields ~2 mg/ml of protein. We then split this crude lysis into three parts and two lots of ~450 μl of lysate used for reciprocal Co-IP (0.9 mg of protein in each sample) with the remaining 100 μl for crude lysate sample for immunoblot validation of protein expression. For each 450 μl of sample immunoprecipitated, we elute in a final volume between 40 and 60 μl, with between 12 and 15 μl of sample analyzed per lane by SDS-PAGE.
10. To minimize disruption to protein–protein interactions, it is recommended that samples for co-immunoprecipitation are not frozen until after the final elution step.
11. The remaining crude lysate can be stored at −20 °C. At the same time stock of crude lysate (15 μl), 4× SDS Loading Dye (40 μl), 1 M DTT (8 μl), and water (12 μl) can be prepared (run 15 μl of this stock per lane by SDS-PAGE) and stored at −20 °C ready to be analyzed by immunoblot.
12. The amount of antibody needed is dependent on the antigen, the antibody, and its concentration. For each antibody–protein combination, the optimal coupling time and conditions need to be determined. The antibody concentration should be decreased/increased until the signal to noise ratio is maximized.
13. If nonspecific binding of polyalanine tract protein persists, the steps to wash the cell lysate blocking the sepharose can be omitted to increase blocking of protein-A sepharose (Fig. 3).
14. Incubation times can vary and if you have many samples, you may choose to continue onto the wash steps by doing the samples in batches while the other sample are still rotating at 4 °C.
15. It is recommended that only one to three samples are done at the same time to ensure the incubation times with 65 °C SDS

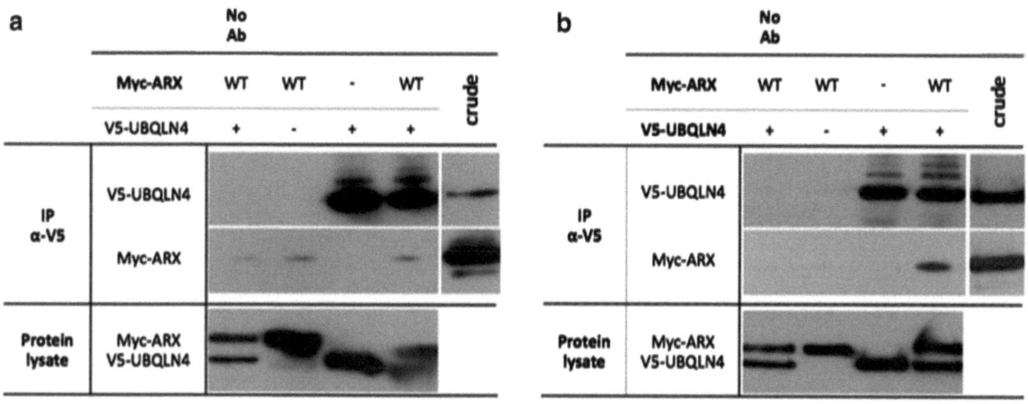

Fig. 3 Pre-blocking of protein-A sepharose with non-transfected HEK293T cell lysate decreases nonspecific binding of PolyA tract containing proteins. HEK293T cells transfected with Myc-ARX-WT and/or V5-UBQLN4 were lysed and protein immunoprecipitated with antibodies against V5 using a 250 mM NaCl wash buffer in both (**a**) and (**b**). Protein-A sepharose is pre-blocked with untransfected HEK293T cell lysate prior to addition of the lysate/antibody mix. Detection of Myc-ARX proteins bound to V5-UBQLN4 by immunoblotting with mouse anti-Myc HRP conjugated antibody. All samples transfected with V5-UBQLN4 showed a protein band of the predicted size upon blotting with anti-V5 HRP-conjugated antibody. V5-UBQLN4 (~64 kDa) and Myc-ARX (~62 kDa) are both present in protein lysates from transfected cells (bottom panel). (**a**) Lysate used to pre-block the protein-A sepharose was removed via washing, and after a 5 min ECL exposure of resulting immunoblot, a nonspecific band in the negative controls can be detected. In an effort to decrease the nonspecific binding of ARX, (**b**) lysate used to pre-block protein-A sepharose was not removed via wash steps. This results in a specific IP of each overexpressed protein but no band present in "no antibody" control of samples from cells transfected with the Myc-ARX negative control and co-transfected with both Myc-ARX and V5-UBQLN4

elution buffer are the same across all samples and to minimize nonspecific elution off the protein-A sepharose.

16. It is good practice to analyze the crude sample first to determine expression of proteins of interest and expected migration of each protein species by SDS-PAGE. When running Co-IP samples, we routinely include the following controls on each blot: (1) no antibody control, (2) no protein of interest control, and (3) a sample of crude lysate expressing the protein(s) of interest. These controls assist in confirming specificity of the immunoprecipitation procedure and ensure the proteins of interest are expressed and are being detected successfully by the immunoblot.

Acknowledgments

This work was supported by grants from the National Health and Medical Research Council of Australia (Project Grant 1002732 to C.S.; Principle Research Fellowship 508043 to J.G.), Australian Research Council Future Fellowship 120100086 to C.S., and MS McLeod Foundation postgraduate scholarship to M.H.T.

References

1. Shoubridge C, Fullston T, Gecz J (2010) ARX spectrum disorders: making inroads into the molecular pathology. Hum Mutat 31(8):889–900. doi:10.1002/humu.21288
2. Albrecht AN, Kornak U, Boddrich A, Suring K, Robinson PN, Stiege AC, Lurz R, Stricker S, Wanker EE, Mundlos S (2004) A molecular pathogenesis for transcription factor associated poly-alanine tract expansions. Hum Mol Genet 13(20):2351–2359
3. Stromme P, Mangelsdorf ME, Scheffer IE, Gecz J (2002) Infantile spasms, dystonia, and other X-linked phenotypes caused by mutations in Aristaless related homeobox gene, ARX. Brain Dev 24(5):266–268. doi: S0387760402000797 [pii]
4. Kitamura K, Yanazawa M, Sugiyama N, Miura H, Iizuka-Kogo A, Kusaka M, Omichi K, Suzuki R, Kato-Fukui Y, Kamiirisa K, Matsuo M, Kamijo S, Kasahara M, Yoshioka H, Ogata T, Fukuda T, Kondo I, Kato M, Dobyns WB, Yokoyama M, Morohashi K (2002) Mutation of ARX causes abnormal development of forebrain and testes in mice and X-linked lissencephaly with abnormal genitalia in humans. Nat Genet 32(3):359–369. doi:10.1038/ng1009 ng1009 [pii]
5. Kato M, Das S, Petras K, Kitamura K, Morohashi K, Abuelo DN, Barr M, Bonneau D, Brady AF, Carpenter NJ, Cipero KL, Frisone F, Fukuda T, Guerrini R, Iida E, Itoh M, Lewanda AF, Nanba Y, Oka A, Proud VK, Saugier-Veber P, Schelley SL, Selicorni A, Shaner R, Silengo M, Stewart F, Sugiyama N, Toyama J, Toutain A, Vargas AL, Yanazawa M, Zackai EH, Dobyns WB (2004) Mutations of ARX are associated with striking pleiotropy and consistent genotype-phenotype correlation. Hum Mutat 23(2):147–159. doi:10.1002/humu.10310
6. Forood B, Perez-Paya E, Houghten RA, Blondelle SE (1995) Formation of an extremely stable polyalanine beta-sheet macromolecule. Biochem Biophys Res Commun 211(1):7–13
7. Shinchuk LM, Sharma D, Blondelle SE, Reixach N, Inouye H, Kirschner DA (2005) Poly-(L-alanine) expansions form core beta-sheets that nucleate amyloid assembly. Proteins 61(3):579–589. doi:10.1002/prot.20536
8. Blondelle SE, Forood B, Houghten RA, Perez-Paya E (1997) Polyalanine-based peptides as models for self-associated beta-pleated-sheet complexes. Biochemistry 36(27):8393–8400. doi:10.1021/bi963015b
9. Albrecht A, Mundlos S (2005) The other trinucleotide repeat: polyalanine expansion disorders. Curr Opin Genet Dev 15(3):285–293. doi:S0959-437X(05)00055-9 [pii] 10.1016/j.gde.2005.04.003
10. Bachetti T, Matera I, Borghini S, Di Duca M, Ravazzolo R, Ceccherini I (2005) Distinct pathogenetic mechanisms for PHOX2B associated polyalanine expansions and frameshift mutations in congenital central hypoventilation syndrome. Hum Mol Genet 14(13):1815–1824. doi:ddi188 [pii] 10.1093/hmg/ddi188
11. Caburet S, Demarez A, Moumne L, Fellous M, De Baere E, Veitia RA (2004) A recurrent polyalanine expansion in the transcription factor FOXL2 induces extensive nuclear and cytoplasmic protein aggregation. J Med Genet 41(12):932–936. doi:41/12/932 [pii] 10.1136/jmg.2004.024356
12. Moumne L, Dipietromaria A, Batista F, Kocer A, Fellous M, Pailhoux E, Veitia RA (2008) Differential aggregation and functional impairment induced by polyalanine expansions in FOXL2, a transcription factor involved in cranio-facial and ovarian development. Hum Mol Genet 17(7):1010–1019. doi:ddm373 [pii] 10.1093/hmg/ddm373
13. Utsch B, McCabe CD, Galbraith K, Gonzalez R, Born M, Dotsch J, Ludwig M, Reutter H, Innis JW (2007) Molecular characterization of HOXA13 polyalanine expansion proteins in hand-foot-genital syndrome. Am J Med Genet A 143A(24):3161–3168. doi:10.1002/ajmg.a.31967
14. Nasrallah IM, Minarcik JC, Golden JA (2004) A polyalanine tract expansion in Arx forms intranuclear inclusions and results in increased cell death. J Cell Biol 167(3):411–416. doi:jcb.200408091 [pii] 10.1083/jcb.200408091
15. Shoubridge C, Cloosterman D, Parkinson-Lawerence E, Brooks D, Gecz J (2007) Molecular pathology of expanded polyalanine tract mutations in the Aristaless-related homeobox gene. Genomics 90(1):59–71. doi:S0888-7543(07)00063-8 [pii] 10.1016/j.ygeno.2007.03.005
16. Fullston T, Finnis M, Hackett A, Hodgson B, Brueton L, Baynam G, Norman A, Reish O, Shoubridge C, Gecz J (2011) Screening and cell-based assessment of mutations in the Aristaless-related homeobox (ARX) gene. Clin Genet 80:510–522. doi:10.1111/j.1399-0004.2011.01685.x
17. Shoubridge C, Gécz J (2011) Polyalanine tract disorders and likely pathogenic mechanisms. In: Tandem repeats in genetics of brain function and disease (AJ Hannan, Ed) Landes Biosciences 2012 Bookshelf ID: NBK51932 ISBN: 978-1-4614-5433-5

18. Shoubridge C, Tan MH, Fullston T, Cloosterman D, Coman D, McGillivray G, Mancini GM, Kleefstra T, Gecz J (2010) Mutations in the nuclear localization sequence of the Aristaless related homeobox; sequestration of mutant ARX with IPO13 disrupts normal subcellular distribution of the transcription factor and retards cell division. Pathogenetics 3:1. doi:10.1186/1755-8417-3-1
19. Shoubridge C, Tan MH, Seiboth G, Gecz J (2012) ARX homeodomain mutations abolish DNA binding and lead to a loss of transcriptional repression. Hum Mol Genet 21(7):1639–1647. doi:10.1093/hmg/ddr601

Chapter 10

Molecular Pathology of Polyalanine Expansion Disorders: New Perspectives from Mouse Models

James N. Hughes and Paul Q. Thomas

Abstract

Disease-causing polyalanine (PA) expansion mutations have been identified in nine genes, eight of which encode transcription factors (TFs) with important roles in development. In vitro and cell overexpression studies have shown that expanded PA tracts result in protein misfolding and the formation of aggregates. This feature of PA proteins is reminiscent of the related polyglutamine (PQ) disease proteins, which have been shown to cause disease via a gain-of-function (GOF) mechanism. However, in sharp contrast to PQ disorders, the disease phenotypes associated with PA mutations are more consistent with a LOF and/or mild GOF mechanism, suggesting that their molecular pathology is inherently different to PQ disorders. Elucidating the cellular impact of PA mutations in vivo has been difficult to address as, unlike the late-onset polyglutamine disorders, all PA disorders associated with TF gene mutations are congenital. However, in recent years, significant advances have been made through the analysis of engineered (knock-in) and spontaneous PA mouse models. Here we review these recent findings and propose an updated model of the molecular and cellular mechanism of PA disorders that incorporates both LOF and GOF features.

Key words Polyalanine (PA) expansion, Protein aggregation, Trinucleotide repeat, Mouse model

1 Introduction

Pathogenic expansions in trinucleotide repeat sequences were first identified in the late 1980s and since then have been associated with a wide variety of human disorders. From a mechanistic point of view, these disorders have attracted considerable interest not least because they have distinct pathological mechanisms depending on the position of the repeat. For example, expansion of noncoding sequences can impair the transcriptional output of neighboring genes, as has been shown to occur in fragile X syndrome [1, 2]. Expansions within expressed noncoding sequences have also been identified, for example, in myotonic dystrophy, and appear to cause disease by a complex RNA gain-of-function (GOF) mechanism that may also include generation of homopolymeric proteins

Danny M. Hatters and Anthony J. Hannan (eds.), *Tandem Repeats in Genes, Proteins, and Disease: Methods and Protocols*, Methods in Molecular Biology, vol. 1017, DOI 10.1007/978-1-62703-438-8_10, © Springer Science+Business Media New York 2013

through non-ATG-initiated translation [3]. Expansions within coding sequences fall into two categories: polyglutamine (PQ) and polyalanine disorders (PA). The former are associated with late-onset dominant-inherited neurological disorders such as Huntington's disease (HD) and the spinocerebellar ataxias and are caused by a GOF mechanism in which formation of cellular aggregates is a hallmark feature [4]. In contrast, all but one of the nine PA disorders is congenital suggesting a different pathological mechanism to the PQ disorders. Biochemical and cell culture studies indicate that PA expansion results in protein misfolding and aggregate formation, although the impact of PA expansion on protein function in vivo is less clear. In recent years, several groups have addressed this issue by generating "knock-in" mouse models that mimic disease-causing mutations for a range of PA disorders. Together, these studies indicate that misfolding of mutant protein is a common feature of PA diseases. However, the fate of the mutant protein (clearance versus aggregation) appears to be cell context- and disease protein-specific and can be manifest as GOF and loss-of-function (LOF) activities.

2 PA Expansion Disorders

To date, nine disorders have been identified that are caused by expansion of PA tracts (Table 1). Apart from *PABPN1* (which encodes a poly(A)-binding protein), all of the known PA disease genes encode developmental transcription factors and are associated with congenital syndromes. Unlike most other disease-causing triplet repeats, PA-encoding tracts are meiotically and mitotically stable and exhibit only a low level of polymorphism. It is generally thought that PA expansion mutations occur via misalignment during meiotic recombination, although other mechanisms have also been proposed [5, 6]. The nine PA disorders are summarized below.

2.1 *HOXD13*

HOXD13 is the most 5′ member of the *HOXD* cluster of homeobox transcription factor genes which play a critical role in assignment of positional information during embryonic development. PA expansion mutations in the *HOXD13* gene of synpolydactyly patients were initially reported in 1996 [7] in a landmark paper that provided the first example of a disease-causing PA expansion mutation. Synpolydactyly is a dominantly inherited distal limb malformation syndrome characterized by incomplete digit separation of the (third and fourth) fingers and (fourth and fifth) toes and supernumerary digits in the syndactylous web. To date, PA expansion mutations in *HOXD13* have been identified in 40 independent families and range from +7 (22Ala) to +14 (29Ala) (reviewed in [8]). Shorter expansions have reduced penetrance and generally manifest as less severe phenotypes, while longer expansions are

Table 1
Summary of polyalanine expansion disorders and associated mouse models

				Alanine tract length		Mouse model	
Gene	Function	Disease	Inheritance	WT	Disease	Null	PA knock-in
HOXD13	TF	Synpolydactyly type II	AD	15	22–29	[68]	[43, 50, 55]
HOXA13	TF	Hand–foot–genital syndrome	AD	14 12 18	24 18 24–30	[69]	[11]
ARX	TF	X-linked mental retardation	XR	16 12	18–23 20	[60]	[57, 58, 70, 71]
PHOX2B	TF	Congenital central hypoventilation syndrome	AD	20	25–33	[20]	[63]
SOX3	TF	X-linked hypopituitarism	XR	15	22–26	[25]	[76]
ZIC2	TF	Holoprosencephaly	AD	15	25	[72]	
RUNX2	TF	Cleidocranial dysplasia	AD	17	27	[73, 74]	
FOXL2	TF	Blepharophimosis–ptosis–epicanthus inversus syndrome	AD	14	22–27	[75]	
PABPN1	Poly(A)-binding protein	Oculopharyngeal muscular dystrophy	AD	10	12–17		

TF transcription factor, *AD* autosomal dominant, *XR* X-linked recessive, *PA* polyalanine, *WT* wild type

more severe and are completely penetrant [9]. Interestingly, frameshift and nonsense mutations that likely result in HOXD13 LOF are associated with a distinct atypical form of synpolydactyly, suggesting that PA mutations do not result in haploinsufficiency and may cause disease via a GOF mechanism.

2.2 HOXA13

HOXA13 is a homeobox gene located at the 5′ end of the *HOXA* cluster that is expressed in the developing limbs and genitourinary tract. HOXA13 has 3 PA tracts, and mutations in each of these have been identified in patients with hand–foot–genital syndrome, an autosomal dominant condition characterized by limb and reproductive tract abnormalities [10–12]. PA tract expansions and presumed null alleles are associated with similar phenotypes suggesting that PA expansions in *HOXA13* result in LOF [11].

2.3 ARX

Aristaless-related homeobox gene (*ARX*) is an X-linked homeodomain transcription factor gene that is broadly expressed in the developing mouse forebrain [13]. The ARX sequence contains four PA tracts and expansion mutations within the first two of these (PA1 and PA2) are associated with a spectrum of neurological disorders of which intellectual disability (ID) is the cardinal feature [14]. In general, the severity of the phenotype correlates with the length of the PA expansion although there is considerable intra- and interfamily variability between individuals with the same mutation. It is important to note that ARX LOF mutations are associated with a more severe disorder (X-linked lissencephaly with abnormal genitalia), suggesting that PA expansion causes partial LOF which is proportional to the PA expansion length.

2.4 *PHOX2B*

Congenital central hypoventilation syndrome (CCHS) is a life-threatening autosomal dominant condition that is characterized by respiratory arrest during sleep accompanied by decreased sensitivity to hypercapnia and hypoxia. More than 90 % of affected individuals have PA mutations in *PHOX2B* [15], which encodes a paired-like homeodomain protein that is expressed in the developing peripheral and central autonomic nervous system. PA expansion mutations in *PHOX2B* range from 25Ala to 33Ala, the latter being the longest known disease-causing PA tract. Independent clinical genetic studies have shown that individuals with longer PA tract expansions generally have more severe phenotypes indicating a genotype–phenotype correlation that is also supported by in vitro functional studies of the mutant protein [16–19]. *Phox2b* heterozygous mice exhibit a similar albeit milder defect indicating that the function of this transcription factor is conserved in mammals [20].

2.5 *SOX3*

PA expansion mutations in the SOX family transcription factor gene *SOX3* have been identified in four families with X-linked hypopituitarism (XH, [21–24]). This X-linked recessive disorder is characterized by GH deficiency, variable deficiencies in other anterior pituitary hormones, and incompletely penetrant ID. Of the two PA expansions that have been identified (22Ala and 26Ala), the latter is associated with a more severe ID phenotype suggesting that the larger expansion has a greater impact on SOX3 protein function. In mice, *Sox3* is expressed in the developing CNS, and mice lacking *Sox3* exhibit a similar phenotype to individuals with XH including hypothalamic–pituitary axis dysfunction [25].

2.6 *ZIC2*

Holoprosencephaly (HPE) is a genetically and phenotypically heterogeneous disorder characterized by malformation of the forebrain midline structures. Many different mutations in the CNS Zn-finger transcription factor gene *ZIC2* have been identified in HPE patients, including a 10Ala PA tract expansion [26]. LOF and PA expansion mutations are associated with similar HPE phenotypes indicating that PA expansion compromises ZIC2 function. This is supported by transactivation assays of ZIC2-25Ala which shows that the mutant protein has impaired activity [27].

2.7 RUNX2

RUNX2 is a Runt family transcription factor gene that is expressed in osteoblasts and early-stage chondrocytes of developing bones in humans [28]. Mutations in *RUNX2* are associated with a dominant skeletal malformation disorder termed cleidocranial dysplasia (CD) which is characterized by osteoporosis, clavicular hypoplasia, and tooth abnormalities [29]. Only one CD family with a PA expansion mutation (+10) has been identified. Interestingly, the phenotype of affected individuals in this family includes brachydactyly which is not found in patients with partial or complete LOF alleles. These data suggest that the PA expansion may cause GOF or dominant negative activity, although additional families are required to support this conclusion.

2.8 FOXL2

Blepharophimosis–ptosis–epicanthus inversus syndrome is an autosomal dominant congenital disorder characterized by malformation of the eyelids which can present as an isolated feature (BPES type I) or in association with premature ovarian failure (BPES type II). Mutations in *FOXL2*, which encodes a forkhead domain transcription factor expressed in the developing eyelids and adult ovary, are associated with both forms of BPES [30]. Approximately 100 different mutations have been identified in BPES patients of which PA expansions ranging between +5 and +13 comprise approximately one-third [31]. It has been suggested that PA expansion mutations are more likely to lead to BPES type II although the evidence is inconclusive. The phenotypic overlap of affected individuals with null and PA expansion alleles is consistent with a LOF mechanism for the latter.

2.9 PABPN1

Poly(A)-binding protein n1 (*PABPN1*) is a ubiquitously expressed gene encoding for a protein involved in several aspects of mRNA polyadenylation. Short expansions of an N-terminal 10-alanine tract to between 12 and 17 alanines result in the disease oculopharyngeal muscular dystrophy (OPMD) [32]. OPMD is a late-onset myopathy that typically presents around 50 years of age with patients experiencing severe swallowing difficulties (dysphagia) and drooping eyelids (ptosis) and can eventually affect all voluntary muscles causing proximal limb weakness. OPMD is generally inherited as a dominant disorder although rare homozygous individuals with more severe symptoms have been also identified [33]. As no other mutations have been described in *PABPN1* that lead to OPMD, it is not currently possible to deduce whether PA expansions are acting as LOF or GOF alleles through phenotype–genotype correlations.

3 Polyalanine Tracts: A Double-Edged Sword

PA tracts of five or more consecutive residues are present in 494 human proteins and are particularly common in nuclear proteins such as transcription factors (which comprise 36 % of human PA proteins [34, 35]). Although the role of PA domains is not well defined, it is thought that they may act as flexible spacer elements that facilitate a range of functional properties including protein–protein interaction and protein–DNA binding [9, 36]. Evolutionary studies in eukaryotes have shown that PA expansions have arisen independently in several TF families (e.g., HOX, GATA, EVX, and SOX; [35, 37]) and that, once established, they become progressively longer, indicating that PA domains confer a selective advantage. However, the absence of any wild-type protein with a PA tract exceeding 20 residues indicates that there is a critical threshold above which PA length expansions are not tolerated. Consistent with this idea, the vast majority of the disease-causing expansions in TF genes (*see* Table 1) are 20 alanine residues or longer.

Selection against longer PA tracts is almost certainly due to the inherent ability of PA tracts to misfold and form aggregates. In vitro studies have shown that PA peptides containing 11–20 residues readily oligomerize and form B-sheets in vitro [38, 39]. Molecular dynamics simulation of PA peptide aggregation has shown that PA proteins initially form amorphous aggregates that ultimately convert to small B-sheets that are stabilized by hydrogen bonds and hydrophobic interactions [40]. Analysis of 7Ala-, 15Ala-, 23Ala-, and 35Ala-containing GST and YFP fusion proteins expressed in COS cells has shown that only the Ala23 and Ala35 proteins oligomerize and form aggregates that are resistant to tryptic digestion [41]. Similarly, expression of GFP–PA tract fusion proteins in heterologous cells has shown that aggregate formation is restricted to proteins containing PA tracts of 19 or longer and that the propensity to aggregate is proportional to the length of the PA tract [42].

In line with the finding that PA tracts cause protein aggregation when inserted into a variety of heterologous peptides, overexpression assays in cultured cells have shown that all nine PA disease-associated proteins can form aggregates [27, 43–47]. Aggregate formation is predominantly cytoplasmic, resulting in a reduction of nuclear protein, and is correlated with the length of the PA expansion [43]. Aggregates for SOX3, HOXD13, ARX, and RUNX2 colocalize with molecular chaperones including HSP40 and HSP70 [43, 46–48]. These chaperones are known to facilitate protein refolding and can rescue protein from aggregates and return it to a soluble form [49]. When cotransfected into COS cells, PA expansion-containing forms of HOXA13, HOXD13, and RUNX2 are able to sequester other PA containing proteins, including their cognate wild-type forms into cytoplasmic aggregates [50], providing scope for a dominant negative mechanism. Of note, the overt aggregates seen in high-level overexpression studies were not observed when mutant PA-expanded (+14) HOXD13 was expressed at lower levels using viral integration. Instead, the mutant protein was virtually absent from the cells but could be restored to cytoplasmic aggregates by inhibition of the proteasome [43]. These differences in cellular response highlight the importance of evaluating disease protein behavior at physiological levels and suggest that aggregate formation may not be a feature of all PA disease proteins in vivo.

4 Mouse Models of PA Disorders

While in vitro studies have provided useful information about the impact of PA expansion mutations on protein structure and function, full understanding of the molecular pathology of PA disorders requires in vivo analysis. This has been a difficult issue to address as all but one of the PA disorders is congenital and therefore results from defects in embryonic processes. The one exception is the late-onset disorder OPMD (Table 1). Interestingly, muscle tissue from OPMD patients exhibits aggregates of mutant protein in the form of intranuclear inclusions, which are a diagnostic feature of this disease. While this appears to support a role for aggregation in PA expansion pathogenesis more generally, it is difficult to extrapolate findings from OPMD to other PA disorders because in many ways OPMD appears to be an outlier among these disorders. Firstly, OPMD is the only late-onset disorder of the group. Secondly, disease-causing PA tracts in PABPN1 are considerably shorter (12–17 alanines) than the rest of the group (18–33 alanines). Indeed, PA tracts of 12 and above are found in several wild-type proteins (including the PA disease-associated TFs, Table 1) where they do not cause aggregation. Interestingly wild-type PABPN1 is also reported to aggregate, suggesting this protein is naturally prone to aggregation [51, 52]. Thirdly, OPMD aggregates form in the nucleus suggesting they may form in a different manner to cytoplasmic aggregates. Finally, OPMD only affects a small subset of the cells despite PABPN1 being a ubiquitously expressed protein and only after many years.

Recent studies using transgenic mice have shown that PABPN1 has additional roles beyond polyadenylation and is involved in selection of alternate polyadenylation signals in the 3′UTR of mRNA. PABPN1 was found to cause a general suppression of proximal polyadenylation signal (PAS) usage in favor of more distal PAS usage [53]. Transgenic mice expressing PA-expanded PABPN1 showed a skewing towards distal PAS usage despite the presence of wild-type protein, suggesting the mutant form acts as a dominant negative, possibly by sequestration of wild-type protein in aggregates. Aggregates do not appear to be toxic per se, as evidenced by the observation that transgenic animals can be partially rescued by the addition of further wild-type protein from a separate transgene, despite having the same amount or more nuclear aggregates [54]. Together these data suggest that interaction of the mutant protein with wild-type protein has the effect of reducing the functional protein level in the nucleus and may cause disease by affecting PAS usage and mRNA stability. The generation of null or hypomorphic alleles would help address whether aggregates are directly pathogenic or whether pathogenesis is simply a result of reduced functional protein levels.

To investigate the functional impact of PA expansions of TFs in vivo, several groups have used targeted mutagenesis of embryonic stem cells to generate "knock-in" mouse models (Table 1). This approach has several advantages. Firstly, modification of the endogenous orthologous gene results in the generation of mutant PA-expanded protein at the appropriate level and cellular context. Secondly, the phenotypic impact of PA expansion alleles and engineered null alleles can be directly compared, thereby allowing conclusions to be drawn regarding the former's activity as LOF or GOF. Thirdly, genetic epistasis of PA alleles can be readily assessed by crossing PA mouse models with other mutant strains and mice from different genetic backgrounds. To date, "knock-in" mouse models for four PA expansion disorders have been published which are summarized below.

4.1 Hoxd13

Synpolydactyly (*spdh*) is a spontaneous mouse mutant that carries a +7Ala expansion in the PA tract [55]. *spdh* homozygous mutants have a severe limb phenotype that includes ectrodactyly (reduced/missing central digits) and synpolydactyly that is similar to humans with heterozygous PA expansions in *HOXD13*. Importantly, this phenotype is much more severe than *Hoxd13* homozygous null mutant mice, indicating that *spdh* is not a simple LOF allele. Indeed, the *spdh* homozygous phenotype is very similar to mice carrying a homozygous triple deletion of the *Hoxd13*, *Hoxd12*, and *Hoxd11* (*Del3/Del3*) as well as a heterozygous mutation in *Hoxa13* [50]. These data, coupled with comprehensive genetic interaction studies of the *spdh*, *Hoxd13* null, *Del3*, and *Hoxa13* alleles, indicate that the *spdh* protein exerts a dominant negative effect over Hoxd11, Hoxd12, Hoxd13, and Hoxa13 which is only fully revealed when *spdh* is in the homozygous state [43, 50, 55]. The cell biological basis for the *spdh* GOF effect does not appear to involve changes in the expression of *Hoxd11*, *Hoxd12*, *Hoxd13*, and *Hoxa13*. In contrast, consistent with cytoplasmic mislocalization of PA-expanded HOXD13 proteins in cell overexpression assays [43], the *spdh* gene product is mainly present in the cytoplasm in vivo with a marked reduction in nuclear levels compared to wild type [50]. While the latter is likely to manifest as LOF, it appears that the cytoplasmic *spdh* protein exerts a GOF effect by sequestration and degradation of other PA containing proteins including HOXA13, HOXD13, and RUNX (but not HOXD12 which lacks a PA tract). An independent spontaneous +7Ala PA expansion allele (*Dyc*) which arose on a different genetic background (Balb/c versus B6) has also been identified [5]. *Dyc* homozygotes have an identical phenotype to *spdh/spdh* mice indicating that genetic background does not contribute significantly to the limb phenotypes of these mutants.

4.2 Hoxa13

Using homologous recombination in embryonic stem cells, Innis et al. [11] engineered a +10Ala knock-in mutation (Ala28) into the third PA-encoding tract of *Hoxa13*. PA expansions in this repeat in humans range from +6Ala to +12Ala and are associated with hand–foot–genital syndrome in the heterozygous state. *Hoxa13*-28Ala heterozygous embryos exhibit defects in forelimb and hind limb buds that are consistent with the human phenotype. Interestingly, in contrast to the *spdh* mutant, heterozygous embryos carrying either the *Hoxa13*-28Ala or *Hoxa13* null alleles exhibit identical phenotypes indicating that *Hoxa13*-28Ala functions as a LOF allele [11]. The molecular defect in the +10 Hoxa13 mice has a posttranslational etiology and is associated with markedly reduced levels of Hoxa13-Ala28 protein in the developing limb buds which likely occurs due to misfolding and degradation [11]. While it is not currently known whether the residual Hoxa13-28Ala protein is mislocalized in the cytoplasm (as occurs in the *spdh* mutant), the absence of any GOF effect would suggest that it is not.

4.3 Arx

Infantile spasms syndrome (ISS, also known as West syndrome) is a pediatric disorder characterized by intellectual disability, seizures and motor spasms. The mutation most commonly associated with ISS is a +7Ala expansion in the first PA tract of the *ARX* gene [56], and ISS mouse models containing a +7Ala PA expansion in the endogenous mouse gene have generated independently by two groups [57, 58]. *Arx* +7Ala mice exhibit the key phenotypic features of ISS including severe seizures, EEG abnormalities, motor spasms, and memory and learning deficits. Interestingly, the phenotype of the +7Ala mice is less severe that *Arx* null mice (which die shortly after birth, [59, 60]) indicating that the +7Ala mutation acts as a partial LOF allele. Consistent with this mechanism, only some of the neuronal migration defects that occur in the *Arx* null mice are present in the +7Ala mice. For example, in the striatum, *Arx* null mice exhibit defects in radial and tangential migration of GABAergic precursors, while only the latter is defective in the +7Ala mutant. Furthermore, in the cortex, tangential migration of the GABAergic precursors from the medial ganglionic eminence is severely affected in the *Arx* null mice but is only slightly impaired in the +7Ala mutants. Like the *Hoxa13* and *Hoxd13* PA expansion mutants, +7Ala PA expansion in *Arx* does not have a detectable impact on steady-state mRNA levels. However, ARX protein in +7Ala embryos is reduced by approximately 50 % by Western blot analysis suggesting that it is misfolded and degraded [57]. A similar reduction in ARX protein is also reportedly observed in a +8Ala expansion mutation in the second PA tract of *Arx* [57]. At the cellular level, Kitamura et al. [57] report that the +7Ala and +8Ala ARX proteins localize to the nucleus and do not form aggregates. Taken together, these in vivo data indicate that PA expansions in *Arx* manifest as partial LOF alleles. This effect is most likely due to a reduction in the level of nuclear protein below a context-dependent threshold that is required for full biological function. It is also worth noting that Price et al. [58] report that wild-type ARX protein is present in the cytoplasm of a minority of ARX-expressing neurons in the adult cortex and that the proportion of neurons with cytoplasmic ARX increases in +7PA mice. However, in the absence of any evidence for GOF associated with *Arx* PA expansion in mice or humans, the significance of this finding is currently unclear.

4.4 Phox2b

The most common PA mutation in patients with CCHS is a heterozygous +7Ala expansion in the 20Ala repeat of the *PHOX2B* gene [19]. It is thought that this mutation acts as a GOF allele although the mechanism is not known [61, 62]. To model this PA disorder, Debreuil et al. [63] introduced a +7Ala mutation into the murine gene (which encodes an identical protein to the human gene) by homologous recombination in embryonic stem cells. Heterozygous pups carrying the +7Ala *Phox2b* mutation died immediately after birth due to severe respiratory difficulties and insensitivity to hypercapnia (high CO_2 levels), thereby phenocopying severe cases of CCHS. Analysis of the retrotrapezoid group/parafacial respiratory group (RTN/pFRG, a region associated with respiratory control and CO_2 sensing) revealed a developmental abnormality in the *Phox2b*-expressing neurons. Other *Phox2b*-positive regions involved in respiratory control including the carotid body, the petrosal/nodose ganglionic complex, and the locus coeruleus appeared normal, suggesting that defective development of the RTN/pFRG was the primary cause of the phenotype. However, selective expression of the PHOX2B-27Ala mutant protein in the RTN/pFRG did not result in the severe CCHS phenotype suggesting that other *Phox2b* expressing regions in the CNS contribute to the phenotype [64]. Regardless of the primary cause of the disorder, it is clear that the phenotype of the *Phox2b*-27Ala heterozygous mice is much more severe than heterozygotes carrying a *Phox2b* null allele indicating that the PA expansion confers a GOF. However, the mechanism that underpins this GOF activity has not been identified. At the cellular level, immunohistochemical analysis of the RTN/pFRG region did not reveal any overt diminution in nuclear protein or the existence of aggregates, although current data cannot exclude the possibility that protein levels are subtly altered in the *Phox2b*-27Ala heterozygous mice.

Given that in vitro cotransfection assays do not support a dominant negative mechanism, it has been proposed that toxic GOF may be responsible [61]. This has some support from in vitro studies but has not been investigated in vivo. Toxic GOF cannot be excluded but seems unlikely given that there is no evidence for aggregation formation in vivo and that other PA expansion models do not excessive levels of apoptosis [55, 57].

5 Conclusions and Future Prospects

One of the goals of studying PA expansion disorders collectively is to understand how the same type of mutation in different genes can cause apparent LOF in some proteins but GOF in others. At a mechanistic level, what are the commonalities and what are the points of divergence that underlie the different outcomes observed across the group? Data from mouse models have provided new insights into these questions (Fig. 1). Clearly protein misfolding

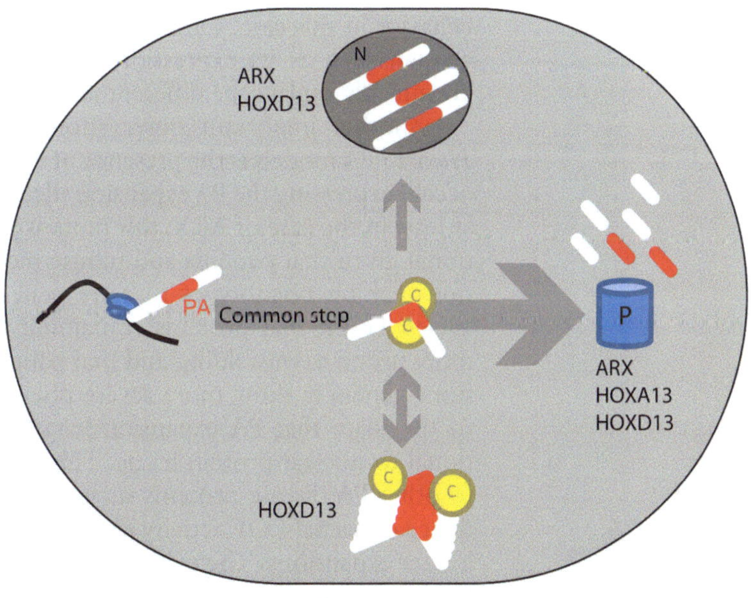

Fig. 1 Model for the molecular pathology of PA disorders adapted from [34] and refined using recent data from PA mouse models. Unlike PQ diseases, there is no evidence that PA expansions affect transcription or translation. We predict that all PA expansion proteins will have a common defect in protein folding. Chaperones (C), particularly heat shock proteins, are likely to be central to the eventual fate of PA-expanded protein. Protein that enters the nucleus (N) is likely to do so because of chaperone-assisted folding. Protein that is not correctly refolded will be targeted for degradation by the proteasome (P). In situations where the degradation machinery is overwhelmed, the misfolded protein will aggregate, where again chaperone-assisted refolding is possible. As described in the text, in vivo evidence for nuclear protein localization is available for ARX and HOXD13, reduced protein levels are seen for ARX, HOXA13, and HOXD13, while cytoplasmic accumulation is observed only in HOXD13 mutant tissue. PA, polyalanine tract

and subsequent aggregation is a potential outcome of PA expansion. However, while this is readily achieved in overexpression assays in vitro, there remains little direct evidence of aggregation in vivo, with the exceptions being the atypical nuclear inclusions of OPMD and cytoplasmic accumulation observed in developing limb buds of mice with Hoxd13+7Ala [32, 43]. On the other hand, a post-transcriptional reduction in protein levels has been reported for three PA disease models, namely, *Arx*, *Hoxa13*, and *Hoxd13* [11, 43, 57]. It seems likely that this reduction in protein reflects the ability of cells to recognize and degrade misfolded mutant protein, thus circumventing the formation of overt aggregates. This would suggest that misfolding is a common first step in the pathogenesis of all the PA diseases, but the question then arises as to why some mutant proteins are effectively degraded and others are not. Of particular interest is the observation that HOXA13 is degraded, while mutant HOXD13 is not (or at least not to the same extent). Both of these similar transcription factors are expressed in the developing limb bud suggesting that potential cell type-specific variables such as chaperone activity are not a major determinant of mutant protein behavior in this case. Other factors that are intrinsic to the disease protein such as its expression level and aggregation propensity presumably explain the different fates of the mutant protein.

Another important observation to arise from the *Arx* and *Hoxd13* PA models is the presence of mutant protein in the nucleus of cells expressing the PA expansion allele [43, 57, 58]. Interestingly, at least in the case of ARX, this mutant protein appears to be functional given that patients and mouse models with *ARX* PA mutations exhibit phenotypes that are generally less severe than null alleles. Given that proteins containing longer PA expansions are more prone to misfolding and that patients with longer PA expansions generally show more severe phenotype, it seems reasonable to speculate that PA expansion length may be inversely proportional to nuclear protein levels. This relationship could also apply to other PA disease proteins such as SOX3 for which there is evidence of partial LOF activity and increasing phenotypic severity of longer expansions. The existence of functional nuclear protein also has implications for the development of therapeutic agents for PA disorders. Many of the drug treatments being investigated for PA disorders are effective at reducing aggregate formation in vitro (reviewed in [65]) and in some instances can alleviate disease phenotypes in PABPN1 transgenic mice [66, 67]. The prevailing model for drug action is that increased chaperone activity leads to reduced aggregate formation. The alternative hypothesis for drug action is that increased chaperone activity increases the efficiency of primary protein folding to increase the amount of functional protein. For the future it would be interesting to assess the effect of these drugs in a PA mouse model where protein aggregates are not seen (such as Hoxa13), thus testing the hypothesis that aggregates are not toxic per se. Finally, another avenue of investigation that has not yet been explored is the use of patient-derived induced pluripotent stem (iPS) cells. This approach is likely to yield further insights into the cellular pathology of PA disorders and provide a useful vehicle for pharmacological screening.

Note Added in Proof

We have recently described the generation and analysis of a mouse model for X-linked hypopituitarism (XH) in which a disease-causing polyalanine expansion mutation in the *Sox3* gene (26Ala) has been introduced into the murine genome through homologous recombination in embryonic stem (ES) cells [76]. We show that the polyalanine expansion mutation has a significant impact on SOX3 protein resulting in a massive reduction in protein levels in vivo and in neurodifferentiated ES cell cultures. We also investigate the developmental consequences of the *Sox3*-26Ala mutation and, through comparison with *Sox3* null embryos, demonstrate that Sox3 polyalanine expansion perturbs development of the hypothalamus and infundibulum, regions that are also abnormal in patients with XH. However, importantly, the residual protein within the nucleus of mutant cells is functional and is sufficient to enable chimeric embryos to undergo gastrulation, a process that is significantly impaired in chimeras generated with *Sox3* null ES cells. Together, these molecular, morphological, and genetic data provide strong evidence for a disease-relevant PA mutant protein that is both nuclear and functional, thereby manifesting as a partial LOF allele.

J. N. H. & P. Q. T.

References

1. Pieretti M, Zhang FP, Fu YH, Warren ST, Oostra BA, Caskey CT, Nelson DL (1991) Absence of expression of the FMR-1 gene in fragile X syndrome. Cell 66(4):817–822. doi:0092-8674(91)90125-I [pii]
2. Sutcliffe JS, Nelson DL, Zhang F, Pieretti M, Caskey CT, Saxe D, Warren ST (1992) DNA methylation represses FMR-1 transcription in fragile X syndrome. Hum Mol Genet 1(6): 397–400
3. Zu T, Gibbens B, Doty NS, Gomes-Pereira M, Huguet A, Stone MD, Margolis J, Peterson M, Markowski TW, Ingram MA, Nan Z, Forster C, Low WC, Schoser B, Somia NV, Clark HB, Schmechel S, Bitterman PB, Gourdon G, Swanson MS, Moseley M, Ranum LP (2011) Non-ATG-initiated translation directed by microsatellite expansions. Proc Natl Acad Sci USA 108(1):260–265. doi:1013343108 [pii] 10.1073/pnas.1013343108
4. Bauer PO, Nukina N (2009) The pathogenic mechanisms of polyglutamine diseases and current therapeutic strategies. J Neurochem 110(6):1737–1765. doi:JNC6302 [pii] 10.1111/j.1471-4159.2009.06302.x
5. Cocquempot O, Brault V, Babinet C, Herault Y (2009) Fork stalling and template switching as a mechanism for polyalanine tract expansion affecting the DYC mutant of HOXD13, a new murine model of synpolydactyly. Genetics 183(1):23–30
6. Trochet D, de Pontual L, Keren B, Munnich A, Vekemans M, Lyonnet S, Amiel J (2007) Polyalanine expansions might not result from unequal crossing-over. Hum Mutat 28(10): 1043–1044

7. Muragaki Y, Mundlos S, Upton J, Olsen BR (1996) Altered growth and branching patterns in synpolydactyly caused by mutations in HOXD13. Science 272(5261):548–551
8. Brison N, Tylzanowski P, Debeer P (2011) Limb skeletal malformations—what the HOX is going on? Eur J Med Genet 55(1):1–7. doi:S1769-7212(11)00073-5 [pii] 10.1016/j.ejmg.2011.06.003
9. Goodman FR, Mundlos S, Muragaki Y, Donnai D, Giovannucci-Uzielli ML, Lapi E, Majewski F, McGaughran J, McKeown C, Reardon W, Upton J, Winter RM, Olsen BR, Scambler PJ (1997) Synpolydactyly phenotypes correlate with size of expansions in HOXD13 polyalanine tract. Proc Natl Acad Sci USA 94(14):7458–7463
10. Debeer P, Bacchelli C, Scambler PJ, De Smet L, Fryns JP, Goodman FR (2002) Severe digital abnormalities in a patient heterozygous for both a novel missense mutation in HOXD13 and a polyalanine tract expansion in HOXA13. J Med Genet 39(11):852–856
11. Innis JW, Mortlock D, Chen Z, Ludwig M, Williams ME, Williams TM, Doyle CD, Shao Z, Glynn M, Mikulic D, Lehmann K, Mundlos S, Utsch B (2004) Polyalanine expansion in HOXA13: three new affected families and the molecular consequences in a mouse model. Hum Mol Genet 13(22):2841–2851
12. Utsch B, Becker K, Brock D, Lentze MJ, Bidlingmaier F, Ludwig M (2002) A novel stable polyalanine [poly(A)] expansion in the HOXA13 gene associated with hand-foot-genital syndrome: proper function of poly(A)-harbouring transcription factors depends on a critical repeat length? Hum Genet 110(5):488–494. doi:10.1007/s00439-002-0712-8
13. Miura H, Yanazawa M, Kato K, Kitamura K (1997) Expression of a novel aristaless related homeobox gene 'Arx' in the vertebrate telencephalon, diencephalon and floor plate. Mech Dev 65(1–2):99–109
14. Shoubridge C, Fullston T, Gecz J (2010) ARX spectrum disorders: making inroads into the molecular pathology. Hum Mutat 31(8):889–900. doi:10.1002/humu.21288
15. Meguro T, Yoshida Y, Hayashi M, Toyota K, Otagiri T, Mochizuki N, Kishikawa Y, Sasaki A, Hayasaka K (2012) Inheritance of polyalanine expansion mutation of PHOX2B in congenital central hypoventilation syndrome. J Hum Genet 57(5):335–337. doi:jhg201227 [pii] 10.1038/jhg.2012.27
16. Di Zanni E, Bachetti T, Parodi S, Bocca P, Prigione I, Di Lascio S, Fornasari D, Ravazzolo R, Ceccherini I (2012) In vitro drug treatments reduce the deleterious effects of aggregates containing polyAla expanded PHOX2B proteins. Neurobiol Dis 45(1):508–518. doi:S0969-9961(11)00317-2 [pii] 10.1016/j.nbd.2011.09.007
17. Matera I, Bachetti T, Puppo F, Di Duca M, Morandi F, Casiraghi GM, Cilio MR, Hennekam R, Hofstra R, Schober JG, Ravazzolo R, Ottonello G, Ceccherini I (2004) PHOX2B mutations and polyalanine expansions correlate with the severity of the respiratory phenotype and associated symptoms in both congenital and late onset Central Hypoventilation syndrome. J Med Genet 41(5):373–380
18. Patwari PP, Carroll MS, Rand CM, Kumar R, Harper R, Weese-Mayer DE (2010) Congenital central hypoventilation syndrome and the PHOX2B gene: a model of respiratory and autonomic dysregulation. Respir Physiol Neurobiol 173(3):322–335. doi:S1569-9048(10)00231-4 [pii] 10.1016/j.resp.2010.06.013
19. Trochet D, Hong SJ, Lim JK, Brunet JF, Munnich A, Kim KS, Lyonnet S, Goridis C, Amiel J (2005) Molecular consequences of PHOX2B missense, frameshift and alanine expansion mutations leading to autonomic dysfunction. Hum Mol Genet 14(23):3697–3708. doi:ddi401 [pii] 10.1093/hmg/ddi401
20. Dauger S, Pattyn A, Lofaso F, Gaultier C, Goridis C, Gallego J, Brunet JF (2003) Phox2b controls the development of peripheral chemoreceptors and afferent visceral pathways. Development 130(26):6635–6642
21. Alatzoglou KS, Kelberman D, Cowell CT, Palmer R, Arnhold IJ, Melo ME, Schnabel D, Grueters A, Dattani MT (2011) Increased transactivation associated with SOX3 polyalanine tract deletion in a patient with hypopituitarism. J Clin Endocrinol Metab 96(4):E685–E690
22. Burkitt Wright EM, Perveen R, Clayton PE, Hall CM, Costa T, Procter AM, Giblin CA, Donnai D, Black GC (2009) X-linked isolated growth hormone deficiency: expanding the phenotypic spectrum of SOX3 polyalanine tract expansions. Clin Dysmorphol 18(4):218–221
23. Laumonnier F, Ronce N, Hamel BC, Thomas P, Lespinasse J, Raynaud M, Paringaux C, Van Bokhoven H, Kalscheuer V, Fryns JP, Chelly J, Moraine C, Briault S (2002) Transcription factor SOX3 is involved in X-linked mental retardation with growth hormone deficiency. Am J Hum Genet 71(6):1450–1455
24. Woods KS, Cundall M, Turton J, Rizotti K, Mehta A, Palmer R, Wong J, Chong WK, Al-Zyoud M, El-Ali M, Otonkoski T, Martinez-Barbera JP, Thomas PQ, Robinson IC, Lovell-Badge R, Woodward KJ, Dattani MT (2005)

Over- and underdosage of SOX3 is associated with infundibular hypoplasia and hypopituitarism. Am J Hum Genet 76(5):833–849

25. Rizzoti K, Brunelli S, Carmignac D, Thomas PQ, Robinson IC, Lovell-Badge R (2004) SOX3 is required during the formation of the hypothalamo-pituitary axis. Nat Genet 36(3): 247–255
26. Paulussen AD, Schrander-Stumpel CT, Tserpelis DC, Spee MK, Stegmann AP, Mancini GM, Brooks AS, Collee M, Maat-Kievit A, Simon ME, van Bever Y, Stolte-Dijkstra I, Kerstjens-Frederikse WS, Herkert JC, van Essen AJ, Lichtenbelt KD, van Haeringen A, Kwee ML, Lachmeijer AM, Tan-Sindhunata GM, van Maarle MC, Arens YH, Smeets EE, de Die-Smulders CE, Engelen JJ, Smeets HJ, Herbergs J (2010) The unfolding clinical spectrum of holoprosencephaly due to mutations in SHH, ZIC2, SIX3 and TGIF genes. Eur J Hum Genet 18(9):999–1005. doi:ejhg201070 [pii] 10.1038/ejhg.2010.70
27. Brown L, Paraso M, Arkell R, Brown S (2005) In vitro analysis of partial loss-of-function ZIC2 mutations in holoprosencephaly: alanine tract expansion modulates DNA binding and transactivation. Hum Mol Genet 14(3):411–420. doi:ddi037 [pii] 10.1093/hmg/ddi037
28. Sugawara M, Kato N, Tsuchiya T, Motoyama T (2011) RUNX2 expression in developing human bones and various bone tumors. Pathol Int 61(10):565–571. doi:10.1111/j.1440-1827.2011.02706.x
29. Hansen L, Riis AK, Silahtaroglu A, Hove H, Lauridsen E, Eiberg H, Kreiborg S (2011) RUNX2 analysis of Danish cleidocranial dysplasia families. Clin Genet 79(3):254–263. doi:CGE1458 [pii] 10.1111/j.1399-0004.2010.01458.x
30. Crisponi L, Deiana M, Loi A, Chiappe F, Uda M, Amati P, Bisceglia L, Zelante L, Nagaraja R, Porcu S, Ristaldi MS, Marzella R, Rocchi M, Nicolino M, Lienhardt-Roussie A, Nivelon A, Verloes A, Schlessinger D, Gasparini P, Bonneau D, Cao A, Pilia G (2001) The putative forkhead transcription factor FOXL2 is mutated in blepharophimosis/ptosis/epicanthus inversus syndrome. Nat Genet 27(2):159–166. doi: 10.1038/84781
31. Beysen D, De Jaegere S, Amor D, Bouchard P, Christin-Maitre S, Fellous M, Touraine P, Grix AW, Hennekam R, Meire F, Oyen N, Wilson LC, Barel D, Clayton-Smith J, de Ravel T, Decock C, Delbeke P, Ensenauer R, Ebinger F, Gillessen-Kaesbach G, Hendriks Y, Kimonis V, Laframboise R, Laissue P, Leppig K, Leroy BP, Miller DT, Mowat D, Neumann L, Plomp A, Van Regemorter N, Wieczorek D, Veitia RA, De Paepe A, De Baere E (2008) Identification of 34 novel and 56 known FOXL2 mutations in patients with blepharophimosis syndrome. Hum Mutat 29(11):E205–E219. doi:10.1002/humu.20819
32. Brais B, Bouchard JP, Xie YG, Rochefort DL, Chretien N, Tome FM, Lafreniere RG, Rommens JM, Uyama E, Nohira O, Blumen S, Korczyn AD, Heutink P, Mathieu J, Duranceau A, Codere F, Fardeau M, Rouleau GA (1998) Short GCG expansions in the PABP2 gene cause oculopharyngeal muscular dystrophy. Nat Genet 18(2):164–167. doi:10.1038/ng0298-164
33. Blumen SC, Brais B, Korczyn AD, Medinsky S, Chapman J, Asherov A, Nisipeanu P, Codere F, Bouchard JP, Fardeau M, Tome FM, Rouleau GA (1999) Homozygotes for oculopharyngeal muscular dystrophy have a severe form of the disease. Ann Neurol 46(1):115–118
34. Albrecht A, Mundlos S (2005) The other trinucleotide repeat: polyalanine expansion disorders. Curr Opin Genet Dev 15(3):285–293
35. Lavoie H, Debeane F, Trinh QD, Turcotte JF, Corbeil-Girard LP, Dicaire MJ, Saint-Denis A, Page M, Rouleau GA, Brais B (2003) Polymorphism, shared functions and convergent evolution of genes with sequences coding for polyalanine domains. Hum Mol Genet 12(22):2967–2979
36. Karlin S, Chen C, Gentles AJ, Cleary M (2002) Associations between human disease genes and overlapping gene groups and multiple amino acid runs. Proc Natl Acad Sci USA 99(26):17008–17013. doi:10.1073/pnas.262658799262658799 [pii]
37. Mojsin M, Kovacevic-Grujicic N, Krstic A, Popovic J, Milivojevic M, Stevanovic M (2010) Comparative analysis of SOX3 protein orthologs: expansion of homopolymeric amino acid tracts during vertebrate evolution. Biochem Genet 48(7–8):612–623. doi:10.1007/s10528-010-9343-2
38. Giri K, Bhattacharyya NP, Basak S (2007) pH-dependent self-assembly of polyalanine peptides. Biophys J 92(1):293–302. doi:S0006-3495(07)70827-5 [pii] 10.1529/biophysj.106.091769
39. Shinchuk LM, Sharma D, Blondelle SE, Reixach N, Inouye H, Kirschner DA (2005) Poly-(L-alanine) expansions form core beta-sheets that nucleate amyloid assembly. Proteins 61(3):579–589. doi:10.1002/prot.20536
40. Nguyen HD, Hall CK (2004) Molecular dynamics simulations of spontaneous fibril formation by random-coil peptides. Proc Natl Acad Sci USA 101(46):16180–16185. doi:0407273101 [pii] 10.1073/pnas.0407273101

41. Nojima J, Oma Y, Futai E, Sasagawa N, Kuroda R, Turk B, Ishiura S (2009) Biochemical analysis of oligomerization of expanded polyalanine repeat proteins. J Neurosci Res 87(10):2290–2296. doi:10.1002/jnr.22052
42. Rankin J, Wyttenbach A, Rubinsztein DC (2000) Intracellular green fluorescent protein-polyalanine aggregates are associated with cell death. Biochem J 348(Pt 1):15–19
43. Albrecht AN, Kornak U, Boddrich A, Suring K, Robinson PN, Stiege AC, Lurz R, Stricker S, Wanker EE, Mundlos S (2004) A molecular pathogenesis for transcription factor associated poly-alanine tract expansions. Hum Mol Genet 13(20):2351–2359
44. Bachetti T, Matera I, Borghini S, Di Duca M, Ravazzolo R, Ceccherini I (2005) Distinct pathogenetic mechanisms for PHOX2B associated polyalanine expansions and frameshift mutations in congenital central hypoventilation syndrome. Hum Mol Genet 14(13):1815–1824. doi:ddi188 [pii] 10.1093/hmg/ddi188
45. Caburet S, Demarez A, Moumne L, Fellous M, De Baere E, Veitia RA (2004) A recurrent polyalanine expansion in the transcription factor FOXL2 induces extensive nuclear and cytoplasmic protein aggregation. J Med Genet 41(12):932–936. doi:41/12/932 [pii] 10.1136/jmg.2004.024356
46. Nasrallah IM, Minarcik JC, Golden JA (2004) A polyalanine tract expansion in Arx forms intranuclear inclusions and results in increased cell death. J Cell Biol 167(3):411–416
47. Wong J, Farlie P, Holbert S, Lockhart P, Thomas PQ (2007) Polyalanine expansion mutations in the X-linked hypopituitarism gene SOX3 result in aggresome formation and impaired transactivation. Front Biosci 12: 2085–2095
48. Abu-Baker A, Messaed C, Laganiere J, Gaspar C, Brais B, Rouleau GA (2003) Involvement of the ubiquitin-proteasome pathway and molecular chaperones in oculopharyngeal muscular dystrophy. Hum Mol Genet 12(20):2609–2623. doi:10.1093/hmg/ddg293 ddg293 [pii]
49. Hartl FU, Hayer-Hartl M (2002) Molecular chaperones in the cytosol: from nascent chain to folded protein. Science 295(5561):1852–1858. doi:10.1126/science.1068408 295/5561/1852 [pii]
50. Villavicencio-Lorini P, Kuss P, Friedrich J, Haupt J, Farooq M, Turkmen S, Duboule D, Hecht J, Mundlos S (2010) Homeobox genes d11-d13 and a13 control mouse autopod cortical bone and joint formation. J Clin Invest 120(6):1994–2004. doi:41554 [pii] 10.1172/JCI41554
51. Raz V, Abraham T, van Zwet EW, Dirks RW, Tanke HJ, van der Maarel SM (2011) Reversible aggregation of PABPN1 pre-inclusion structures. Nucleus 2(3):208–218. doi:10.4161/nucl.2.3.15736 1949-1034-2-3-8 [pii]
52. Tavanez JP, Calado P, Braga J, Lafarga M, Carmo-Fonseca M (2005) In vivo aggregation properties of the nuclear poly(A)-binding protein PABPN1. RNA 11(5):752–762
53. Jenal M, Elkon R, Loayza-Puch F, van Haaften G, Kuhn U, Menzies FM, Oude Vrielink JA, Bos AJ, Drost J, Rooijers K, Rubinsztein DC, Agami R (2012) The poly(A)-binding protein nuclear 1 suppresses alternative cleavage and polyadenylation sites. Cell 149(3):538–553. doi:S0092-8674(12)00398-4 [pii] 10.1016/j.cell.2012.03.022
54. Davies JE, Sarkar S, Rubinsztein DC (2008) Wild-type PABPN1 is anti-apoptotic and reduces toxicity of the oculopharyngeal muscular dystrophy mutation. Hum Mol Genet 17(8):1097–1108. doi:ddm382 [pii] 10.1093/hmg/ddm382
55. Bruneau S, Johnson KR, Yamamoto M, Kuroiwa A, Duboule D (2001) The mouse Hoxd13(spdh) mutation, a polyalanine expansion similar to human type II synpolydactyly (SPD), disrupts the function but not the expression of other Hoxd genes. Dev Biol 237(2):345–353
56. Poirier K, Eisermann M, Caubel I, Kaminska A, Peudonnier S, Boddaert N, Saillour Y, Dulac O, Souville I, Beldjord C, Lascelles K, Plouin P, Chelly J, Bahi-Buisson N (2008) Combination of infantile spasms, non-epileptic seizures and complex movement disorder: a new case of ARX-related epilepsy. Epilepsy Res 80(2–3):224–228. doi:S0920-1211(08)00085-5 [pii] 10.1016/j.eplepsyres.2008.03.019
57. Kitamura K, Itou Y, Yanazawa M, Ohsawa M, Suzuki-Migishima R, Umeki Y, Hohjoh H, Yanagawa Y, Shinba T, Itoh M, Nakamura K, Goto Y (2009) Three human ARX mutations cause the lissencephaly-like and mental retardation with epilepsy-like pleiotropic phenotypes in mice. Hum Mol Genet 18(19):3708–3724
58. Price MG, Yoo JW, Burgess DL, Deng F, Hrachovy RA, Frost JD Jr, Noebels JL (2009) A triplet repeat expansion genetic mouse model of infantile spasms syndrome, Arx(GCG)10+7, with interneuronopathy, spasms in infancy, persistent seizures, and adult cognitive and behavioral impairment. J Neurosci 29(27): 8752–8763
59. Collombat P, Mansouri A, Hecksher-Sorensen J, Serup P, Krull J, Gradwohl G, Gruss P (2003) Opposing actions of Arx and Pax4 in endocrine pancreas development. Genes Dev

17(20):2591–2603. doi:10.1101/gad.269003 17/20/2591 [pii]

60. Kitamura K, Yanazawa M, Sugiyama N, Miura H, Iizuka-Kogo A, Kusaka M, Omichi K, Suzuki R, Kato-Fukui Y, Kamiirisa K, Matsuo M, Kamijo S, Kasahara M, Yoshioka H, Ogata T, Fukuda T, Kondo I, Kato M, Dobyns WB, Yokoyama M, Morohashi K (2002) Mutation of ARX causes abnormal development of forebrain and testes in mice and X-linked lissencephaly with abnormal genitalia in humans. Nat Genet 32(3):359–369. doi:10.1038/ng1009 ng1009 [pii]
61. Amiel J, Dubreuil V, Ramanantsoa N, Fortin G, Gallego J, Brunet JF, Goridis C (2009) PHOX2B in respiratory control: lessons from congenital central hypoventilation syndrome and its mouse models. Respir Physiol Neurobiol 168(1–2): 125–132. doi:S1569-9048(09)00060-3 [pii] 10.1016/j.resp.2009.03.005
62. Benailly HK, Lapierre JM, Laudier B, Amiel J, Attie T, De Blois MC, Vekemans M, Romana SP (2003) PMX2B, a new candidate gene for Hirschsprung's disease. Clin Genet 64(3):204–209. doi:105 [pii]
63. Dubreuil V, Ramanantsoa N, Trochet D, Vaubourg V, Amiel J, Gallego J, Brunet JF, Goridis C (2008) A human mutation in Phox2b causes lack of CO2 chemosensitivity, fatal central apnea, and specific loss of parafacial neurons. Proc Natl Acad Sci USA 105(3):1067–1072
64. Ramanantsoa N, Hirsch MR, Thoby-Brisson M, Dubreuil V, Bouvier J, Ruffault PL, Matrot B, Fortin G, Brunet JF, Gallego J, Goridis C (2011) Breathing without CO(2) chemosensitivity in conditional Phox2b mutants. J Neurosci 31(36):12880–12888. doi:31/36/12880 [pii] 10.1523/JNEUROSCI.1721-11.2011
65. Di Zanni E, Bachetti T, Parodi S, Bocca P, Prigione I, Di Lascio S, Fornasari D, Ravazzolo R, Ceccherini I (2011) In vitro drug treatments reduce the deleterious effects of aggregates containing polyAla expanded PHOX2B proteins. Neurobiol Dis 45(1):508–518. doi:S0969-9961(11)00317-2 [pii] 10.1016/j.nbd.2011.09.007
66. Davies JE, Wang L, Garcia-Oroz L, Cook LJ, Vacher C, O'Donovan DG, Rubinsztein DC (2005) Doxycycline attenuates and delays toxicity of the oculopharyngeal muscular dystrophy mutation in transgenic mice. Nat Med 11(6):672–677. doi:nm1242 [pii] 10.1038/nm1242
67. Davies JE, Sarkar S, Rubinsztein DC (2006) Trehalose reduces aggregate formation and delays pathology in a transgenic mouse model of oculopharyngeal muscular dystrophy. Hum Mol Genet 15(1):23–31. doi:ddi422 [pii] 10.1093/hmg/ddi422
68. Dolle P, Dierich A, LeMeur M, Schimmang T, Schuhbaur B, Chambon P, Duboule D (1993) Disruption of the Hoxd-13 gene induces localized heterochrony leading to mice with neotenic limbs. Cell 75(3):431–441. doi:0092-8674(93)90378-4 [pii]
69. Fromental-Ramain C, Warot X, Messadecq N, LeMeur M, Dolle P, Chambon P (1996) Hoxa-13 and Hoxd-13 play a crucial role in the patterning of the limb autopod. Development 122(10):2997–3011
70. Beguin S, Crepel V, Aniksztejn L, Becq H, Pelosi B, Pallesi-Pocachard E, Bouamrane L, Pasqualetti M, Kitamura K, Cardoso C, Represa A (2012) An epilepsy-related ARX polyalanine expansion modifies glutamatergic neurons excitability and morphology without affecting GABAergic neurons development. Cereb Cortex. doi:bhs138 [pii] 10.1093/cercor/bhs138
71. Nasrallah MP, Cho G, Simonet JC, Putt ME, Kitamura K, Golden JA (2012) Differential effects of a polyalanine tract expansion in Arx on neural development and gene expression. Hum Mol Genet 21(5):1090–1098. doi:ddr538 [pii] 10.1093/hmg/ddr538
72. Warr N, Powles-Glover N, Chappell A, Robson J, Norris D, Arkell RM (2008) Zic2-associated holoprosencephaly is caused by a transient defect in the organizer region during gastrulation. Hum Mol Genet 17(19):2986–2996. doi:ddn197 [pii] 10.1093/hmg/ddn197
73. Komori T, Yagi H, Nomura S, Yamaguchi A, Sasaki K, Deguchi K, Shimizu Y, Bronson RT, Gao YH, Inada M, Sato M, Okamoto R, Kitamura Y, Yoshiki S, Kishimoto T (1997) Targeted disruption of Cbfa1 results in a complete lack of bone formation owing to maturational arrest of osteoblasts. Cell 89(5):755–764. doi:S0092-8674(00)80258-5 [pii]
74. Otto F, Thornell AP, Crompton T, Denzel A, Gilmour KC, Rosewell IR, Stamp GW, Beddington RS, Mundlos S, Olsen BR, Selby PB, Owen MJ (1997) Cbfa1, a candidate gene for cleidocranial dysplasia syndrome, is essential for osteoblast differentiation and bone development. Cell 89(5):765–771. doi:S0092-8674(00)80259-7 [pii]
75. Uda M, Ottolenghi C, Crisponi L, Garcia JE, Deiana M, Kimber W, Forabosco A, Cao A, Schlessinger D, Pilia G (2004) Foxl2 disruption causes mouse ovarian failure by pervasive blockage of follicle development. Hum Mol Genet 13(11):1171–1181. doi:10.1093/hmg/ddh124 ddh124 [pii]
76. Hughes J, Piltz S, Rogers N, McAninch D, Rowley L, Thomas P (2013) Mechanistic insight into the pathology of polyalanine expansion disorders revealed by a mouse model for x linked hypopituitarism. PLoS Genet. 9(3):e1003290. doi:10.1371/journal.pgen.1003290. Epub 2013 Mar 7

Chapter 11

Yeast as a Platform to Explore Polyglutamine Toxicity and Aggregation

Martin L. Duennwald

Abstract

Protein misfolding is associated with many neurodegenerative diseases, including neurodegenerative diseases caused by polyglutamine expansion proteins, such as Huntington's disease. The model organism baker's yeast (*Saccharomyces cerevisiae*) has provided important general insights into the basic cellular mechanisms underlying protein misfolding. Furthermore, experiments in yeast have identified cellular factors that modulate the toxicity and the aggregation associated with polyglutamine expansion proteins. Notably, many features discovered in yeast have been proven to be highly relevant in other model organisms and in human pathology. The experimental protocols depicted here serve to reliably determine polyglutamine toxicity and polyglutamine aggregation in yeast.

Key words Protein misfolding, Neurodegeneration, Polyglutamine toxicity, Polyglutamine aggregation, Yeast model

1 Introduction

Protein misfolding describes a process in which proteins attain a nonfunctional, often cytotoxic three-dimensional conformation instead of their proper, functional three-dimensional conformation. Many human diseases, including neurodegenerative diseases, are associated with protein misfolding [1, 2]. Research in model organisms including mice, flies, worms, and yeast has provided important new insights into biochemical, genetic, and cellular factors that modulate protein misfolding and its ensuing toxicity. We and others have established, characterized, and utilized yeast as a platform to study the misfolding of polyglutamine (polyQ) expansion proteins, which cause neurodegenerative disease, such as Huntington's disease (HD) [3–7]. These yeast polyQ models recapitulate fundamental aspects of polyQ biology, above all, polyQ length-dependent protein aggregation and toxicity [4]. Further, yeast polyQ models have revealed many intricate aspects of polyQ

Danny M. Hatters and Anthony J. Hannan (eds.), *Tandem Repeats in Genes, Proteins, and Disease: Methods and Protocols*, Methods in Molecular Biology, vol. 1017, DOI 10.1007/978-1-62703-438-8_11,

misfolding, polyQ aggregation, and polyQ toxicity [8–10]. Many of these results from experiments in yeast have been recapitulated in other model organism and even in samples from human patients, thus demonstrating the high relevance of polyQ studies in yeast.

Here, I outline experimental protocols to determine polyQ toxicity and polyQ aggregation in yeast. For clarification in the context of this chapter, I define polyQ toxicity as a growth defect of yeast cells expressing polyQ expansion proteins, i.e., a defect in dividing at normal rates. Consequently, polyQ toxicity can be determined by measuring the growth of yeast cells either in liquid cultures or on solid media (plates). Further, I define polyQ aggregation as the formation of highly insoluble protein aggregates. PolyQ aggregation can therefore be determined by documenting the ratio of the insoluble portion of polyQ proteins to the total amount of polyQ proteins in yeast cells.

2 Materials

2.1 PolyQ Expansion Proteins in Yeast

The protocols below describe experiments that have been carried out with specific polyQ fusion proteins [4]. These fusion proteins contain an amino-terminal FLAG-tag, followed by the first 17 amino-terminal amino acids of the human huntingtin (Htt) protein, followed by a pure polyQ region of different length (25, 46, 72, and 103 Qs), followed by a fluorescent protein, such as GFP. Please note that the sequences that flank the polyQ region both at the carboxy and at the amino-terminus have a decisive impact on polyQ toxicity in yeast [4]. Also, it seems essential to use DNA constructs that utilize mixed codons (CAG and CAA) for the expression of polyQ proteins because pure CAG repeats (as they occur in the human disease genes) greatly reduce the expression levels of polyQ expansion proteins (our unpublished observation).

The expression levels of polyQ proteins strongly influence the levels of polyQ toxicity and polyQ aggregation: higher polyQ protein expression results in greater levels of polyQ toxicity and aggregation [4]. Therefore, we routinely use the GAL1 promoter for the expression of polyQ proteins in yeast. This promoter is strongly induced when yeast cells are grown in media with galactose as carbon source. Also, the GAL1 promoter can be repressed by the presence of glucose in growth media. The GAL1 promoter system thus allows tightly controlled expression of polyQ proteins for defined periods of time.

Furthermore, we have engineered yeast strains that express polyQ protein from DNA constructs that are stably integrated into the yeast genome. These strains have reproducible expression levels and thus are ideally suited for studies of polyQ toxicity and aggregation.

2.2 Yeast Strains

We and others have found that the presence of yeast prions, foremost the yeast prion [RNQ+] (also known as [PIN+]), strongly modulates polyQ aggregation and polyQ toxicity [3, 6]. Briefly, yeast prions are proteins that can exist either in a soluble, non-prion conformation or in an insoluble, often aggregated, prion conformation [11]. For instance, yeast cells can harbor the protein Rnq1p either in its prion conformation, [RNQ+], or in its non-prion conformation, [rnq−] [12].

In [RNQ+] cells, polyQ toxicity and aggregation are far greater than in [rnq−] cells [3]. While the protocols below have been carried out mostly in the commonly used laboratory yeast strain W303, other [RNQ+] yeast strains also work well.

2.3 Yeast Media

The protocols below use standard yeast media. We have obtained highly reproducible results with premixed media from Sunrise Media (San Diego, CA, USA). For optimal results, we recommend using fresh media (i.e., prepared not earlier than 2 weeks before starting experiments).

2.4 Spotter for PolyQ Toxicity Assays on Solid Media

For uniform and reproducible transfer of liquid yeast cultures onto solid media we use spotters (V&P Scientific, San Diego, CA, USA) with either 48 pins or 96 pins.

2.5 BioscreenC for Monitoring PolyQ Toxicity in Liquid Culture

The BioscreenC instrument (Oy Growthcurves, Helsinki, Finland) determines polyQ toxicity by measuring the growth of liquid yeast cultures expressing polyQ proteins. Specifically, the BioscreenC measures the absorption of light at a wavelength of 600 nm (OD_{600}) by yeast cultures grown in wells of honey-well plates (Oy Growthcurves) at a constant temperature with continuous agitation. The instrument allows the simultaneous growth measurements of up to 200 different yeast cultures with volumes of up to 300 μl for each culture.

2.6 Membranes for Solubility Assays

1. Cellulose acetate membranes (from Schleicher and Schuell, Goettingen, Germany) with a pore size of 2 μm. Dot blots are carried out using PVDF membranes or nitrocellulose membranes (from BioRad, Hercules, CA, USA). For filter papers use thin 3 M Whatman filter papers.

2.7 Reagents for Solubility Assays

1. Lysis buffer: 50 mM Hepes pH 7.5; 150 mM NaCl; 5 mM EDTA; 1 % (v/v) Triton X 100. Add protease inhibitors directly before use: 50 mM *N-ethyl*-maleimide (NEM) from 2 M stock solution in DMSO and 1 mM phenylmethylsulfonyl fluoride (PSMF) from 300 mM stock solution in ethanol.
2. Glass beads: For protein lysis, use acid-washed glass beads with a diameter range between 425 and 600 μm (Sigma-Aldrich).

3. Manifold: For easy application of protein lysates to membranes we use a 96-well manifold (from Schleicher and Schuell) for both filter retardation assays and dot blots.
4. Blocking solution: 5 % (w/v) solution of nonfat milk powder in PBS (137 mM NaCl; 2.7 mM KCl; 10 mM $Na_2HPO_4 \cdot 2H_2O$; 2 mM KH_2PO_4; pH = 7.4).
5. Wash solution: PBS buffer with 0.01 % (v/v) Tween-20.
6. Primary antibodies: Anti-GFP.
7. Secondary antibodies: Anti-mouse horseradish peroxidase (HRP)-coupled antibodies.
8. ECL solution: An ECL kit (e.g., from Pierce/Thermo Scientific (Rockford, IL, USA)).
9. 10 % (w/v) SDS solution.

2.8 Other Miscellaneous Items

1. Sterile tubes.
2. Petri dishes.
3. 96-Well plates.
4. 30 °C Incubator.
5. Digital camera.
6. Centrifuge.
7. Spectrophotometer.
8. Vortex.
9. Heating block.
10. Multichannel pipette.

3 Methods

Yeast cultures should be handled using sterile microbiology practices, i.e., working close to a flame on a bench top or under a sterile hood. Sterile tubes, Petri dishes, and reagents should be used for all experiments described below (*see* **Note 1**).

3.1 PolyQ Toxicity: Starting the Yeast Culture

1. To start polyQ toxicity experiments, retrieve yeast strains freshly from frozen stocks or use freshly made yeast strains. Use a minimum of three different colonies per experiment for each condition you wish to test (*see* **Note 2**). Restreak these three colonies on fresh plates for following repeat experiments.
2. Grow yeast cells in liquid cultures overnight (e.g., 15 h) at 30 °C with constant agitation to avoid flocculation. Use selective yeast media appropriate for the yeast strains that you use. When using polyQ constructs under the control of the GAL1 promoter, use media that contain glucose as carbon source to repress polyQ expression in the overnight cultures.

3.2 Spotting Assays on Plates

These spotting assays are relatively facile and can detect strong to medium levels of polyQ toxicity.

1. Prepare dilutions of yeast cultures in 96-well plates. First, dilute the yeast cultures such that the most concentrated sample of each culture has an $OD_{600}=0.2$. Use a volume of 200 μl for the most concentrated yeast culture in wells of the first row in the 96-well plate. Dilute these cultures 1:5 with sterile water in wells of the other rows of the 96-well plate. The volume of each of these dilutions should be no less than 150 μl (*see* **Note 3**).
2. Before use, let plates dry at room temperature until there is no superfluous water (e.g., from condensation) on the plates. Mix the yeast cultures thoroughly with the spotter before transferring to plates. When using the GAL1 promoter, spot diluted yeast cultures on plates that contain galactose as sole carbon source for induction of the expression of polyQ proteins, and on plates with glucose as sole carbon source for polyQ repression, and full media (YPD) as controls. Transfer the diluted yeast cultures from the 96-well plates with a spotter. Ensure even application of the spotter to the surface of the plates for optimal results (*see* **Note 4**).
3. Allow the plates to dry after spotting until the spotted liquid drops have completely dried. Incubate plates at 30 °C for 2–3 days (*see* **Note 5**).
4. Document results by taking pictures of the plates with a digital camera.

3.3 Growth Curves Determined Using BioscreenC

Determining growth curves with the BioscreenC instrument allows the detection of even subtle levels of polyQ toxicity (*see* **Note 6**).

1. Yeast cultures are started as described under Subheading 3.1.
2. Wash overnight cultures three times by spinning them in a centrifuge (not harder than 3,000 × *g*), discarding the supernatant, and adding 3 ml of sterile water in each washing cycle.
3. Dilute washed overnight cultures to an $OD_{600}=0.2$. Then dilute these cultures further 1:200 into galactose-containing media.
4. Fill the wells of the honey-well plates with 300 μl of the diluted yeast cultures. Use multiple wells for each sample (minimum of three) as technical replicas. Insert plate into the BioscreenC instrument.
5. Start BioscreenC instrument by opening the Easy Bioscreeen Experiment window on the computer attached to the BioscreenC. Determine the number of samples to be measured (maximum of 200 different samples); set the temperature to 30 °C; set the experiment length to 72 h; set the measurement intervals to 5 min; set the filter to 600 nm (Brown); and set the shaking mode to medium and to 15-s duration before each

measurement. Start the BioscreenC run. The measured ODs will be displayed in a Microsoft Excel sheet and can be viewed during the experiment.

3.4 PolyQ Aggregation

PolyQ aggregation is best assessed by determining the ratio between insoluble to total polyQ protein (*see* **Note 7**). To this end, we perform filter retardation assays to detect insoluble polyQ protein species and, in parallel, dot blots to detect total polyQ proteins.

3.4.1 Protein Lysis

1. Start yeast cultures as described under Subheading 3.1, **step 1**.
2. Wash the overnight yeast cultures (as described under Subheading 3.2, **step 2**).
3. Dilute washed cultures in 10 ml galactose media to an $OD_{600} = 0.2$.
4. Grow 10 ml yeast cultures to an $OD_{600} = 0.6$ (*see* **Note 8**).
5. Spin down the cells and wash with sterile water.
6. Spin down the cells and add 150 μl of cold lysis buffer containing NEM and PMSF.
7. Add 200 μg acid-washed glass beads.
8. Vortex six times for 30 s each, keeping on ice for 30 s between vortexing.
9. Spin down for 10 min at 4 °C in tabletop microcentrifuge at maximum speed.
10. Take 80 μl of the supernatant, transfer to fresh tubes, and keep on ice.

3.4.2 Sample Preparation for Filter Retardation

1. Add 2 % (final concentration) SDS from 10 % (w/v) stock solution.
2. Boil samples for 5 min in a heating block.
3. Store samples on ice.
4. Prepare dilutions (e.g., six serial 1/5 dilutions) in 96-well plates, use a multichannel pipette, and use sterile water for dilutions.

3.4.3 Filter Retardation Assay

1. Soak cellulose acetate membrane in 2 % (w/v) SDS solution (*see* **Note 9**), and then wash cellulose acetate membrane in 0.2 % (w/v) SDS.
2. Soak filter paper in water.
3. Put cellulose acetate membrane on top of filter paper in manifold.
4. Close manifold (*see* **Note 10**).
5. Apply 100 μl of each sample/dilution using a multichannel pipette, and apply vacuum to manifold.

6. Wash five times with 200 μl of a 0.2 % SDS solution.
7. Open manifold carefully (*see* **Note 11**).
8. Use cellulose acetate membrane for immunodetection.

3.4.4 Dot Blot

1. Take samples from protein lysis (Subheading 3.3, **step 1**). Do not add SDS and do not boil samples.
2. Continue precisely as described for filter retardation assays (Subheading 3.3, **step 3**) except use PVDF or nitrocellulose membrane (*see* **Note 12**).
3. Immunodetection (same for filter retardation assay and dot blot).
4. Standard protocols for immunodetection using the listed antibodies and reagents work well for polyQ dot blots and filter retardation assays (*see* **Note 13**).

4 Notes

1. It is particularly important to avoid cross-contaminations of different yeast cultures. Use fresh sterile tips, inoculation loops, and tubes for each experimental step.
2. Yeast cells can spontaneously lose polyQ toxicity [6, 13]. It is therefore important to always carry out multiple technical and biological repeats for each experimental condition.
3. The dilutions of yeast cultures are most easily prepared using a multichannel pipette. It is important to thoroughly mix the cultures for each dilution step, e.g., by pipetting them up and down several times.
4. The transfer of yeast cultures to plates using a spotter may require experience and a steady hand. Prepare enough plates for each experiment for multiple attempts to ensure generation of presentable results.
5. The incubation times may vary greatly depending on specific experimental conditions. Take pictures after multiple incubation times (2, 3, and 4 days) and document the incubation times carefully.
6. Note that experiments assessing polyQ toxicity by monitoring the growth of liquid yeast cultures can be flawed by the appearance of spontaneous suppressors of polyQ toxicity [6, 13]. Therefore, perform multiple technical and biological repeats for each experimental condition.

   PolyQ toxicity using liquid yeast cultures can also be determined without the BioscreenC instrument. Use cultures of a larger volume (at least 5 ml) and grow them at 30 °C with constant agitation. Every 2 h, take out aliquots of 200 μl volumes, dilute with 800 μl water, and measure the OD_{600} every 2 h with a spectrophotometer. Prepare growth curves using Excel or similar programs.

7. Note that the described protocols cannot detect more soluble polyQ protein species, such as oligomeric species.
8. This might take a longer period of time. Plan your experiments accordingly.
9. Cellulose acetate membranes are delicate and break easily. Handle them with caution.
10. Ensure that the manifold is properly sealed and no liquid leaks out. If necessary, use an additional soaked filter paper. If possible, avoid the wells positioned at the perimeter of the manifold because these can be leaky.
11. Ensure that there is no remaining liquid in the wells of the manifold before opening.
12. We typically use dry nitrocellulose membranes to avoid excessive diffusion of liquids.
13. Longer blocking (e.g., overnight at 4 °C with mild shaking) can greatly reduce background so that even weaker signals can be detected.

Acknowledgments

Research in the Duennwald lab is supported by grants from the American Federation for Aging Research (AFAR), the Hereditary Disease Foundation (HDF), and the William Wood Foundation.

References

1. Soto C (2003) Unfolding the role of protein misfolding in neurodegenerative diseases. Nat Rev Neurosci 4(1):49–60. doi:10.1038/nrn1007 nrn1007 [pii]
2. Soto C, Estrada LD (2008) Protein misfolding and neurodegeneration. Arch Neurol 65(2):184–189. doi:65/2/184 [pii] 10.1001/archneurol.2007.56
3. Duennwald ML, Jagadish S, Giorgini F, Muchowski PJ, Lindquist S (2006) A network of protein interactions determines polyglutamine toxicity. Proc Natl Acad Sci USA 103(29):11051–11056. doi:0604548103 [pii] 10.1073/pnas.0604548103
4. Duennwald ML, Jagadish S, Muchowski PJ, Lindquist S (2006) Flanking sequences profoundly alter polyglutamine toxicity in yeast. Proc Natl Acad Sci USA 103(29):11045–11050. doi:0604547103 [pii] 10.1073/pnas.0604547103
5. Krobitsch S, Lindquist S (2000) Aggregation of huntingtin in yeast varies with the length of the polyglutamine expansion and the expression of chaperone proteins. Proc Natl Acad Sci USA 97(4):1589–1594. doi:97/4/1589 [pii]
6. Meriin AB, Zhang X, He X, Newnam GP, Chernoff YO, Sherman MY (2002) Huntington toxicity in yeast model depends on polyglutamine aggregation mediated by a prion-like protein Rnq1. J Cell Biol 157(6):997–1004. doi:10.1083/jcb.200112104 jcb.200112104 [pii]
7. Muchowski PJ, Schaffar G, Sittler A, Wanker EE, Hayer-Hartl MK, Hartl FU (2000) Hsp70 and hsp40 chaperones can inhibit self-assembly of polyglutamine proteins into amyloid-like fibrils. Proc Natl Acad Sci USA 97(14):7841–7846. doi:10.1073/pnas.140202897 140202897 [pii]

8. Giorgini F, Muchowski PJ (2009) Exploiting yeast genetics to inform therapeutic strategies for Huntington's disease. Methods Mol Biol 548:161–174. doi:10.1007/978-1-59745-540-4_9
9. Sherman MY, Muchowski PJ (2003) Making yeast tremble: yeast models as tools to study neurodegenerative disorders. Neuromolecular Med 4(1–2):133–146. doi:NMM:4:1–2:133 [pii] 10.1385/NMM:4:1–2:133
10. Winderickx J, Delay C, De Vos A, Klinger H, Pellens K, Vanhelmont T, Van Leuven F, Zabrocki P (2008) Protein folding diseases and neurodegeneration: lessons learned from yeast. Biochim Biophys Acta 1783(7):1381–1395. doi:S0167-4889(08)00046-3 [pii] 10.1016/j.bbamcr.2008.01.020
11. Shorter J, Lindquist S (2005) Prions as adaptive conduits of memory and inheritance. Nat Rev Genet 6(6):435–450. doi:nrg1616 [pii] 10.1038/nrg1616
12. Sondheimer N, Lindquist S (2000) Rnq1: an epigenetic modifier of protein function in yeast. Mol Cell 5(1):163–172. doi:S1097-2765(00)80412-8 [pii]
13. Duennwald ML (2011) Monitoring polyglutamine toxicity in yeast. Methods 53(3): 232–237. doi:S1046-2023(10)00284-7 [pii] 10.1016/j.ymeth.2010.12.001

Chapter 12

Immuno-based Detection Assays to Quantify Distinct Mutant Huntingtin Conformations in Biological Samples

Gregor P. Lotz and Andreas Weiss

Abstract

A pathological hallmark of many protein-misfolding diseases is the formation of insoluble aggregates. Quantitative methods are needed to better resolve and define the formation, aggregation, and temporal dynamics of soluble misfolded proteins in native settings. In this book chapter we describe simple and sensitive detection methods to characterize high ordered aggregates (AGERA) and subsets of distinct soluble aggregates (SEC-FRET) of mutant huntingtin protein in biological samples.

Key words Endogenous mutant huntingtin, Oligomers, Soluble fragments, Insoluble aggregates, Agarose gel electrophoresis, Size-exclusion chromatography

1 Introduction

An important hallmark of many protein-misfolding diseases is the accumulation of protein aggregates. There is much evidence that misfolding and aggregation cause pathogenesis [1] and the prevention of the formation of these aggregates in neurons represents an attractive therapeutic strategy to ameliorate protein-misfolding diseases. Recent studies suggest that protein aggregation is a complex process that adopts multiple misfolded monomeric and higher order soluble and insoluble conformations [2]. Immunohistochemical methods provide qualitative information of aggregate morphology, number, and regional localization but do not provide quantitative characterization. Standard methods for protein analysis such as polyacrylamide gel electrophoresis can poorly resolve protein aggregates because of common biochemical features of protein aggregates like insolubility and resistance to chemical extraction [3]. The filter-retardation assay for aggregates [4] provides more quantitative information about size and SDS-solubility of protein aggregates but is limited on size of the filter pores of the membrane used. Aggregate growth or composition also cannot be

Danny M. Hatters and Anthony J. Hannan (eds.), *Tandem Repeats in Genes, Proteins, and Disease: Methods and Protocols*, Methods in Molecular Biology, vol. 1017, DOI 10.1007/978-1-62703-438-8_12, © Springer Science+Business Media New York 2013

determined with this method. Therefore, quantitative methods are needed to better resolve and define the formation, aggregation, and temporal dynamics of soluble misfolded proteins forming insoluble aggregates. In this chapter we introduce sensitive and quantitative methods (agarose gel electrophoresis for resolving aggregates (AGERA) and SEC-FRET) to monitor aggregate formation and progression in Huntington's disease models. Both methods provide opportunities to assess the efficacy of potential drug candidates or to identify biomarkers of disease onset or progression. We also discuss some caveats and challenges of the described methods by adding notes to individual protocol steps.

2 Materials

Prepare all solutions using ultrapure water and analytical grade reagents. Prepare and store all reagents at room temperature (unless indicated otherwise). Diligently follow all waste disposal regulations when disposing waste materials. We do not add sodium azide to the reagents.

2.1 Agarose Gel Electrophoresis for Resolving Aggregates

1. Agarose.
2. Stock-Buffer: 100 mL 375 mM Tris–HCl, pH 8.83. 20 % (w/v) SDS stock solution.
3. Non-reducing Laemmli sample buffer (150 mM Tris–HCl pH 6.8, 33 % (w/v) glycerol, 1.2 % SDS, and bromophenol blue).
4. High-molecular-weight size markers: Purified catalase (232 kDa), ferritin (440 kDa), and thyroglobulin (669 kDa) (Amersham Bioscience HMW Gel Filtration Kit, #17-0441-01).
5. Laemmli running buffer: 192 mmol/L glycine, 25 mM Tris base, 0.1 % (w/v) SDS.
6. Transfer buffer: 192 mM glycine, 25 mM Tris base, 0.1 % (w/v) SDS, 15 % (v/v) methanol.
7. MW8 antibody was developed by Paul Patterson [7] and obtained from the Developmental Studies Hybridoma Bank developed under the auspices of the NICHD and maintained by The University of Iowa, Department of Biological Sciences, Iowa City, IA 52242.
8. DNA Sub CellTM trays (Biorad). *See* **Note 1**.
9. Semi-dry electroblotter model B (Ancos, Højby, Denmark). *See* **Note 1**.
10. PDVF membranes (Millipore, Zug, Switzerland; Immobilon-P, #IPVH00010). *See* **Note 1**.

2.2 Resolving Aggregate State in Brain Samples by Size-Exclusion Chromatography Combined with TR-FRET Assay

1. Lysis buffer: Phosphate-buffered saline, 1 % (w/v) Triton X-100, anti-proteases (Roche, cOmplete, Mini Protease Inhibitor Cocktail Tablets, #04693124001), and anti-phosphatases (Roche, PhosStop Phosphatase Inhibitor Tablets, #04906845001).
2. Protein detection reagents (e.g., BCA-kit).
3. Running buffer: Phosphate-buffered saline, 0.05 % (w/v) Triton X-100.
4. Terbium donor fluorophore (Tb) and d2 acceptor fluorophore (d2) (CisBio, France).
5. Tris-buffered saline: 20 mM Tris–HCl, pH 7.4, 0.9 % (w/v) NaCl.
6. ECL reagent: Enhanced chemiluminescence blotting substrate (e.g., Pierce ECL Western Blotting substrate).
7. High-speed ultra-centrifuge with polycarbonate tubes (Beckman Coulter #343778).
8. Ceramic beads (Precellys, Bertin Technologies).
9. Filters with 0.45 μm membrane.
10. Gelfiltration column: Superdex 200 10/300 (GE Healthcare #17-5175-01).
11. Chromatography system (e.g., AKTA-FPLC (GE-Healthcare)).
12. Fraction collector with multi-well format capability.
13. Low-volume 96-well plates (Greiner Bio-one, #784080).
14. Plate reader with filter for emission and excitation profiles of terbium and d2 fluorophores.

3 Methods

3.1 Agarose Gel Electrophoresis for Resolving Aggregates

The first simple method for qualitative and quantitative characterization of SDS-resistant mutant huntingtin (HTT) aggregates in mouse models is AGERA (Figs. 1 and 2). AGERA relies on the formation of large gel pores in the agarose gel that allow for migration of bulk protein complexes or aggregates through the gel by applying an electric current. Following electrophoresis, the protein aggregates are transferred by semidry blotting to a PDVF membrane where they can be visualized by HTT-specific antibodies. Similar techniques have previously been described for detection of larger amyloid prion protein complexes [5–7]. The following protocol describes the procedure to determine the load of mutant HTT aggregates in mouse brain samples by AGERA.

1. For short agarose gels (1.5 or 2 % (w/v)), dissolve 1.5 or 2 g agarose in 100 mL 375 mM Tris–HCl, pH 8.8 and bring the mixture to its boiling point in a microwave oven (*see* **Note 2**).

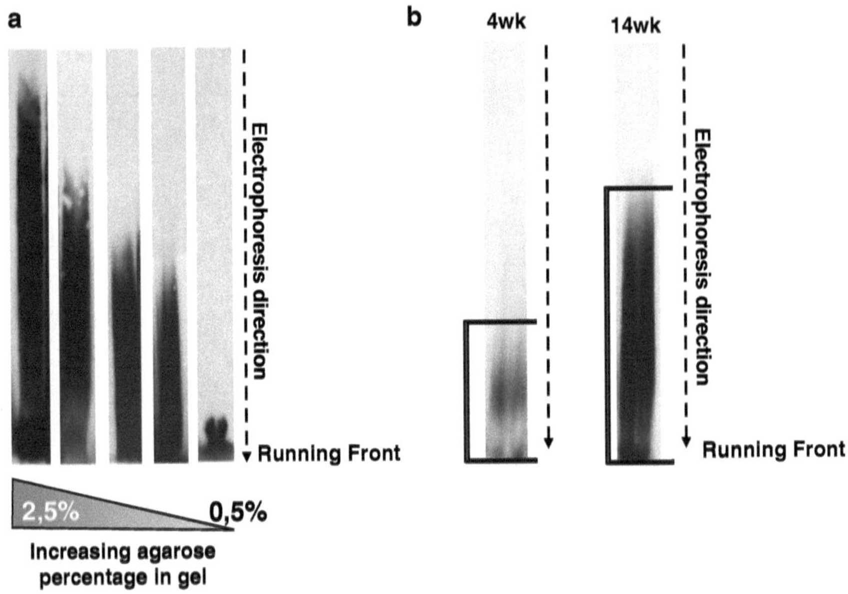

Fig. 1 Mutant huntingtin protein aggregates in R6/2 mouse brains detected by AGERA. (**a**) Running behavior of aggregates from a brain sample of a 14-week R6/2 mouse in gels with increasing percentages. Resolution of mutant HTT aggregates increases with increased agarose percentage. (**b**) An increase in aggregate size and load in the brains of 4- vs. 14-week-old R6/2 mice can be visualized with AGERA. Blots developed with MW8 antibody

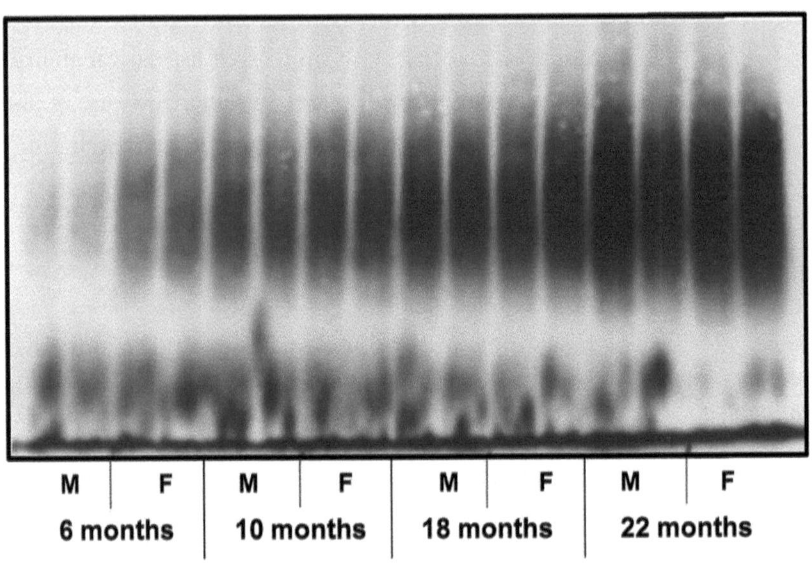

Fig. 2 Mutant HTT aggregate increase in the brains of aging male (M) and female (F) Hdh150Q knock-in mice visualized by AGERA. Blot developed with MW8 antibody

After the agarose is melted, add SDS to a final concentration of 0.1 % (*see* **Note 3**). Pour gels on short Biorad DNA Sub CellTM trays with a gel thickness of 8 mm.

2. For long agarose gels (1 or 1.5 %), dissolve 2.5 or 3.75 g agarose in 250 mL 375 mmol/L Tris–HCl, pH 8.8, adjusting

the final SDS concentration to 0.1 % (*see* **Note 3**), and pour the gels on long Biorad DNA Sub CellTM trays (gel thickness = 8 mm).

3. Dilute biological samples 1:1 into non-reducing Laemmli sample buffer and incubate for 5 min at 95 °C. For mouse brain tissue samples, load 0.15 mg of total protein per AGERA lane (*see* **Note 4**).
4. Run gels in Laemmli running buffer at 100 V, 2 A until the bromophenol blue running front reaches the bottom of the gel.
5. Use semidry electroblotter to blot the gels on PDVF membranes at 200 mA for 1 h. Use transfer buffer to wet the membranes and gel (*see* **Note 5**).
6. Block membrane with 10 % (w/v) milk powder in Tris-buffered saline for 1 h.
7. Incubate with primary HTT antibody (e.g., MW8 [7] or mEM48 (Millipore #5374)) overnight at 4 °C.
8. Wash membrane 3× with Tris-buffered saline with Tween-20.
9. Incubate with secondary anti-mouse antibody coupled to HRP.
10. Wash membrane 3× with Tris-buffered saline with Tween-20.
11. Develop immunoblot with ECL reagent.

3.2 Resolving Aggregate State in Brain Samples by Size-Exclusion Chromatography Combined with TR-FRET Assay

The second simple method for qualitative and quantitative characterization of mutant HTT aggregates in biological samples is a combination of size-exclusion chromatography (SEC) and time-resolved fluorescence resonance energy transfer (TR-FRET) assay. The SEC separates mutant HTT protein and complexes according to size and shape and the fractions collected are further characterized using a TR-FRET-based immunoassay that specifically detects soluble mutant HTT species [8–10]. This methodology resolves and defines the formation, aggregation, and temporal dynamics of subsets of native soluble mutant HTT aggregates in brain samples of HD mouse models (Figs. 3 and 4).

1. Add 500 μl of complete ice-cooled lysis buffer to your brain tissue sample (e.g., half brain of HD R/2 mouse) and homogenize with ceramic beads according to the manufacturer's instructions. Transfer 500 μl homogenate extract to 1 ml polycarbonate tubes.
2. Sonicate homogenate-extracts for 10 s (optional).
3. Centrifuge the 500 μl extracts at 100,000 × *g* for 30 min using a high-speed fixed-angle benchtop ultracentrifuge with polycarbonate tubes (*see* **Note 6**).
4. Filter supernatant through a 0.45 μm membrane and determine protein concentration by BCA assay (*see* **Note 7**).

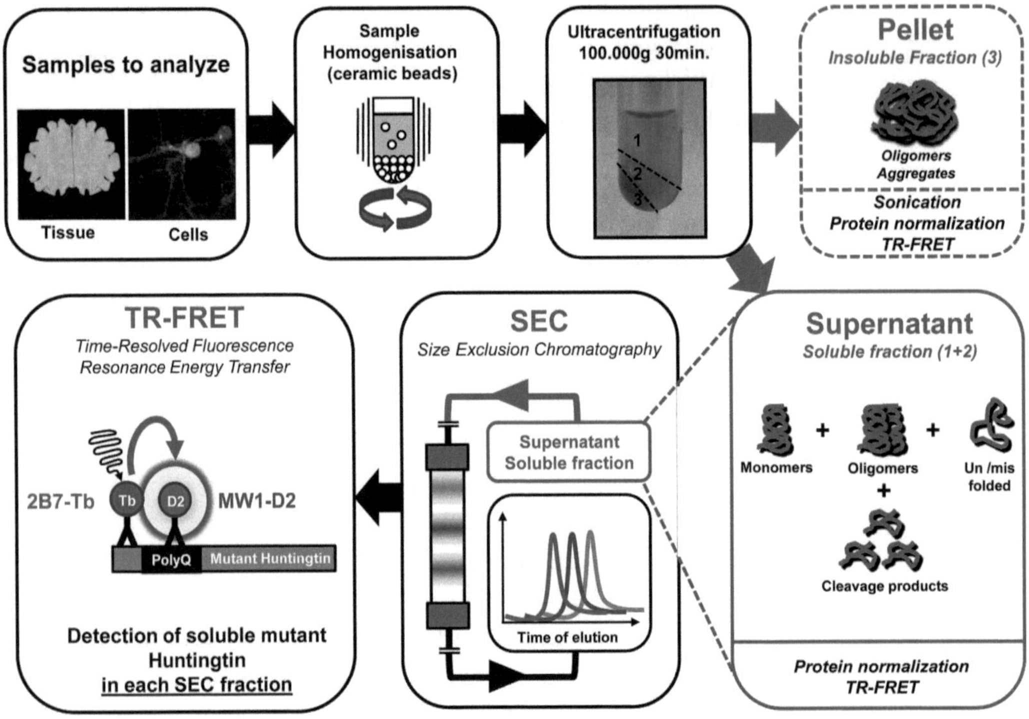

Fig. 3 Schematic overview of the SEC-FRET methodology. The SEC-FRET technology is a useful tool to detect multiple mHTT species in biological samples (adapted from ref. 10)

5. Equilibrate the Superdex 200 10/300 column with several column volumes of running buffer until baseline is stable by using chromatography system (*see* **Note 8**).
6. Load and inject protein samples (brain) on the Superdex 200 10/300 column in a sample volume of 500 μl (*see* **Note 9**).
7. Perform the separation of the protein complexes by eluting one column volume (~24 ml), at 4 °C, with a flow rate of 0.5 ml/min and in running buffer (*see* **Note 10**).
8. Elution is collected by fraction collector in 96-well plate format with 250 μl per fraction.
9. 96-Well plates can be stored at −80 °C and used for further analysis when needed. However, several freezing-thawing cycles will impact measurements.
10. Thaw plate, transfer 10 μl to a low-volume plate, and add 1 μl mix of HTT antibodies labeled with terbium (Tb) or d2 fluorophores. The final antibody amount per well for soluble mutant huntingtin detection should be 1 ng 2B7-Tb and 10 ng MW1-d2. TR-FRET detection of aggregated mHTT should be performed with MW8 antibody labeled with terbium (donor) or d2 fluorophores (acceptor) [8–10] (*see* **Note 11**).

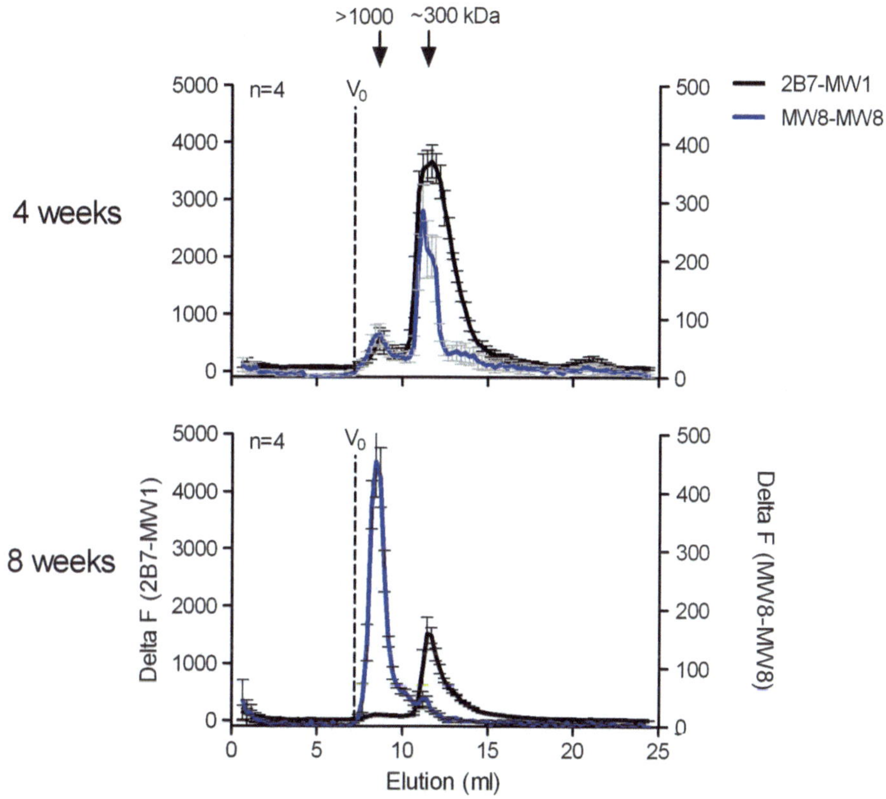

Fig. 4 Detection of a subset of soluble mHTT aggregates in R6/2 brain lysate at different age with a combination of size-exclusion chromatography and TR-FRET. Supernatant obtained from centrifuged R6/2 brain homogenate was loaded onto a Superdex 200 column and the fractions were analyzed by TR-FRET with indicated antibody combination. Graphs show TR-FRET signal profiles of SEC-eluted supernatant of R6/2 brain tissue homogenate at 4 weeks and 8 weeks of age ($n = 4$ per age). *Black line* corresponds to 2B7/MW1 and blue to MW8/MW8 TR-FRET signal profile. *Arrows* indicate estimated size of main peaks in kDa by using protein standards in the same running buffer. Error bars depict standard deviations (adapted from ref. 9)

11. Incubate antibody mix with fraction samples overnight at 4 °C.
12. Measure TR-FRET signal by using an EnVision Reader (Perkin Elmer, excitation 320 nm, time delay 100 μs; integration time 400 μs) (*see* **Note 12**).
13. Plot Delta F values (percentage signal intensities over lysis buffer background signal normalized to total protein concentration) to elution (in ml) to obtain mutant HTT elution profile of your brain samples (Fig. 4) (*see* **Note 13**).

4 Notes

1. The instrumentation does not have to be these specific models. These represent those used in our laboratory. Other external labs have in the meantime adapted the method using other brands of semidry blotters and trays to pour the gels.

2. When melting the agarose, care has to be taken that the mixture does not boil over. Additionally, to avoid evaporation and thereby variation in the final gel percentage, a loose cover such as a petri dish should be placed above the flask while the mixture is heated in the microwave.

3. SDS should always be added after melting the agarose to avoid excess formation of air bubbles in the poured gel.

4. Purified catalase (232 kDa), ferritin (440 kDa), and thyroglobulin (669 kDa) were taken as high-molecular-weight size markers (all proteins included in Amersham Bioscience HMW Gel Filtration Kit, #17-0441-01). Ferritin is an especially useful marker protein as it is readily detectable on the blotted PDVF membrane due to its brownish color, thereby providing a quick and easy size and protein transfer control.

5. Thickness of the lower percentage gels (<1.5 %) decreases during the immunoblot transfer step. Therefore, it is advised to center a ~500 g weight on the electroblotter's top. This will facilitate constant and even contact between the gel and the electroblotter when blotting these gels.

6. Centrifugation at 100,000 × *g* or higher results in three phases of the homogenate (Fig. 3). The two top soluble phases are collected for the SEC-FRET analysis. The third insoluble pellet phase is processed further for TR-FRET analysis only. Pellet from R6/2 brain homogenate is processed further by washing once (PBS, 1 % (w/v) TritonX), then resuspended in 2 % (w/v) Triton X-100 in PBS, and sonicated. 10 μl of the suspension is measured by TR-FRET assay with labeled antibody combinations (e.g., MW8-Tb/MW-8-D2) [9, 10].

7. Several standard protein detection assays (Lowry or Bradford assays) can be used to measure protein concentration. However, bicinchoninic acid (BCA) assay is most useful for samples that include nonionic detergents such as Triton X-100.

8. If no chromatography system is available, a combination of peristaltic pump, chromatography column, and fraction collector would work, too.

9. Do not load your protein samples into high volume to avoid loss of resolution by broadening the peak. Concentration of the protein load depends on estimated HTT concentration in the brain (determined by total soluble protein Coomassie staining related to total soluble HTT level determined by western blotting). For instance, the R6/2 HD mouse model expressed high level of mutant HTT exon1 fragments, whereas other HD models have less mutant HTT protein expressed. To obtain reasonable signal for TR-FRET measurement, we typically load around 300 μg total protein of mutant HTT-overexpressing cells, 1.2 mg protein load taken from R6/2 half brains, and 3 mg from HD150Q knock-in mouse model [9, 10].

10. The Superdex 200 10/30 column can be calibrated with Gel Filtration Marker Kit for protein molecular weights 29,000–700,000 Da (Sigma # MWGF1000). Dilute proteins in PBS and inject into the column. Normalize the volume of elution for each protein marker, *Ve*, to the elution volume obtained with Blue dextran 2000 (void volume = *V0*). Plot the ratio *Ve*/*V0* with their respective molecular weight logarithm to obtain a linear regression.
11. MW1 and MW8 antibodies were developed by Paul Patterson [7] and obtained from the Developmental Studies Hybridoma Bank developed under the auspices of the NICHD and maintained by The University of Iowa, Department of Biological Sciences, Iowa City, IA 52242. Generation and characterization of 2B7 were described previously [8].
12. Antibodies were labeled by CisBio Bioassays (Parc Marcel Boiteux, France). Soluble mutant HTT was measured with 2B7-Tb/MW1-D2. Aggregated mutant HTT was measured with MW8-Tb/MW8-D2 [8–10].
13. Values are expressed as the ratio between the emission at 665 and 620 nm. Each ratio is expressed in Delta F: First, the sample ratio is reduced with the background signal ratio (antibodies in lysis buffer), then divided by the background signal ratio, and finally multiplied by 100. Delta F values are then normalized to protein concentration of the brain tissues.

References

1. Chiti F, Dobson CM (2006) Protein misfolding, functional amyloid, and human disease. Annu Rev Biochem 75:333–366
2. Sarah L, Hands SL, Wyttenbach A (2010) Neurotoxic protein oligomerisation associated with polyglutamine diseases. Acta Neuropathol 120:419–437
3. Hazeki N, Tukamoto T, Gogo J et al (2000) Formic acid dissolves aggregates of an N-terminal huntingtin fragment containing an expanded polyglutamine tract: applying to quantification of protein components of the aggregates. Biochem Biophys Res Commun 277(2):386–393
4. Scherzinger E, Lurz R, Turmaine M et al (1997) Huntingtin-encoded polyglutamine expansions form amyloid-like protein aggregates in vitro and in vivo. Cell 90(3):549–558
5. Kryndushkin DS, Alexandrov IM, Ter-Avanesyan MD et al (2003) Yeast [PSI+] prion aggregates are formed by small Sup35 polymers fragmented by Hsp104. J Biol Chem 278(49):49636–49643
6. Bagriantsev SN, Kushnirov VV, Liebman SW (2006) Analysis of amyloid aggregates using agarose gel electrophoresis. Methods Enzymol 412:33–48
7. Ko J, Ou S, Patterson PH (2001) New anti-huntingtin monoclonal antibodies: implications for huntingtin conformation and its binding proteins. Brain Res Bull 56:319–329
8. Weiss A, Abramowski D, Bibel M et al (2009) Single-step detection of mutant Huntingtin in animal and human tissues: a bioassay for Huntington's disease. Anal Biochem 395(1): 8–15
9. Baldo B, Paganetti P, Grueninger S et al (2012) TR-FRET based duplex immunoassay reveals an inverse correlation of soluble and aggregated mutant huntingtin in Huntington's disease. Chem Biol 19(2):264–275
10. Marcellin D, Abramowski D, Young D et al (2012) Fragments of HdhQ150 mutant huntingtin form a soluble oligomer pool that declines with aggregate deposition upon aging. PLoS One 7(9):e44457

Chapter 13

Modeling and Analysis of Repeat RNA Toxicity in *Drosophila*

S.E. Samaraweera, L.V. O'Keefe, C.L. van Eyk, K.T. Lawlor, D.T. Humphreys, C.M. Suter, and R.I. Richards

Abstract

Expansion of repeat sequences beyond a pathogenic threshold is the cause of a series of dominantly inherited neurodegenerative diseases that includes Huntington's disease, several spinocerebellar ataxias, and myotonic dystrophy types 1 and 2. Expansion of repeat sequences occurring in coding regions of various genes frequently produces an expanded polyglutamine tract that is thought to result in a toxic protein. However, in a number of diseases that present with similar clinical symptoms, the expansions occur in untranslated regions of the gene that cannot encode toxic peptide products. As expanded repeat-containing RNA is common to both translated and untranslated repeat expansion diseases, this repeat RNA is hypothesized as a potential common toxic agent.

We have established *Drosophila* models for expanded repeat diseases in order to investigate the role of multiple candidate toxic agents and the potential molecular pathways that lead to pathogenesis. In this chapter we describe methods to identify candidate pathogenic pathways and their constituent steps. This includes establishing novel phenotypes using *Drosophila* and developing methods for using this system to screen for possible modifiers of pathology. Additionally, we describe a method for quantifying progressive neurodegeneration using a motor functional assay as well as small RNA profiling techniques, which are useful in identifying RNA intermediates of pathogenesis that can then be used to validate potential pathogenic pathways in humans.

Key words Dynamic mutations, Repeat expansion diseases, Polyglutamine, RNA dominant pathogenesis, Neurodegeneration, Huntington's disease, Drosophila models of human disease

1 Introduction

There are over 20 dominantly inherited neurodegenerative diseases caused by expanded repeat sequences including Huntington's disease (HD), a series of related movement disorders (ataxias), and myotonic dystrophy types 1 and 2 [1–3]. Recently, some subtypes of amyotrophic lateral sclerosis (ALS) and frontotemporal lobar degeneration (FTLD) have also been shown to be caused by this type of mutation [4]. These diseases typically manifest as late onset, progressive neurodegenerative, and/or muscular diseases, some

Danny M. Hatters and Anthony J. Hannan (eds.), *Tandem Repeats in Genes, Proteins, and Disease: Methods and Protocols*, Methods in Molecular Biology, vol. 1017, DOI 10.1007/978-1-62703-438-8_13, © Springer Science+Business Media New York 2013

with a high degree of clinical overlap despite involving distinct proteins. Despite the fact that the causative mutation is known in each of these diseases, the molecular basis of pathogenesis is unclear. Since pathology typically involves cell death, the cells of interest are rarely available in affected patient tissue. Furthermore cell death could be the result of numerous cellular processes, rendering the identification of specific pathogenic pathways problematic.

Where expanded repeats are translated, the repeat RNA generally codes for polyglutamine [1]. Therefore much attention has been focused on expanded polyglutamine as the common basis of pathology [5–7]. However, in some of these diseases, the repeat expansions are located within untranslated RNA [1, 2], yet they exhibit similar symptoms. Hence an alternative hypothesis is that one or more RNA-based pathogenic pathways are responsible for most (if not all) of the dominant neurodegenerative diseases caused by expanded repeat sequences. That is, repeat expansion leads to alteration in RNA structure and/or form that initiates and/or perturbs a pathway resulting in loss of cellular function and ultimately cell death. One means of testing this hypothesis is to define and validate model systems of repeat RNA pathogenesis. These models can be used to address key questions—What types of repeat RNA cause pathogenesis? What are the hallmarks of RNA pathogenic pathways? Do human diseases exhibit these hallmarks?—thus providing validation of these candidate pathways in the pathogenesis of the human disease.

With extensive resources and over 100 years of experimental development, *Drosophila* offers not only clear advantages for the genetic dissection of complex biological pathways, but there is also a growing appreciation of the benefits and opportunities that *Drosophila* genetics brings to therapeutic drug discovery [8]. Furthermore, *Drosophila* has been extensively used to model neurodegenerative diseases [9] including ALS and FTLD [10, 11]. Substantial and significant progress has now been made in the use of *Drosophila* to identify and characterize distinct expanded repeat-based pathogenic pathways. In this chapter we discuss methods that have been used to explore repeat RNA pathogenesis in *Drosophila* [12–15]. Disruption to the patterning of the eye and abdomen (i.e., tergite disruption) can be used as qualitative and quantitative assays of expanded repeat-induced toxicity. These systems can be used as the basis for genetic modification analyses to identify potential pathogenic pathways. We also discuss methods for the analysis of in vivo repeat-containing RNA localization. In addition, we describe a functional measure for progressive neurodegeneration using a motor function assay and methods for small RNA profiling which we have used to identify intermediates of repeat RNA pathogenesis.

2 Materials

2.1 *Drosophila* Stocks

1. *UAS-repeat* stocks (generated in our lab and described in [12–14]).
2. Driver stocks (obtained from Bloomington stock center): *GMR-GAL4* (#9146), *da-GAL4* (*#8641*), *and P{GAL4-elav.L}2/CyO* (*#8765*, referred to as *elavII-GAL4*).
3. Double balancer stock (obtained from Bloomington stock center): *Bl/CyO; TM2/TM6B* (#3704).

2.2 Standard Fly Food Media

1. 11 % sucrose, 5.2 % polenta, 5 % baker's yeast, 5 % treacle, 1.8 % tegosept (10 % methyl parahydroxybenzoate), and 0.9 % J-grade agar in reverse osmosis purified water.

2.3 Phenotypic Analysis and Screening Using the Eye

1. Dissecting microscope with camera.
2. Collections to be screened are available from Bloomington (*flystocks.bio.indiana.edu*) and Vienna *Drosophila* RNAi Center (*stockcenter.vdrc.at*) stock centers.

2.4 Visualizing RNA Localization

1. Cryomolds.
2. Freezing medium (Tissue-Tek® O.C.T.™).
3. Cryostat for sectioning specimens.
4. Lysine-coated slides and coverslips.
5. 4 % paraformaldehyde (in PBS).
6. 1× PBS: 7.5 mM Na_2HPO_4, 2.5 mM NaH_2PO_4, 145 mM NaCl.
7. Ethanol.
8. Fluorescently labeled oligonucleotide probes.
9. SSC: 150 mM NaCl, 15 mM sodium citrate.
10. Denhardt's reagent: 0.02 % (w/v) Ficoll 400, 0.02 % (w/v) polyvinylpyrrolidone, 0.02 % (w/v) bovine serum albumin dissolved in H_2O, filtered, and sterilized.
11. Hybridization buffer: 4× SSC, 0.2 g/mL dextran sulfate, 50 % (v/v) formamide, 0.25 mg/mL poly(A) RNA, 0.25 mg/mL single-stranded DNA, 0.1 M DTT, 0.5× Denhardt's reagent.
12. Humid chamber.
13. Mounting medium.
14. DAPI or Hoechst stain.

2.5 Climbing Assays

1. 500 mL measuring cylinders.
2. Parafilm.
3. Small funnel.
4. Lab timer.

2.6 Small RNA Sequencing and RNA Extraction

1. Plastic pestles (that fit microcentrifuge tubes).
2. 20 G 1 mL syringes.
3. Parafilm.
4. RNAse-free water.
5. RNAse-free microcentrifuge tubes.
6. 10 % SDS: 10 % w/v lauryl sulfate in water.
7. Phenol without chloroform (e.g., Trizol or similar).
8. Chloroform.
9. Isopropanol.
10. Ethanol.
11. 3 M sodium acetate, pH 5.2, pH adjusted with glacial acetic acid.

2.7 Small RNA Library Preparation and Sequencing

1. SOLiD RNA sequencing kit containing the following: adaptors, primers, ligation enzyme, reverse transcriptase, polymerase, dNTPs, and all related buffers (Applied Biosystems).
2. Emulsion PCR kit (Applied Biosystems).
3. Glycerol.
4. 10 bp dsDNA ladder.
5. 25 bp ladder.
6. 19:1 acrylamide/bis-acrylamide mix with urea (500 g/L).
7. Scalpels.
8. Isopropanol.
9. RNAse-free microcentrifuge tubes.
10. SOLiD™ System Analysis Pipeline Tool (Corona Lite) (http://solidsoftwaretools.com/gf/project/corona/).

3 Methods

3.1 Expression of Transgenic Expanded Repeat Constructs in Drosophila

Ectopic expression of expanded repeat constructs is carried out using the GAL4/UAS system [16]. In this model, flies carrying expanded repeat sequences downstream of the yeast upstream activation sites (UAS) are crossed to GAL4 driver lines [13]. The resulting progeny will express the GAL4 yeast transcriptional activator which binds the UAS and activates expression of the repeat transgenes in a tissue specific manner (Fig. 1) (*see* **Note 1**).

3.2 Phenotypic Analysis and Modification Screens Using the Adult Eye

The precise patterning events of the *Drosophila* eye during development result in a highly ordered array of approximately 800 hexagonally shaped ommatidia on the external surface of the adult eye [17]. This organized array of ommatidia can be visualized under a dissecting light microscope (Fig. 2a) and can be disrupted by

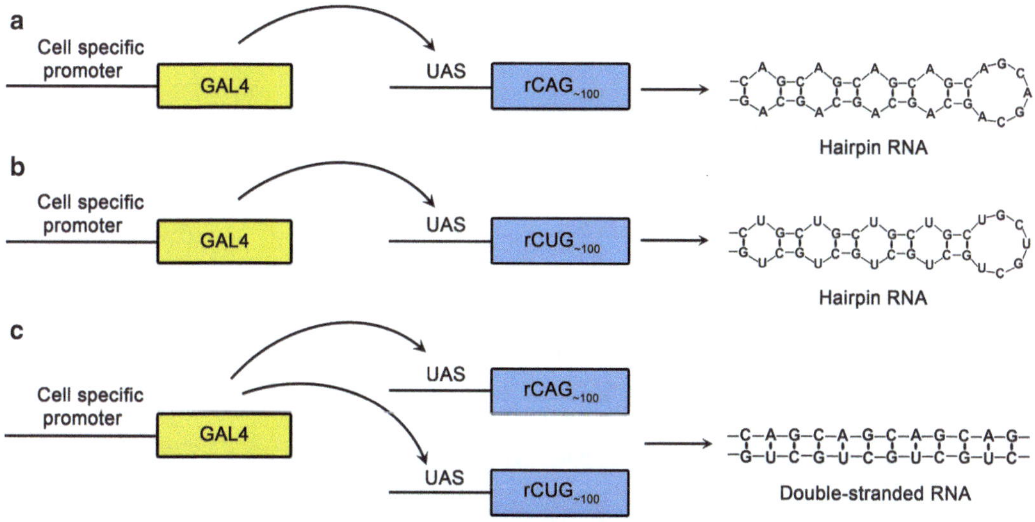

Fig. 1 The GAL4/UAS expression system. Flies express the yeast transcriptional activator GAL4 under the regulation of a cell-specific promoter. GAL4 protein binds and activates the yeast UAS activation sequence to drive expression of (**a**) rCAG$_{\sim100}$ repeat forming hairpin RNA, (**b**) rCUG$_{\sim100}$ repeat forming hairpin RNA, and (**c**) both rCAG$_{\sim100}$ and rCUG$_{\sim100}$ repeats which can form perfectly paired double-stranded RNA

ectopic expression of various repeat sequences using the *GAL4/UAS* system (Fig. 1). *GMR-GAL4* expresses in all differentiating cells (both non-neuronal and neuronal) of the adult eye [18] and can be used for ectopic expression of repeat sequences that are either translated (CAG and CUG) or untranslated (rCAG and rCUG). The resulting phenotypes can be visualized as "loss of pigment" where the amount of color pigment in the eye (due to the expression of the *white* gene in each of the *GAL4/UAS* ectopic expression constructs) is decreased. A "loss of pigment" phenotype is observed with ectopic expression of a transgene-encoding translated expanded CAG repeats (*UAS-CAG*$_{\sim100}$) under the control of *GMR-GAL4* (*GMR > CAG*$_{\sim100}$, Fig. 2b). In contrast, a phenotype may be observed as disruption to the ordered rows of ommatidia leading to a "roughening" of the external surface of the eye. This is observed with ectopic expression of translated expanded CUG repeat sequence (*GMR > CUG*$_{\sim100}$, Fig. 2c). Phenotypes can also be generated showing both of these characteristics. Although ectopic expression of multiple copies of either untranslated expanded *rCAG* (*GMR > 4xrCAG*$_{\sim100}$) or *rCUG* (*GMR > 4xrCUG*$_{\sim100}$) results in no obvious phenotype (Fig. 2d, e), co-expression of untranslated expanded *rCAG* and *rCUG* together (*GMR > 2xrCAG.2xCUG*$_{\sim100}$) gives a phenotype characterized by both disruption to ommatidial patterning and some loss of pigment (Fig. 2f).

Phenotypes such as these can form the basis of genetic modification analyses to determine whether altered expression of any candidate interactor is able to modify (i.e., enhance or suppress)

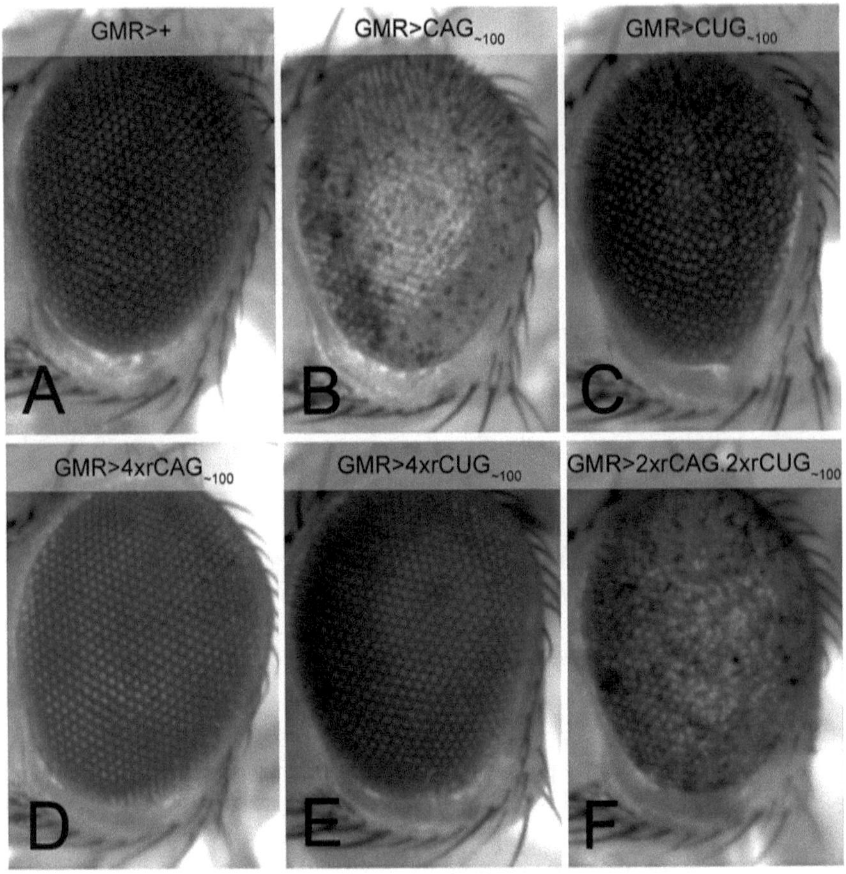

Fig. 2 Expression of expanded repeat sequences in the eye. (**a**) Control eyes show the organized array of ommatidia and the presence of pigment. (**b**) Expression of $CAG_{\sim100}$ demonstrates a strong "loss of pigment" phenotype whereas (**c**) expression of $CUG_{\sim100}$ disturbs the ordered array of ommatidia leading to "roughening" of the eye. The expression of (**d**) $4xrCAG_{\sim100}$ and (**e**) $4xrCUG_{\sim100}$ results in no obvious phenotype. (**f**) $2xrCAG.2xrCUG_{\sim100}$ double-stranded RNA results in both "loss of pigment" and "roughening" of the eye. Expression of all UAS-repeat constructs is under the *GMR-GAL4* driver. All experiments are at 25 °C [12, 14]

the original repeat-induced phenotype in the eye. A genetic interaction in this way is an indicator of a functional interaction between the repeat sequence and the candidate interacting gene. For example, ectopic expression of *GMR > 2xrCAG.2xCUG*$_{\sim100}$ using a stronger line than shown in Fig. 2f results in an eye phenotype characterized by roughening, loss of pigment, as well as some black necrotic regions (Fig. 3a). This *GMR > 2xr CAG.2xCUG*$_{\sim100}$ phenotype was tested for a genetic interaction with the well-characterized double-strand RNA-binding protein *Dicer-2* [12]. Decreased dosage of *Dicer-2* on its own has no effect on patterning of the adult eye (Fig. 3b) but results in a significant suppression of the *GMR > 2xrCAG.2xCUG*$_{\sim100}$ rough eye phenotype (Fig. 3c). This genetic interaction is supportive of a functional role for Dicer-2 processing of expanded double-strand rCAG. $rCUG_{\sim100}$ RNA in pathogenesis.

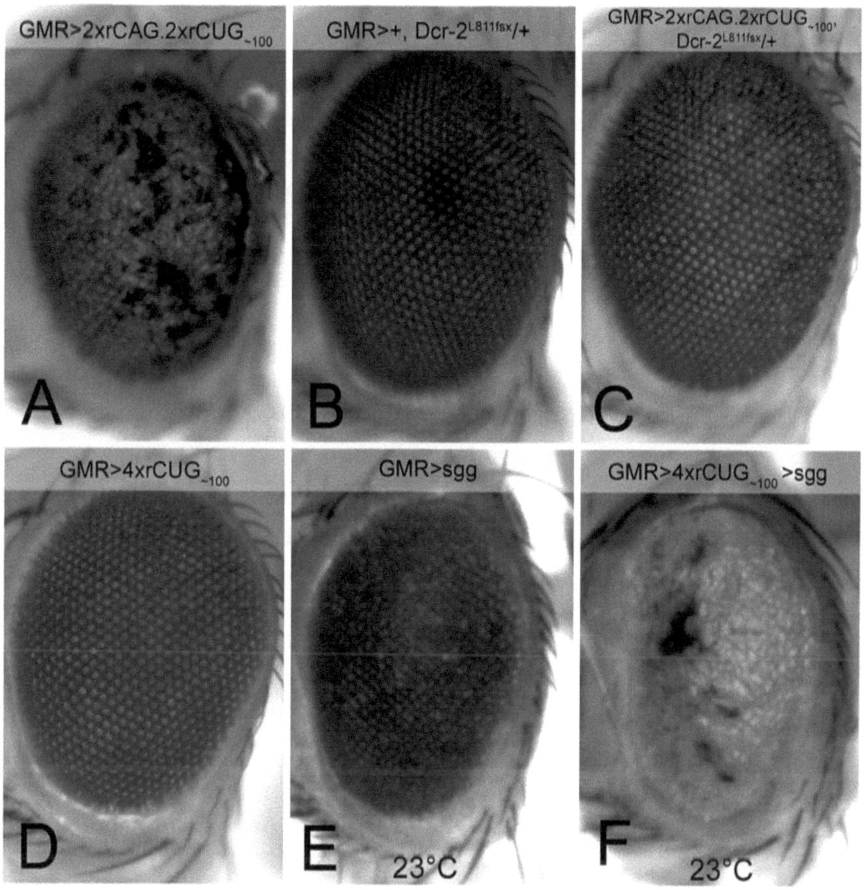

Fig. 3 Modification of eye phenotypes identifies potential pathogenic pathways. (**a**) Expression of 2xrCAG.2xrCUG$_{\sim 100}$ caused roughening of the eye, loss of pigmentation, and some necrotic regions. (**b**) *Dcr-2*^{L811fsx}/+ heterozygous loss of function mutation has no phenotype alone but (**c**) significantly suppresses the 2xrCAG.2xrCUG$_{\sim 100}$ phenotype [12]. (**d**) Expression of 4xrCUG$_{\sim 100}$ gives no phenotype in the eye. (**e**) *UAS-sgg* has a mild rough eye phenotype alone which is (**f**) significantly enhanced when co-expressed with 4xrCUG$_{\sim 100}$ resulting in strong loss of pigment, roughening, and necrotic regions [15]. Expression of all UAS-repeat constructs is under the *GMR-GAL4* driver. All experiments performed at 25 °C unless specified

A second example of genetic modification is the phenotype observed with expression of multiple copies of rCUG$_{\sim 100}$ (*GMR* > *4xrCUG*$_{\sim 100}$) which results in no obvious phenotype in the eye (Fig. 3d). This *GMR* > *rCUG* phenotype was tested for an interaction with Shaggy, the *Drosophila* orthologue of GSK3β [15]. Expression of Shaggy together with rCUG$_{\sim 100}$ (*GMR* > *rCUG*$_{\sim 100}$ > *sgg*) is lethal at 25 °C, and hence these experiments were performed at the lower temperature of 23 °C. While ectopic expression of Shaggy alone at 23 °C (*GMR* > *sgg*) results in a rough eye phenotype (Fig. 3e), a significant enhancement is observed in the presence of ectopic untranslated *rCUG*$_{\sim 100}$ (*GMR* > *rCUG*$_{\sim 100}$ > *sgg*), with the eyes showing a more significant disruption to ommatidial patterning, loss of pigment, as well as some black patches suggestive of necrotic regions

(Fig. 3f). This genetic interaction is supportive of a functional role for Shaggy in single-strand expanded CUG RNA pathogenesis.

In addition to testing individual candidates in this way, one of the major strengths of the *Drosophila* system is the possibility for genome-wide screens, where an unbiased approach can be taken to determine significant and often novel genetic modifiers of the gene or pathway of interest. Mutant collections available to be screened include the Bloomington Deficiency Kit, which is comprised of chromosomal deletions that cover 98.4 % of the *Drosophila* genome [19]; the Gene Disruption Project, which now has mutant stocks available for 9,440 genes (approximately two-thirds of the total number of genes in *Drosophila*) [20]; and collections of RNAi ectopic expression constructs for knockdown of 93 % of all known genes in *Drosophila* [21].

1. Collect virgins from the *UAS-repeat* stock to be tested and cross to males from the *GMR-GAL4* driver stock. Keep cross at 25 °C and turn to new vial (or discard parents) after 3–4 days (*see* **Notes 2** and **3**).
2. After 10–12 days, adult flies will start to emerge from cross. Clear all adults and collect age-matched females the following day to score phenotypes (i.e., they will be 0–1-day-old).
3. Determine any phenotype by close examination of the external surface of the eye on a dissecting microscope to reveal any disruption to the organized rows of ommatidia and/or any loss of pigment in regions of the eye (*see* **Notes 3–5**).
4. Photography with camera attached to the dissecting microscope is usually sufficient to record any phenotype (*see* **Note 6**).
5. To perform a genetic screen, flies carrying *GAL4* and *UAS-repeat* constructs are crossed to double balancer flies to generate double balanced stocks (i.e., *GMR-GAL4/CyO, TM2/TM6B* and *Bl/CyO, UAS-CAG*$_{\sim 100}$*/TM6B*). These stocks can then be crossed together to generate screening males (i.e., *GMR-GAL4/CyO, UAS-CAG*$_{\sim 100}$*/TM6B*) (*see* **Note 7**).
6. These screening males can be crossed separately to virgins of all stocks in the mutant collection being tested, and all non-balancer-containing progeny will have both the *GMR-GAL4* driver and the *UAS-repeat* construct in the altered genetic background of the potential modifier.

3.3 Quantitative Scoring of Adult Drosophila Morphological Phenotypes

In our *Drosophila* model, expression of single-stranded expanded repeat RNAs (4xrCAG$_{\sim 100}$ or 4xrCUG$_{\sim 100}$ alone) has no phenotype in the eye (Fig. 2d, e). Nevertheless other morphological changes can be examined in the presence of these repeat sequences. *Drosophila* provides an ideal system in which to undertake quantitative analysis of morphological phenotypes caused by specific

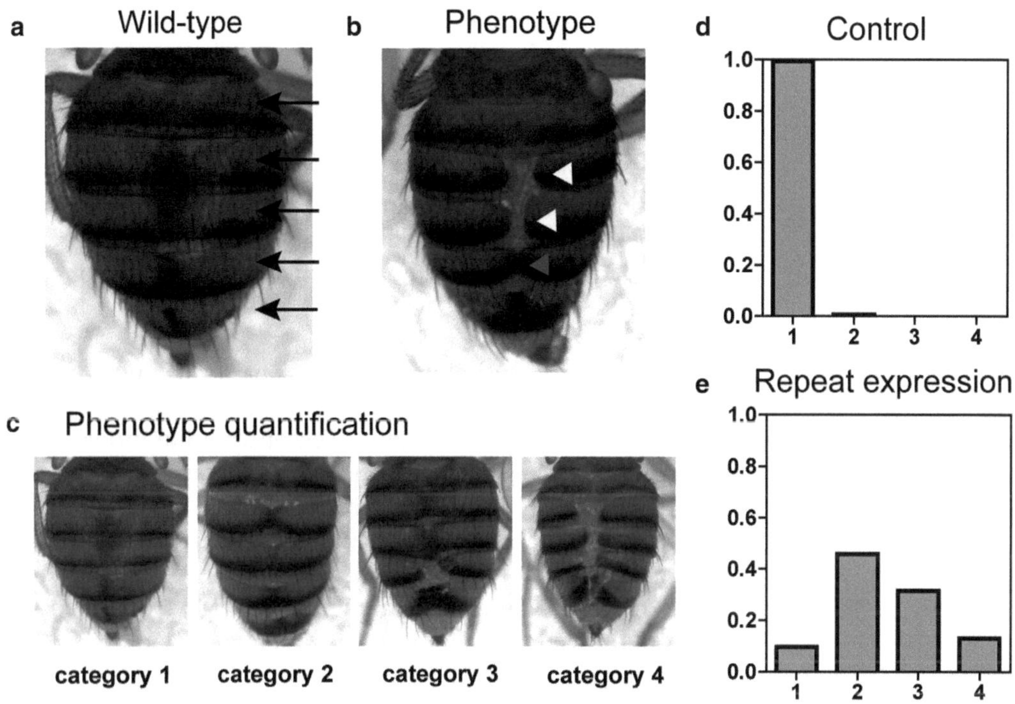

Fig. 4 Scoring disruption of tergite patterning in flies. (**a**) Wild-type flies show a regular arrangement of tergite bands along the dorsal abdomen (*black arrows*). (**b**) Ubiquitous expression of expanded repeat RNA gives rise to disruption of the adult abdominal tergites, where tergites do not fuse correctly (*gray arrow*) or sometimes do not fuse at all (*white arrows*). (**c**) Phenotype severity can be scored on a scale of 1–4, where category 1 is like wild type, category 2 has tergites disrupted but still fused, category 3 has one tergite entirely split, and category 4 has two or more tergites entirely split. (**d** and **e)** Graphs showing the proportion of progeny within each scoring category can be generated from this data, giving a quantitative measure of the tergite disruption phenotype which can be used as the basis for genetic modification [23]

genetic manipulations. In our system, repeat RNA-containing transgenes can be expressed ubiquitously, giving rise to specific perturbation of the adult abdominal tergites (Fig. 4a, b). This phenotype is manually scored in each individual fly, using a scale based on the severity of disruption (Fig. 4c). Scores for an entire population are tallied to give an overall phenotypic severity for each genotype (Fig. 4d, e). Establishing methods to quantify such phenotypes subsequently enables genetic modifier screening, using genetic tools available in *Drosophila* (as described in Subheading 3.2), to determine potential pathways of pathogenesis.

1. Cross flies carrying *UAS-repeat* transgenes to flies carrying the ubiquitous *daughterless-GAL4* (*da-GAL4*) [22] driver (*see* **Note 8**). Care should be taken to include appropriate controls (as described in **Note 3**).
2. Collect 0–1-day-old female flies. Crosses should be set up to obtain sufficient progeny to enable statistical comparison.

3. Score the disruption to the dorsal abdominal tergites in each individual based on the number of tergites disrupted and severity of disruption (according to the range shown in Fig. 4c).
4. Tally results to obtain an overall phenotypic strength for each repeat expression genotype (Fig. 4d, e).
5. Modifier genes can be examined by undertaking an identical experiment where a specific mutation or ectopic expression construct is present within the genetic background of repeat expression flies.

3.4 Examination of Repeat RNA Cellular Localization Using Fluorescent In Situ Hybridization

Expanded repeat RNA in human patients has been shown to form discrete nuclear foci that co-localize with the RNA-binding protein MBNL-1 [24, 25]. Foci have also been observed in model organisms, including *Drosophila*, identifying a conserved hallmark of repeat RNA-mediated pathology (reviewed in [26]). The ability to perform precise genetic manipulations and fluorescent imaging techniques makes *Drosophila* an ideal system in which to examine the role of repeat RNA localization in pathology. Combined with immunofluorescence techniques, in situ hybridization methods may also form the basis for studies examining co-localization of repeat RNA with candidate binding proteins.

Methods presented here are adapted from previous studies in human tissue and *Drosophila* [25, 27]. We and others have successfully examined RNA localization in frozen sections of whole third instar larvae, which provides a convenient way to examine multiple tissues at once (Fig. 5). This protocol has been optimized for use with sections; however, with modified conditions, the visualization of repeat RNA localization within whole-mount tissue may also be achieved.

Design of Probes: Previous studies have made use of complementary oligo probes that are directly labeled with a fluorescent molecule such as a cyanine dye (Cy3, Cy5), rhodamine, or FITC. Using a short oligo probe (generally 30 nt) complementary to the repeat of interest, it is possible to visualize the localization of repeat RNA independent of transcript context (Fig. 5). However, this approach may allow probe binding to repeat sequences present in endogenous transcripts. An alternative approach involves the use of fluorescently labeled oligo probes, or labeled RNA probes, complementary to a flanking sequence. This has the advantage of enabling the use of the same probe to make comparisons between different repeat sequence-containing transgenes.

1. Collect wandering third instar larvae and place in cryomolds filled with OCT medium. Quickly freeze on dry ice and store at −80 °C.
2. Cut 10–20 µm sections using a cryostat. Cutting conditions should be optimized to obtain intact, reproducible sections.

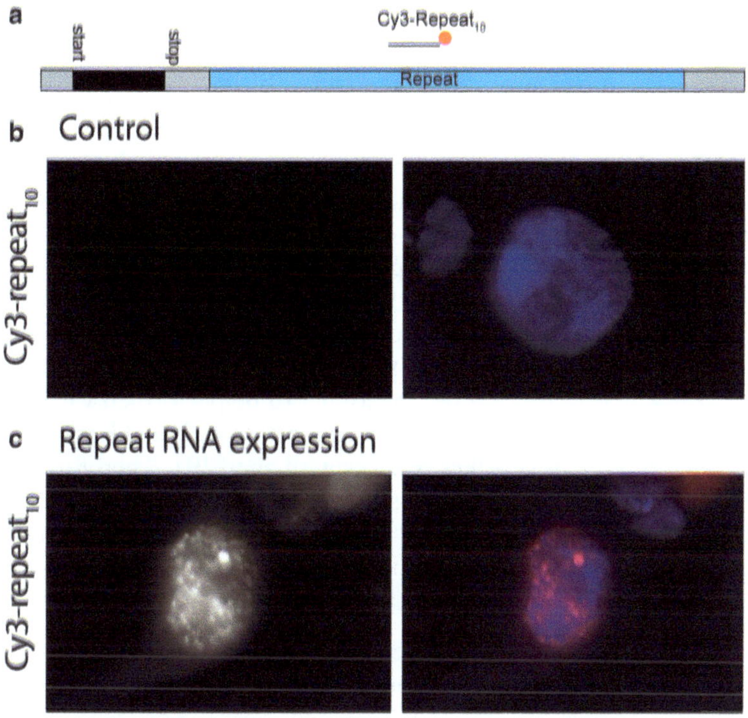

Fig. 5 Visualization of RNA localization in *Drosophila* tissue sections. (**a**) Fluorescently labeled oligonucleotide probes can be designed which are complementary to the expanded repeat sequence. (**b**) A cell from a larva not expressing the expanded repeat construct shows no signal in the nucleus (DAPI, *blue staining*). (**c**) A cell from a larva expressing expanded repeat RNA, showing multiple foci (*red*) throughout the nucleus (DAPI, *blue staining*) [23]

3. Transfer sections immediately to lysine-coated slides and quickly transfer slides to dry ice to proceed with the protocol, or store slides at −80 °C.
4. Fix for 15 min in cold, freshly thawed 4 % paraformaldehyde in 1× PBS.
5. Wash three times for 15 min with PBS and rinse quickly in 100 % ethanol.
6. Dilute probe to a final concentration of 0.5 ng/μL in hybridization buffer. Probe concentration can be adjusted to obtain optimum signal with minimal nonspecific binding.
7. Add sufficient diluted probe in buffer to just cover the sample, and cover with a coverslip. For the remainder of the protocol, care should be taken to avoid exposure to excess light that may reduce probe fluorescence. Hybridize 2 h to overnight at 37 °C in a sealed humid chamber (*see* **Note 9**).
8. Wash slides four times for 15 min each in SSC (*see* **Note 10**).

9. Air dry and mount in mounting medium containing a DNA stain such as DAPI or Hoechst to visualize the nucleus.
10. Visualize RNA localization using an epifluorescence microscope.

3.5 Functional Analysis of Neurodegeneration-Climbing Assays

Climbing or negative geotaxis measures the startle-induced vertical movement of flies up the inside wall of a container. Negative geotaxis is a robust response in which over 90 % of flies from most wild-type strains will instantly respond when the flies are tapped to the bottom of a container [28]. Climbing assays are a common measure of neurodegeneration in flies as the measure of mobility is similar to ataxia in humans [9]. To document degeneration over time, the assay can be performed weekly or fortnightly until up to 4–6 weeks after eclosion. Notably, the climbing ability of wild-type flies reduces with age [29]. Therefore with the use of appropriate controls, degeneration caused by the expanded repeat sequences beyond that of merely "aged" flies can be determined. In our experiments we use the *elavII-GAL4* [30] driver for pan-neuronal expression of the expanded repeat sequence.

1. Age-matched (e.g., 0–3-day-old) adult male flies of the desired genotype must be selected. In order to ensure that the flies are age-matched, clear the progeny from each vial on day zero and then collect flies that eclose in the next 3 days. Separate the progeny into vials of 20–25 flies. At least three replicates for each genotype must be used to eliminate biological and technical variation (*see* **Note 11**).
2. Transfer one replicate of flies (≤25 flies) to a 500 mL (~48 mm diameter) measuring cylinder and seal the top of the cylinder with Parafilm. Similarly set up multiple identical measuring cylinders to test "replicate 1" of all genotypes in one session (*see* **Note 12**).
3. Mark 50 mL (~27 mm) on the cylinder (Fig. 6).
4. To score "replicate 1" of the first genotype, gently tap the cylinder on the bench top to knock the flies to the bottom and immediately start a timer (*see* **Note 13**).
5. After 25 s, count and record the numbers of flies remaining below the 50 mL mark to determine flies that are unable to climb (*see* **Note 14**).
6. Score and record "replicate 1" for all other genotypes.
7. Return to the first genotype and score again. Perform a total of five consecutive trials per replicate allowing for at least 3 min rest between trails for a given replicate.
8. At the end of the experiment, transfer the flies back into regular vials with the aid of a small funnel. Take care to not let any escape (*see* **Note 15**).
9. Repeat for the remaining replicates for all genotypes.

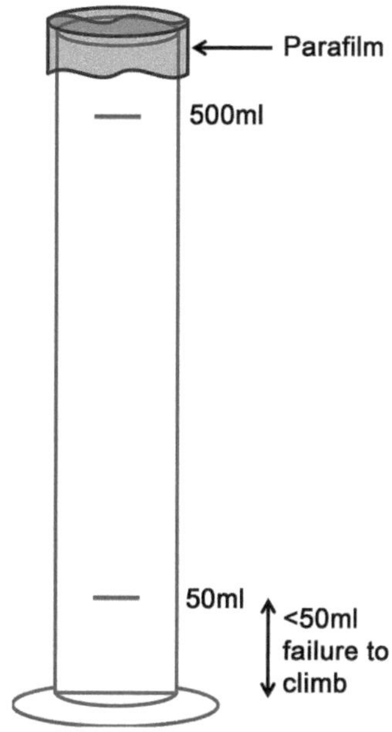

Fig. 6 Apparatus for climbing assays. A 500 mL measuring cylinder is marked at 50 mL. Flies are transferred into the cylinder and sealed with Parafilm. The climbing ability of flies is measured over a period of 25 s, and those remaining below 50 mL are scored as "failure to climb"

10. To maintain flies over a period of time for aging experiments, the flies will need to be transferred into fresh vials every 2–3 days. This is also a good opportunity to score any dead flies and adjust the *n*-value of the climbing assays accordingly. The assay can be performed at any time point.
11. To analyze the data, use the five trials of each genotype to determine the mean climbing ability. These scores can then be compared to independent replicates of data (biological replicates) and to aged experiments of the same replicate (for age-dependent motor degeneration) (Fig. 7).

3.6 Small RNA Analyses

Small noncoding RNA populations, particularly microRNAs (miRNA), small-interfering RNAs (siRNA), and piwi RNAs (piRNA), are important regulators of gene expression in development, apoptosis, and survival [31, 32]. Small RNA deep sequencing has been useful in determining alterations to particularly the endogenous miRNA profile of *Drosophila* with mutations in proteins of the small RNA pathway [33–35]. In *Drosophila* models of expanded repeat diseases, analysis of small RNA populations can be used to determine processing of repeat-containing RNAs and alterations to endogenous miRNA profiles [12]. In order to profile changes in small RNAs that may be relevant to neurodegeneration, we expressed the expanded repeats using the *elav-GAL4* [30] driver and profiled newly eclosed flies.

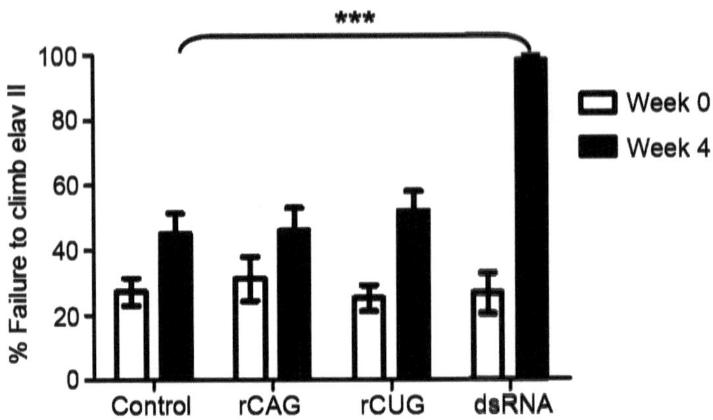

Fig. 7 dsRNA climbing phenotype. Pan-neuronal expression of expanded rCAG, rCUG, or dsRNA with *elavII-GAL4* does not affect the climbing ability of flies at week 0 compared to control. Similarly no effect is observed with either rCAG or rCUG alone at week 4. Flies expressing dsRNA show a significant increase in the percentage of flies that fail to climb ($n=70$, $p=7.6\times10^{-8}$, 2-tailed Student's *t*-test) at week 4 compared to the control ($n=70$). ***$p<0.001$; error bars show s.d. At week 0 $n\geq74$ and at week 4 $n\geq47$ (rCAG = $4xrCAG_{\sim100}$, rCUG = $4xrCUG_{\sim100}$, dsRNA = $2xrCAG.2xrCUG_{\sim100}$) [12]

3.6.1 RNA Extraction

1. Use flies that are age-matched for this experiment as expression patterns change with age. Collect thirty whole flies or one hundred heads from adult flies at the desired age (e.g., 0–1-day-old) (*see* **Note 16**).
2. Freeze the specimen immediately using liquid nitrogen and store at −80 °C until ready for use.
3. All steps from this point must be performed *on ice* unless specified. Wipe down bench tops with 10 % SDS. Use RNAse-free water and RNAse-free tubes.
4. To extract total RNA, in a fume hood, homogenize sample in 200 μL Trizol or similar phenol-based product using a plastic pestle (*see* **Note 17**).
5. Add a further 800 μL Trizol and homogenize with a 20G syringe. If required, a volume greater than 200 μL can be used in the first homogenizing step and Trizol added to a total of 1 mL.
6. Incubate at room temperature for 5 min and then centrifuge at 16 rcf for 10 min at 4 °C.
7. Transfer the supernatant to a fresh microcentrifuge tube. If any pelleted tissue is transferred, spin again.
8. In the fume hood add 200 μL of chloroform to each tube and shake vigorously by hand.
9. Incubate at room temperature for 3 min and then spin at 16 rcf for 15 min at 4 °C.

10. Transfer the upper aqueous phase (~500 μL) to a fresh RNAse-free microcentrifuge tube (*see* **Note 18**).
11. Add an equal volume isopropanol, mix and incubate at room temperature for 10 min, and then spin at 16 rcf for 10 min at 4 °C.
12. Remove the supernatant and wash the pellet in 1 mL 75 % ethanol.
13. Discard the supernatant. Take care to remove all traces of ethanol and air dry the pellet.
14. Resuspend the pellet in 100 μL RNAse-free water. The RNA concentration can be determined at this point using a spectrophotometer.
15. To further purify the RNA, add 0.1 volume (10 μL) 3 M sodium acetate pH 5.2 and 2.5 volume (250 μL) ice-cold 100 % ethanol and precipitate at −80 °C for at least 30 min.
16. Spin at 16 rcf for 5 min at 4 °C.
17. Remove the supernatant and wash the pellet in 250 μL of 75 % ethanol to remove remaining salts.
18. Add 250 μL fresh 75 % ethanol and store the RNA at −80 °C (*see* **Note 19**).

3.6.2 Small RNA Library Preparation for SOLiD Sequencing

Libraries for SOLiD deep sequencing are prepared from total RNA using the Small RNA Expression Kit (SREK; Ambion, as per the Applied Biosystems protocol). Briefly:

1. Total RNA (250–1,000 ng) is annealed and ligated to SOLiD P1 and P2 adaptors at 16 °C for 16 h.
2. The resulting ligated RNAs are then reverse transcribed with Array script™ at 42 °C for 30 min and then treated with RNase H (30 min at 37 °C) (*see* **Note 20**).
3. The resulting cDNA (1 μL) is used in a 100 μL PCR reaction for 5–18 cycles of 95 °C, 30 s; 62 °C, 30 s; and 72 °C, 30 s. Primers for this PCR step introduce a unique five-nucleotide barcode/index to each sample. Barcoding/indexing of samples allows for multiple samples to be sequenced on the one slide to reduce costs.
4. PCR products are then resolved on an 8 % polyacrylamide TBE gel, and the 105–150-bp fraction (corresponding to small RNAs 16–61 nucleotides in length) is excised from the gel.

3.6.3 Emulsion PCR and SOLiD Sequencing

1. Amplified PCR products are purified by incubating the gel slice overnight in buffer containing 5 mM Tris–HCl, 0.5 mM EDTA, and 2.5 M ammonium acetate with pH 8.
2. The PCR products are then precipitated using isopropanol to yield a SOLiD small RNA library for emulsion PCR. Emulsion

PCR generates clonally templated beads for sequencing. A single emulsion PCR reaction is used to couple up to ten barcoded libraries to P1-coated beads as per the standard Applied Biosystems protocol (*see* **Note 21**).

3. After emulsion PCR, templated beads are enriched in a glycerol gradient and deposited onto the surface of a glass slides for SOLiD sequencing.

3.6.4 Sequence Analysis

1. The number of trinucleotide repeat-containing small RNAs can be calculated directly from the SOLiD .csfasta file from sequences using the grep command line tool; here only exact matches can be accepted. For example, the number of CAG repeat reads of length 21 can be determined via grep T21231 231231231231231233302010303131 bc5.fasta | wc |.
2. For microRNA analyses, the SOLiD™ System Analysis Pipeline Tool is used to identify adapters (25 seed, maximum two color-space mismatches) and subsequently to map cloned inserts to miRNA precursors (*D. melanogaster* miRBase version 14) with a maximum of two color-space errors. If a read maps to multiple locations, the longest alignment should be selected; reads with multiple maximally long alignments are discarded.

Reads mapping to microRNA precursors in miRBase are considered further if they match the mature miR sequence or reported (or predicted) star sequence. To allow for isomiRs the match is accepted if it occurs within ±3 nt at either the 5′ or 3′ ends of the annotated sequence. Reads mapping to the precursor that could be considered neither mature nor star (e.g., including the loop sequence) should be discarded from further analyses. The total number of reads aligning to miRBase is used as the normalizing factor to allow for comparisons between libraries.

4 Notes

1. Specific driver lines can be chosen depending on the desired tissue or developmental time. Expression of transgenes with the ubiquitous *da-GAL4* [22] driver enables observation of effect on development and viability as well as adult morphological changes. The *GMR-GAL4* [18] driver is used to model repeat toxicity in the *Drosophila* eye which is useful in screening for potential modifiers of toxicity. Additionally the pan-neuronal *elav-GAL4* [30] driver is also useful in modeling functional neurodegenerative effects of repeat expression.
2. Changing temperature from the standard 25 °C may affect the severity of the phenotype. Crosses performed at 18 °C or 23 °C result in a significantly reduced phenotype, while 29 °C gives

significantly enhanced phenotypes (however, caution must be taken since *GMR-GAL4* alone gives a phenotype at this higher temperature).

3. The use of appropriate controls is essential to score and compare phenotypes. *GMR* > *CAG* can be compared to *GMR* > +; however, a better control is *GMR* > *EV* where ectopic expression lines are generated with the empty vector used to generate the repeat constructs (i.e., *UAS-EV*). Where multiple repeat constructs are expressed, they should be compared to a control with the same number of *UAS-EV* insertions.
4. If the phenotype observed within a cross is consistent and fully penetrant, then recording six eyes is usually sufficient. However, if the phenotype is variable within a cross, then it may be necessary to design a scoring strategy (i.e., mild, medium, and strong) where larger numbers of eyes need to be recorded for statistical analyses.
5. Phenotypes may vary between males and females. They are often stronger in females and these are the progeny scored in all of these analyses.
6. Scanning electron microscopy may be required for quantification of rough eye phenotypes as the number of ommatidia can be accurately determined.
7. When performing genetic modification screens with ectopic repeat-containing constructs, it is desirable to maintain *UAS*-repeat stocks in the absence of any *GAL4* driver stocks.
8. In our system robust phenotypes are obtained using the ubiquitous *da-GAL4* driver. Other ubiquitous drivers are also available, including *Act5c-GAL4* [36] and *tub-GAL4* [37].
9. A humid chamber can be created by placing moist tissue in a sealed plastic container. Slides can be placed on a platform, such as an upturned tube rack to prevent contact with any moisture.
10. Wash conditions for each probe should be optimized by adjusting temperature and salt concentration. Typical starting conditions in our hands are two washes for 15 min in 1× SSC at 37 °C, followed by two washes for 15 min in 0.5× SSC at 37 °C.
11. In our experience 25 flies in a set is ideal. Using a larger number of flies makes the transfer of flies to and from the measuring cylinder difficult and often results in escapees. In addition counting flies in each section of the cylinder is more difficult.
12. To prevent any complications in interpretation due to the use of anesthetic, do not anesthetize flies in order to transfer them to the cylinder. Simply tap the vial and swiftly transfer flies directly into the cylinder, taking care to cover the cylinder with one hand to prevent escapees. This may take some practice.

13. Set the apparatus up against a white wall or a similar background. This makes it easier to count the flies.
14. Scoring flies that remain below 50 mL will account for flies that fail to "right" themselves after the cylinder is tapped as well as those which have no climbing ability. In addition to this, the experiment can be expanded to count flies at different levels of the cylinder. Scoring flies in different levels of the cylinder may be useful in distinguishing between those that fail to "right" themselves and the degree of mobility. For example, count the flies above 500 mL to determine the number of "fast climbers." Increasing the number of sections scored within the cylinder gives more information for interpretation but makes accurate counting of flies difficult.
15. Do not leave flies in the cylinders for extended periods of time as low humidity could be detrimental and often makes the flies lethargic which can lead to false results. In our experience six replicates can be score in approximately 30 min, i.e., six cylinders scored five times each.
16. Collecting one hundred age-matched flies from a single cross is often not possible. Therefore, set up multiple crosses for each genotype and collect the progeny. If reaching the desired number takes a number of days, simply take care to ensure the flies are age-matched (collecting flies at the same time of day will ensure this) and freeze separately. The samples can be pooled at RNA extraction. If the desired tissue is adult fly heads, use a scalpel to separate heads and freeze appropriately.
17. Homogenizing with a plastic pestle in large volumes of liquid is difficult as the flies/fly heads tend to float in the liquid. If fly heads are spread over multiple tubes, split the volume across the tubes and homogenize the tissue. Then use a pipette to pool the tissue into one tube of one hundred fly heads.
18. To perform RNA extractions for quantitative PCR or microarray analysis, following isolation of the upper aqueous phase from the chloroform extraction, proceed to use a commercial RNA column purification kit. If small RNA fractionation is required, be wary of column kits that are not specific for small RNA purification as the desired RNAs may be lost in the flow through.
19. RNA can be transported under 75 % (v/v) ethanol with cold (wet ice) packs. Use Parafilm to cover the lid to prevent evaporation of ethanol. Alternatively use screw cap tubes.
20. In an update to this protocol, Step 20 is replaced by the following: The sample is heat denatured (70 °C for 5 min) and then RT primer added. RT performed as previously described. RT products are purified using the Qiagen mini-elute column and half of the reaction is heat denatured (95 °C for 3 min)

and loaded onto a denaturing 10 % PAGE gel. The RT products in the gel between 60 and 80 nt are sliced and divided into four subslices. One subslice is loaded directly into PCR reaction mixture and PCR performed as in Step 21.

21. There are now 16 different 5 nt barcodes and 100 different 10 nt barcodes available to be mixed into emulsion PCR. An emulsion PCR must contain at least four barcodes that are "balanced" to minimize risk of color miscalling.

References

1. La Spada AR, Taylor JP (2010) Repeat expansion disease: progress and puzzles in disease pathogenesis. Nat Rev Genet 11:247–258
2. Richards RI (2001) Dynamic mutations: a decade of unstable expanded repeats in human genetic disease. Hum Mol Genet 10: 2187–2194
3. Richards RI, Sutherland GR (1992) Dynamic mutations: a new class of mutations causing human disease. Cell 70:709–712
4. Orr HT (2011) FTD and ALS: genetic ties that bind. Neuron 72:189–190
5. Marsh JL, Walker H, Theisen H et al (2000) Expanded polyglutamine peptides alone are intrinsically cytotoxic and cause neurodegeneration in *Drosophila*. Hum Mol Genet 9:13–25
6. Olshina MA, Angley LM, Ramdzan YM et al (2010) Tracking mutant huntingtin aggregation kinetics in cells reveals three major populations that include an invariant oligomer pool. J Biol Chem 285:21807–21816
7. Ordway JM, Tallaksen-Greene S, Gutekunst CA et al (1997) Ectopically expressed CAG repeats cause intranuclear inclusions and a progressive late onset neurological phenotype in the mouse. Cell 91:753–763
8. Pandey UB, Nichols CD (2011) Human disease models in *Drosophila* melanogaster and the role of the fly in therapeutic drug discovery. Pharmacol Rev 63:411–436
9. Lessing D, Bonini NM (2009) Maintaining the brain: insight into human neurodegeneration from *Drosophila melanogaster* mutants. Nat Rev Genet 10:359–370
10. Estes PS, Boehringer A, Zwick R et al (2011) Wild-type and A315T mutant TDP-43 exert differential neurotoxicity in a *Drosophila* model of ALS. Hum Mol Genet 20:2308–2321
11. Lin MJ, Cheng CW, Shen CK (2011) Neuronal function and dysfunction of *Drosophila* dTDP. PLoS One 6:e20371
12. Lawlor KT, O'Keefe LV, Samaraweera SE et al (2011) Double-stranded RNA is pathogenic in *Drosophila* models of expanded repeat neurodegenerative diseases. Hum Mol Genet 20: 3757–3768
13. McLeod CJ, O'Keefe LV, Richards RI (2005) The pathogenic agent in *Drosophila* models of 'polyglutamine' diseases. Hum Mol Genet 14:1041–1048
14. van Eyk CL, McLeod CJ, O'Keefe LV et al (2012) Comparative toxicity of polyglutamine, polyalanine and polyleucine tracts in *Drosophila* models of expanded repeat disease. Hum Mol Genet 21:536–547
15. van Eyk CL, O'Keefe LV, Lawlor KT et al (2011) Perturbation of the Akt/Gsk3-beta signalling pathway is common to *Drosophila* expressing expanded untranslated CAG, CUG and AUUCU repeat RNAs. Hum Mol Genet 20:2783–2794
16. Brand AH, Perrimon N (1993) Targeted gene expression as a means of altering cell fates and generating dominant phenotypes. Development 118:401–415
17. Dickson BJ, Hafen E (1993) Genetic dissection of eye development in *Drosophila*. In: Bate M, Martinex AA (eds) The development of *Drosophila melanogaster*, vol II. Cold Spring Harbour Laboratory Press, New York, pp 1327–1362
18. Freeman M (1996) Reiterative use of the EGF receptor triggers differentiation of all cell types in the *Drosophila* eye. Cell 87:651–660
19. Cook RK, Christensen SJ, Deal JA et al (2012) The generation of chromosomal deletions to provide extensive coverage and subdivision of the *Drosophila melanogaster* genome. Genome Biol 13:R21
20. Bellen HJ, Levis RW, He Y et al (2011) The *Drosophila* gene disruption project: progress using transposons with distinctive site specificities. Genetics 188:731–743
21. Dietzl G, Chen D, Schnorrer F et al (2007) A genome-wide transgenic RNAi library for conditional gene inactivation in *Drosophila*. Nature 448:151–156

22. Wodarz A, Hinz U, Engelbert M et al (1995) Expression of crumbs confers apical character on plasma membrane domains of ectodermal epithelia of *Drosophila*. Cell 82:67–76
23. Lawlor KT, O'Keefe LV, Samaraweera SE et al (2012) Ubiquitous Expression of CUG or CAG Trinucleotide Repeat RNA Causes Common Morphological Defects in a *Drosophila* Model of RNA-Mediated Pathology. PLoS One 7:e38516
24. Miller JW, Urbinati CR, Teng-Umnuay P et al (2000) Recruitment of human muscleblind proteins to (CUG)(n) expansions associated with myotonic dystrophy. EMBO J 19:4439–4448
25. Taneja KL, McCurrach M, Schalling M et al (1995) Foci of trinucleotide repeat transcripts in nuclei of myotonic dystrophy cells and tissues. J Cell Biol 128:995–1002
26. Wojciechowska M, Krzyzosiak WJ (2011) Cellular toxicity of expanded RNA repeats: focus on RNA foci. Hum Mol Genet 20:3811–3821
27. Houseley JM, Wang Z, Brock GJ et al (2005) Myotonic dystrophy associated expanded CUG repeat muscleblind positive ribonuclear foci are not toxic to *Drosophila*. Hum Mol Genet 14:873–883
28. Rhodenizer D, Martin I, Bhandari P et al (2008) Genetic and environmental factors impact age-related impairment of negative geotaxis in *Drosophila* by altering age-dependent climbing speed. Exp Gerontol 43:739–748
29. Gargano JW, Martin I, Bhandari P et al (2005) Rapid iterative negative geotaxis (RING): a new method for assessing age-related locomotor decline in *Drosophila*. Exp Gerontol 40: 386–395
30. Luo L, Liao YJ, Jan LY et al (1994) Distinct morphogenetic functions of similar small GTPases: *Drosophila* Drac1 is involved in axonal outgrowth and myoblast fusion. Genes Dev 8:1787–1802
31. Ambros V (2004) The functions of animal microRNAs. Nature 431:350–355
32. Erson AE, Petty EM (2008) MicroRNAs in development and disease. Clin Genet 74: 296–306
33. Czech B, Malone CD, Zhou R et al (2008) An endogenous small interfering RNA pathway in *Drosophila*. Nature 453:798–802
34. Hartig JV, Forstemann K (2011) Loqs-PD and R2D2 define independent pathways for RISC generation in *Drosophila*. Nucleic Acids Res 39:3836–3851
35. Marques JT, Kim K, Wu PH et al (2010) Loqs and R2D2 act sequentially in the siRNA pathway in *Drosophila*. Nat Struct Mol Biol 17: 24–30
36. Ito K, Awano W, Suzuki K et al (1997) The *Drosophila* mushroom body is a quadruple structure of clonal units each of which contains a virtually identical set of neurones and glial cells. Development 124:761–771
37. Lee T, Luo L (1999) Mosaic analysis with a repressible cell marker for studies of gene function in neuronal morphogenesis. Neuron 22:451–461

Chapter 14

Analyzing Modifiers of Protein Aggregation in *C. elegans* by Native Agarose Gel Electrophoresis

Mats Holmberg and Ellen A.A. Nollen

Abstract

The accumulation of specific aggregation-prone proteins during aging is thought to be involved in several diseases, most notably Alzheimer's and Parkinson's disease as well as polyglutamine expansion disorders such as Huntington's disease. *Caenorhabditis elegans* disease models with transgenic expression of fluorescently tagged aggregation-prone proteins have been used to screen for genetic modifiers of aggregation. To establish the role of modifying factors in the generation of aggregation intermediates, a method has been developed using native agarose gel electrophoresis (NAGE) that enables parallel screening of aggregation patterns of fluorescently labeled aggregation-prone proteins. Together with microscopy-based genetic screens this method can be used to identify modifiers of protein aggregation and characterize their molecular function. Although described here for analyzing aggregates in *C. elegans*, NAGE can be adjusted for use in other model organisms as well as for cultured cells.

Key words *C. elegans*, Protein aggregation, Native agarose gel electrophoresis, Aggregation intermediates, Polyglutamine

1 Introduction

Proper protein folding and correct protein–protein interactions are vital for the maintenance of cellular function. Disturbances in these processes have been associated with a wide range of disorders, including Alzheimer's disease, Parkinson's disease, and polyglutamine (polyQ) expansion disorders such as Huntington's disease [1]. These processes are likely to be affected by a broad spectrum of intra- and extracellular factors, many of which have yet to be identified.

Genome-wide genetic screens in *C. elegans* using RNA interference and forward mutagenesis induced by ethyl methane sulfonate (EMS) have been used to identify genes that either enhance or suppress the aggregation of aggregation-prone proteins [2–6]. Methods for performing such genetic screens in *C. elegans* have been described elsewhere [7–11]. To further establish the function of modifying factors in aggregation, a native agarose gel

Danny M. Hatters and Anthony J. Hannan (eds.), *Tandem Repeats in Genes, Proteins, and Disease: Methods and Protocols*, Methods in Molecular Biology, vol. 1017, DOI 10.1007/978-1-62703-438-8_14, © Springer Science+Business Media New York 2013

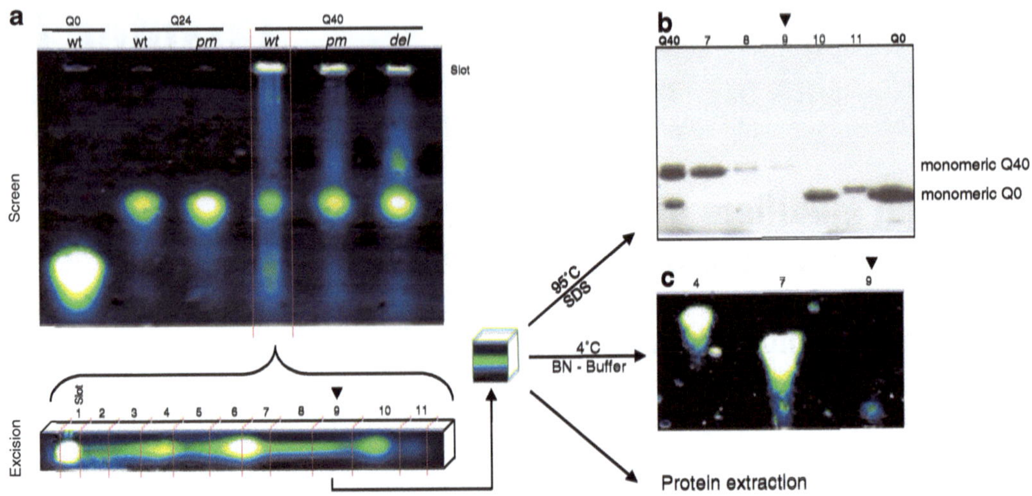

Fig. 1 (**a**) Native agarose gel electrophoresis (NAGE) of polyQ-YFP constructs (0, 24, and 40 glutamines) in different genetic backgrounds showing the detection of protein folding and aggregation intermediates only in lysates of worms expressing aggregation-prone polyQ proteins with 40 glutamine residues. Inactivation of a gene, *moag-4*, either by point mutation (pm) or deletion (del), reduces the number of aggregates in living worms and changes the relative distribution of aggregation intermediates such as excisions 8 and 9. (**b**) Excised gel fragments marked by *dashed red lines* were further analyzed by second-dimension SDS-PAGE showing that the species is a full-length polyglutamine-YFP. (**c**) Regions 4, 7, and 9 were also analyzed using blue-native (BN) PAGE to study the role of the charge effects of the migration of the intermediates. These experiments suggest that wild-type MOAG-4 is involved in the formation of a compact aggregation intermediate

electrophoresis (NAGE) method has been developed to visualize the distribution of aggregation intermediates in multiple lysates simultaneously. This method complements other biochemical methods for studying aggregation, such as size-exclusion chromatography, which provide detailed information about the size and distribution of the intermediates but which require specialized equipment and analyze samples one by one.

This NAGE protocol can be used to compare the relative distribution of different aggregation intermediates in multiple samples. The intermediates are visualized using a fluorescence laser scanner. This enables fast and quantitative analysis of non-denatured aggregation intermediates and requires neither transfer to a membrane nor antibody detection (Fig. 1a). High-molecular-weight species such as aggregates are retained in the gel slots and can be quantified. For visualization of SDS-insoluble material only, NAGE analysis can be complemented by other aggregation assays, such as a filter retardation assay [12]. To further characterize intermediates of interest, selected bands can be excised from the gel and analyzed using other methods, such as denaturing SDS-polyacrylamide gel electrophoresis (PAGE), blue-native PAGE, or mass spectrometry (Fig. 1b, c).

It should be noted that although we developed this protocol for samples from *C. elegans*, the protocol has also been used for mammalian cells. In addition, all reagents and equipment needed for our protocol are usually available in a standard microbiology or biochemistry laboratory, with the possible exception of the fluorescent laser scanner.

Current methods for analyzing protein aggregates in agarose gels usually involve protein denaturation and western analyses [13–17]. However, the use of fluorescence instead of blotting allows for not only quantification of the relative amounts of each aggregation intermediate but also comparison of their distribution in different *C. elegans* strains. The methods described here can also be used to study other processes, such as protein complex formation, and to screen for other types of modifiers, such as biologically active compounds.

2 Materials

2.1 Sample Preparation and Lysis (Optimized for *C. elegans*)

1. Nematode growth media (NGM) plates: 3 g NaCl, 7.5 g casein digest, 17.5 g agar. Add distilled H_2O to a volume of 950 ml. Autoclave and cool to 55 °C. Add 1 ml 5 g/ml cholesterol in ethanol, 1 ml of 1 M $MgSO_4$, 1 ml 1 M $CaCl_2$, and 25 ml 1 M sodium phosphate buffer, pH 6.0. Aseptically pour 25–30 ml into 9 cm petri dishes and let cool. Apply 300 μl overnight culture of OP50 bacteria per plate (1 ml 5× concentrated culture to fluorodeoxyuridine (FUDR) plates, see below) and spread to cover plate using a sterile glass or metal rod.
2. FUDR: 200 mg/ml.
3. M9 buffer: 3 g KH_2PO_4, 6 g Na_2HPO_4 (anhydrous), 5 g NaCl, 1 ml 1 M $MgSO_4$, add H_2O to 1 L. Sterilize by autoclaving.
4. Hypochlorite solution: 4 M NaOH, sodium hypochlorite (13 % stock) (ratio 2:3). This should be prepared fresh before use.
5. Sample preparation system: Fastprep-24 (MP Biomedicals).
6. Mechanically durable microcentrifuge tubes.
7. Silica beads: 1 mm.
8. PBS: 4 g NaCl, 1 g KCl, 7.2 g Na_2HPO_4 $2H_2O$, and 0.27 g KH_2PO_4 per liter, pH 7.4.
9. 1× Complete protease inhibitors (Roche).
10. Liquid nitrogen for snap-freezing.

2.2 Native Agarose and Secondary Gel Electrophoresis

1. Melt 1 % w/v agarose in Tris–glycine buffer, pH 8.3. Pour into a horizontal gel tray appropriate for a gel of 15 cm in length. Insert a well comb with a slot volume of 50–70 μL. The thickness of the gel should be less than 8 mm to fit in the Bio-Rad molecular imager (*see* **Note 1**).

2. Native loading buffer: 62.5 mM Tris–HCl, 10 % w/v glycerol, bromophenol blue, pH 8.7.
3. SDS gel-loading buffer (5×): 250 mM Tris–Cl pH 6.8, 8 % SDS, 0.1 % bromophenol blue, 40 % v/v glycerol, 100 mM dithiothreitol, H_2O.
4. Anode buffer: 50 mM Tris–HCl, pH 7.0.
5. Blue native loading buffer: 5 % (w/v) Coomassie, 100 mM Tris–HCl, pH 7.0.
6. Laser Scanner: Molecular Imager FX, Bio-Rad.
7. Image analysis software: ImageJ 1.42q [18].
8. Standard protein gel electrophoresis setup.

3 Methods

3.1 Sample Preparation and Lysis (Optimized for *C. elegans*)

1. Prepare NGM agar plates. For analysis of worms beyond young adult stage, prepare additional plates with 12.5 mg/l FUDR.
2. Age-synchronize worms using hypochlorite treatment and grow on NGM agar plates using appropriate bacterial strains. Culturing and synchronization are described in detail elsewhere (*see* ref. 19). Add approximately 2,500 worms to each 9 cm NGM plate. While one plate is enough for screening, we recommend using three or more plates depending on expression. Ensure that the animals have not been starved for at least two generations.
3. (Optional) For analysis of aggregation in aged adult worms, rinse plates containing young adults in M9 buffer and transfer to 15 ml tubes. Let worms settle and wash at least once with M9 to remove any young worms already hatched.
4. Harvest plates by rinsing with PBS. Wash worms at least once with additional PBS to remove residual bacteria, which will influence measurement of protein concentration. A yellow color in samples generally indicates significant bacterial contamination.
5. Reduce sample volume as much as possible using a pipette, add protease inhibitor, and snap freeze using liquid nitrogen. Store at −80 °C. Samples processed directly do not need to be frozen, but doing so will improve lysis.

3.2 Lysis

1. Transfer samples to mechanically durable microcentrifuge tubes.
2. Add an equal volume of 1 mm silica beads. Keep on ice.
3. To lyse samples, shake the tubes in a bead beater for 45 s at five movements per second (*see* **Note 2**). Note that numbers can vary depending on the model of bead beater. Return the tubes to ice.

4. Repeat shaking for 20 s at least once. If possible, visualize lysis by light microscopy by pipetting a drop of 0.5 μl onto a microscope slide. Repeat shaking if any unlysed worms or multiple intact heads can be seen in the sample. Aggregate integrity can be checked at this step by fluorescence microscopy.
5. Transfer samples away from beads with a pipette and measure protein concentration.

3.3 Native Agarose Gel Electrophoresis

1. Place a gel electrophoresis tank in a cold room or a refrigerator (4 °C) and fill it with 1× Tris–glycine buffer to fully cover the gel.
2. Equilibrate the gel in the gel electrophoresis tank for 1 h at 50 V. Check the temperature of the buffer, which should be 4–8 °C (*see* **Note 3**).
3. Add native loading buffer 1:5 to protein lysate samples. Keep the samples on ice until loading. We used 20 μg (screening) to 200 μg (excision and analysis) of starting material.
4. Pipette the liquid up and down a few times to resuspend any insoluble material from the bottom of the tube before loading the samples onto the gel. Avoid air bubbles.
5. Start the gel running at constant 50 V overnight for 14 h. The running time can be extended by lowering the voltage.
6. Scan the gel at a resolution of at least 100 μm using the gel scanner. If the samples are to be excised from the gel, keep the gel on ice for as much of the procedure as possible and include the borders of the gel in the gel scan for orientation (*see* **Note 4**).
7. Analyze the samples using ImageJ or a similar software. The following ImageJ macro can be used for convenient quantifications: http://rsbweb.nih.gov/ij/macros/BackgroundCorrectedDensity.txt.

3.4 Gel Excision

1. Print the gel image 1:1. Align the gel borders with the image and keep the gel on ice while cutting.
2. Cut out the desired bands using a clean scalpel and remove any excess agarose. Transfer the bands to a microcentrifuge vial or similar. Clean the scalpel with ethanol between each sample (*see* **Note 5**).

3.5 Secondary Analysis of Intermediates

3.5.1 SDS-PAGE

1. Pour a standard SDS-PAGE gel of desired concentration with a stacking gel of a polyacrylamide concentration of at least 5 % (w/v).
2. Add 1:5 v/w SDS gel-loading buffer (5×) to the samples.
3. Heat at 95 °C for 10 min.
4. Keep the samples warm and pipette the sample mixture into the gel wells.
5. Run and blot the gel as a regular SDS-PAGE.

3.5.2 Blue-Native PAGE

1. Pour a standard polyacrylamide gel without SDS. Set up a refrigerated gel electrophoresis tank in a cold room or a refrigerator. Include gel and anode buffer.
2. While keeping the samples on ice, add approximately 1:5 v/w blue-native loading buffer to the sample and mechanically disrupt the agarose with a spatula.
3. Leave samples on ice for 10 min.
4. Transfer the samples to the dry wells with a spatula and use a pipette to transfer any remaining liquid. From this point on, run as any blue-native PAGE (see for instance Wittig et al., ref. 20). The gel can be scanned for fluorescent signals and also be probed using immunoblotting (*see* **Note 5**).

4 Notes

1. Diagonally running samples or horizontal stripes throughout the gel image indicate that the reader is in contact with the gel. Try removing the slightly elevated edges of the gel or rerun with a thinner gel.
2. Lysis should be done by mechanical (bead) lysis. Sonication is possible but in our experience this disrupts aggregates, resulting in a smeared out mixture of different protein species.
3. This ensures uniformity, with the gel remaining at 4 °C throughout the experiment.
4. For mass-spectrometry analysis, bands can be excised as described under Subheading 3.3 and extracted as reported [21].
5. Possible pause point in protocol: The vials can now be stored at −80 °C for later use.
6. The Coomassie dye in blue-native gels can interfere with the excitation of certain fluorophores. For example, GFP and YFP are both excited by blue light. In our experience, however, although most bands detected in the agarose gel can still be detected with this method, higher sensitivity can be reached using blotting and antibody detection.

Acknowledgements

We thank Sally Hill for editing the manuscript, and Renée Sienstra and Olga Sin for critical reading of the manuscript. This project was funded by grants from the Prinses Beatrix Fonds Foundation, an NWO Meervoud grant (E.A.A.N), and a UMCG Rosalind Franklin Fellowship (E.A.A.N.).

References

1. Chiti F, Dobson CM (2006) Protein misfolding, functional amyloid, and human disease. Annu Rev Biochem 75:333–366
2. Nollen EA, Garcia SM, van Haaften G, Kim S, Chavez A, Morimoto RI, Plasterk RH (2004) Genome-wide RNA interference screen identifies previously undescribed regulators of polyglutamine aggregation. Proc Natl Acad Sci USA 101:6403–6408
3. van Ham TJ, Thijssen KL, Breitling R, Hofstra RM, Plasterk RH, Nollen EA (2008) C. elegans model identifies genetic modifiers of alpha-synuclein inclusion formation during aging. PLoS Genet 4:e1000027
4. Wang J, Farr GW, Hall DH, Li F, Furtak K, Dreier L, Horwich AL (2009) An ALS-linked mutant sod1 produces a locomotor defect associated with aggregation and synaptic dysfunction when expressed in neurons of Caenorhabditis elegans. PLoS Genet 5:e1000350
5. Silva MC, Fox S, Beam M, Thakkar H, Amaral MD, Morimoto RI (2011) A genetic screening strategy identifies novel regulators of the proteostasis network. PLoS Genet 7:e1002438
6. van Ham TJ, Holmberg MA, van der Goot AT, Teuling E, Garcia-Arencibia M, Kim HE, Du D, Thijssen KL, Wiersma M, Burggraaff R, van Bergeijk P, van Rheenen J, Jerre van Veluw G, Hofstra RM, Rubinsztein DC, Nollen EA (2010) Identification of MOAG-4/SERF as a regulator of age-related proteotoxicity. Cell 142:601–612
7. Kamath RS, Ahringer J (2003) Genome-wide RNAi screening in Caenorhabditis elegans. Methods 30:313–321
8. Pothof J, van Haaften G, Thijssen K, Kamath RS, Fraser AG, Ahringer J, Plasterk RHA, Tijsterman M (2003) Identification of genes that protect the C. elegans genome against mutations by genome-wide RNAi. Genes Dev 17:443–448
9. van Haaften G, Vastenhouw NL, Nollen EA, Plasterk RH, Tijsterman M (2004) Gene interactions in the DNA damage-response pathway identified by genome-wide RNA-interference analysis of synthetic lethality. Proc Natl Acad Sci USA 101:12992–12996
10. Jorgensen EM, Mango SE (2002) The art and design of genetic screens: Caenorhabditis elegans. Nat Rev Genet 3:356–369
11. Gidalevitz T, Krupinski T, Garcia S, Morimoto RI (2009) Destabilizing protein polymorphisms in the genetic background direct phenotypic expression of mutant SOD1 toxicity. PLoS Genet 5:e1000399
12. Wanker EE, Scherzinger E, Heiser V, Sittler A, Eickhoff H, Lehrach H (1999) Membrane filter assay for detection of amyloid-like polyglutamine-containing protein aggregates. Methods Enzymol 309:375–386
13. Bagriantsev SN, Kushnirov VV, Liebman SW (2006) Analysis of amyloid aggregates using agarose gel electrophoresis. Methods Enzymol 412:33–48
14. Halfmann R, Lindquist S (2008) Screening for amyloid aggregation by Semi-Denaturing Detergent-Agarose Gel Electrophoresis. J Vis Exp (17). pii: 838. doi: 10.3791/838
15. Weiss A, Klein C, Woodman B, Sathasivam K, Bibel M, Regulier E, Bates GP, Paganetti P (2008) Sensitive biochemical aggregate detection reveals aggregation onset before symptom development in cellular and murine models of Huntington's disease. J Neurochem 104:846–858
16. Yerkes S, Vesenka J, Kmiec EB (2010) A stable G quartet binds to a huntingtin protein fragment containing expanded polyglutamine tracks. J Neurosci Res 88:335–345
17. Kryndushkin DS, Alexandrov IM, Ter-Avanesyan MD, Kushnirov VV (2003) Yeast [PSI+] prion aggregates are formed by small Sup35 polymers fragmented by Hsp104. J Biol Chem 278:49636–49643
18. Schneider CA, Rasband WS, Eliceiri KW (2012) NIH Image to ImageJ: 25 years of image analysis. Nat Methods 9:671–675
19. Brenner S (1974) The genetics of Caenorhabditis elegans. Genetics 77:71–94
20. Wittig I, Braun HP, Schagger H (2006) Blue native PAGE. Nat Protoc 1:418–428
21. Kim R, Yokota H, Kim SH (2000) Electrophoresis of proteins and protein-protein complexes in a native agarose gel. Anal Biochem 282:147–149

Chapter 15

Kinetic Analysis of Aggregation Data

Regina M. Murphy

Abstract

Aggregation of repeat-containing proteins is associated with neurodegenerative disorders; a specific example is the established link between expansion of the polyglutamine domain in huntingtin and the appearance of nuclear inclusions in Huntington's disease. This connection between aggregation and pathology has motivated numerous investigations into the kinetics of aggregation. Quantitative analysis of kinetic data is needed both for comparative purposes (e.g., to compare the effect of different compounds on aggregation kinetics) and for mechanistic insight. Here we describe some analytical equations that can be used to model aggregation data and demonstrate appropriate and simple methods for extracting valid model parameters by fitting equations to kinetic data.

Key words Amyloid, Polyglutamine, Parameter estimation, Thioflavin T, Aggregation kinetics, Monomer loss kinetics

1 Introduction

Aggregation of peptides and proteins containing homo-amino acid repeats is a well-known phenomenon, and there have been numerous reports in which the kinetics of aggregation have been measured (for a few examples, see ref. [1–5]). Quantitative analysis of the data is sometimes attempted, not only specifically in the case of repeat-containing proteins but also for other aggregation-prone proteins (for a few examples, see ref. [6–9]). In some cases these kinetic data have been used to fit empirical or semiempirical equations; this analysis can be particularly useful when comparisons are of interest, for example, in explorations of the effect of repeat length on aggregation or the potency of various compounds at inhibiting aggregation. In other cases the goal has been to extract mechanistic information, a far more challenging objective. Complications arise for several reasons, such as the following: There are many possible species in solution (monomers of both native and nonnative conformation, reversibly formed oligomers, amorphous aggregates, fibrillar aggregates, etc.); measurements

Danny M. Hatters and Anthony J. Hannan (eds.), *Tandem Repeats in Genes, Proteins, and Disease: Methods and Protocols*, Methods in Molecular Biology, vol. 1017, DOI 10.1007/978-1-62703-438-8_15, © Springer Science+Business Media New York 2013

obtained by available experimental techniques do not necessarily directly report on the concentration of desired species, or with sufficient sensitivity; and multiple mechanisms can equally well capture the observed data within experimental error. Furthermore, the language can obfuscate—when a researcher reports that a specific treatment reduced aggregation, what does that *exactly* mean? By fitting aggregation kinetic data to equations and reporting parameter values, we can be more exact and descriptive.

Before analysis of data can begin, one must collect experimental data; knowledge of what is required for reliable kinetic modeling should influence and direct the experimental design. First, it is critical to know what the experimental technique is, and is not, measuring and to be attuned to the sensitivity of each technique and the suitability of that technique for detecting small concentrations of key intermediates. Second, it is important to pay attention to whether the quantity measured is in mass or molar units (and if in moles, moles of aggregate or monomer equivalents) and to establish that the signal is linear with quantity. Third, experimental data over the entire time course of aggregation is needed.

We list some of the more common techniques and briefly summarize the information that can be obtained from each:

1. *Circular dichroism or FTIR*: These methods report on the secondary structure, averaged over all species and conformations, and are used to detect changes in conformation that may accompany aggregation. However, one must keep in mind that it cannot be presumed a priori that conformational change equals aggregation. Association into oligomers can precede, or occur in the absence of, conformational changes; and conversely, monomers may undergo conformational change without concomitant association.
2. *Size-exclusion chromatography* (SEC): The monomer concentration (both native and nonnative conformations), as well as the concentration and size of smaller oligomers, can be determined. However, dissociation of reversible aggregates is likely to occur due to dilution and shear. Sizing of aggregates can also be challenging if the aggregates are non-spherical and/or interact with the column.
3. *Native gel electrophoresis*: Similar to SEC, what appears as monomer may be a combination of monomers (potentially of a variety of conformations) as well as any reversibly associated oligomers.
4. *Dye binding*: Thioflavin T (ThT) binding is widely used to detect fibrillar aggregates; fluorescence intensity is generally assumed to be linear to the mass of β-sheet aggregates. Amorphous aggregates are not detected, and a minimum size of aggregates to obtain ThT-positive signal is not well established.

5. *Micro- and ultrafiltration*: Separation is based on size, but is not a sharp cutoff, and reversibly associated oligomers are likely to dissociate. This technique is most useful for measuring mass concentration of large, irreversible aggregates.
6. *Centrifugation*: Aggregates that have sufficiently higher densities and sufficiently large size will precipitate upon centrifugation. Since conformational change from a disordered to a highly β-sheet conformation can be accompanied by dehydration and density increase, species that are measured by centrifugation are most likely, but not necessarily, β-sheet aggregates.
7. *Turbidity*: Measurements of turbidity using a nephelometer, spectrophotometer (through apparent decrease in absorbance), or spectrofluorimeter (through forward scattering) are simple and convenient. Turbidity is proportional to the mass concentration of aggregates above ~1 μm; amorphous and fibrillar aggregates both contribute to the signal.
8. *Laser light scattering*: A laser light scattering apparatus with multi-angle intensity detection yields an average molecular weight, averaged over all particles in solution, and the angular dependence can under favorable conditions be used to interpret aggregate morphology. Dynamic light scattering yields information on the average hydrodynamic diameter of the aggregates (through a measurement of diffusion coefficient) and is an excellent method for determining if there are small amounts of aggregates in a sample. These techniques measure aggregate size, not concentration. Size data can be useful in fitting kinetic models, as complementary to the measurement of monomer concentration.

The most comprehensive approach requires that all possible species in an aggregating system are included: monomers, dimers, reversibly formed oligomers, and irreversible aggregates. Further, the number of monomers in each species, and the conformation of each species, should be tracked. Comprehensive kinetic models are available in the literature, for example, in refs. [10, 11]. At their most general level, these models include folding/unfolding transitions in the monomer, reversible association into oligomers, conformational transitions within aggregates, monomer addition to preexisting aggregates, and aggregate-aggregate condensation. This approach is only appropriate if the data are sufficiently complete and multifaceted to support this level of detail. Such models can be greatly simplified if one considers only monomers and fibrils and lumps all fibrils into a single category. Even with this simplification, there are several different reaction schemes imaginable, resulting in a system of coupled differential and algebraic equations that are generally nonlinear [6, 8, 12]. These must be solved numerically, and parameter estimation requires the use of more advanced software.

In this chapter, we limit ourselves to the analysis of monomer loss data and focus specifically on fitting data where there are explicit analytical expressions for the concentration of monomer as a function of time. We show how this analysis can be readily carried out in Microsoft Excel.

2 Method

2.1 Select Model Equation

1. *Exponential growth*. The simplest model for aggregation postulates exponential growth of aggregates, or equivalently that monomer loss is

$$\frac{[M]}{[M]_0} = 1 - \delta \times \exp(k_1 t) \tag{1}$$

where $[M]$ = the monomer concentration at any time t, $[M]_0$ = initial monomer concentration, k_1 is a rate constant, and δ is a fitted parameter (*see* **Note 1**). Equation 1 and its variations have been used on occasion to model protein aggregation kinetics, including huntingtin kinetics [13, 14] (*see* **Notes 2** and **3**).

2. *Logistic equation*. The *logistic* equation is used frequently to describe sigmoidal growth and decay curves such as those that appear in population growth studies where growth is initially proportional to population but becomes limited by available resources. For protein aggregation, the rate expression is

$$\frac{d[M]}{dt} = -k[M]([M]_0 - [M])$$

An integrated solution that has been used by many researchers to describe protein aggregation is [15, 16]

$$\frac{[M]}{[M]_0} = \left[1 + \exp\left(k_{\text{app}}\left(t - t_{\text{m}}\right)\right)\right]^{-1} \tag{2}$$

where k_{app} is an apparent rate constant and t_{m} is the time to 50 % maximum signal. (Note that this equation does not go identically to 1 at $t = 0$.) This equation has been used in slightly modified form [17]:

$$\frac{[M]}{[M]_0} = \left[1 + \nu \times \exp\left(k_{\text{app}}\left(t - t_{\text{m}}\right)\right)\right]^{\frac{-1}{\nu}} \tag{3}$$

where ν is an empirical parameter that accounts for asymmetry of the sigmoidal curve.

Equation 2 can be used mechanistically if we postulate (a) that only monomer and aggregate exist, (b) that the rate of monomer loss is proportional to the monomer concentration times the monomer-equivalent concentration of all aggregated species (*see* **Note 4**), (c) that at some time t_{lag} there is a small concentration of aggregates $\delta[M]_0$ that appears by an unknown mechanism, and (d) that aggregation is irreversible. Then we can write:

$$\frac{\mathrm{d}[M]}{\mathrm{d}t} = -k_1[M]\sum_{i=2}^{\infty} i[A]_i = -k_1[M]\left([M]_0 - [M]\right)$$

Then $[M] = [M]_0$ at $t < t_{\text{lag}}$, $[M] = (1-\delta)[M]_0$ at $t = t_{\text{lag}}$ with $\delta \ll 1$, and for $t > t_{\text{lag}}$,

$$\frac{[M]}{[M]_0} = \left[1 + \delta \times \exp\left(k_1[M]_0\left(t - t_{\text{lag}}\right)\right)\right]^{1} \tag{4}$$

There are three parameters, t_{lag}, k_1, and δ, but only two can be determined independently. We can see this by rearranging the expression and noticing that the term in the wavy brackets is a constant:

$$\frac{[M]}{[M]_0} = \left[1 + \left\{\delta \times \exp\left(-k_1[M]_0 t_{\text{lag}}\right)\right\}\exp\left(k_1[M]_0 t\right)\right]^{-1} \tag{5}$$

Equation 5 is formally equivalent to Eq. 2 with $k_{\text{app}} = k_1[M]_0$ and $t_{\text{m}} = t_{\text{lag}} - \ln\delta / k_1[M]_0$.

3. *Autocatalytic model.* The autocatalytic model postulates a two-step mechanism: (1) conversion of monomer M to an aggregation-competent monomer A_1, with rate constant k_0, and (2) further loss of monomer by reaction of M with A_1 to produce more A_1, at a rate constant k_1. The analytical solution is [12]

$$\frac{[M]}{[M]_0} = \left[\frac{k_1[M]_0}{k_0 + k_1[M]_0} + \frac{k_0}{k_0 + k_1[M]_0}\exp\left(\left(k_0 + k_1[M]_0\right)t\right)\right]^{-1} \tag{6}$$

which is formally almost equivalent to Eq. 5 if $k_0 << k_1[M]_0$. Equation 6 was used to fit an extensive set of aggregation kinetic data [18].

4. *Gompertz model.* Another well-known function for describing sigmoidal growth and decay curves is the Gompertz equation. When applied to monomer loss kinetics, this model postulates that the rate of monomer loss is proportional to the monomer concentration times the log of the monomer concentration [19, 20]. Monomer loss can be expressed as

$$\frac{[M]}{[M]_0} = 1 - \delta \times \exp\left[-\exp\left(-k_{\text{app}}\left(t - t_{\text{m}}\right)\right)\right] \tag{7}$$

where δ, k_{app}, and t_{m} are fitted parameters.

5. *Monomer partitioning*. Another example of a kinetic model where an analytic solution is obtained is "monomer partitioning," where it is assumed that the monomer population initially contains two different conformers, M_A and M_B, where M_A initiates aggregation but does not participate in growth and M_B adds to extant aggregates but cannot start new ones [12] (*see* **Note 5**). This is a 3-parameter model, with f = the fraction of total monomer that initiates aggregates and k_0 and k_1 the rate constants of initiation and growth, respectively.

$$\frac{[M]}{[M]_0} = f\exp(-k_0 t) + (1-f)\exp\left[\frac{k_1 f[M]_0}{k_0}(1 - k_0 t - \exp(-k_0 t))\right] \quad (8)$$

2.2 Estimate Parameters

Once the experimental data have been collected and a model equation has been chosen, the data are fitted to the model, and model parameters are estimated. Suppose we have N data points (y_1, y_2,… y_n), taken at time t_1, t_2,…t_n. (In our case, these are the monomer concentration measurements, or equivalently the aggregation measurements, taken at initial monomer concentration $[M]_0$.) (*see* **Note 6**) Suppose our model has K parameters (a_1, a_2,…a_k). We guess values for these parameters, and we use the chosen model equation to calculate the value f_i at each time t_i as a function of the parameter values ($f_1 = f_1(t_1, a_1, a_2,\ldots a_k)$, $f_2 = f_2(t_2, a_1, a_2,\ldots a_k)$,… $f_n = f_n(t_n, a_1, a_2,\ldots a_k)$). The difference between experimental data and model calculation, $y_i - f_i$, is called the residual. We calculate the residual at each time point, square it, and then sum over all time points to get the sum of squares SS as a function of the chosen values of the parameters a:

$$SS(a) = \sum_{i=1}^{N} (y_i - f_i)^2 \quad (9)$$

The goal is to find the set of K parameter values that minimizes the sum of squares SS of the difference between the experimental data points and the value calculated from the model at the same time point. There are a number of different algorithms for efficiently finding the minimum parameter set, and there are software packages that carry out these algorithms. We will demonstrate how parameter estimation with analytical model equations can be done easily in Microsoft Excel. Refer to Fig. 1 for screenshots illustrating some of these steps.

1. Set up cells to contain parameters, one for each parameter, and name the parameters, by highlighting the cell that will contain the parameter value and using the INSERT/NAME command (*see* **Note 7**). Insert initial guessed values of the parameters in these cells.

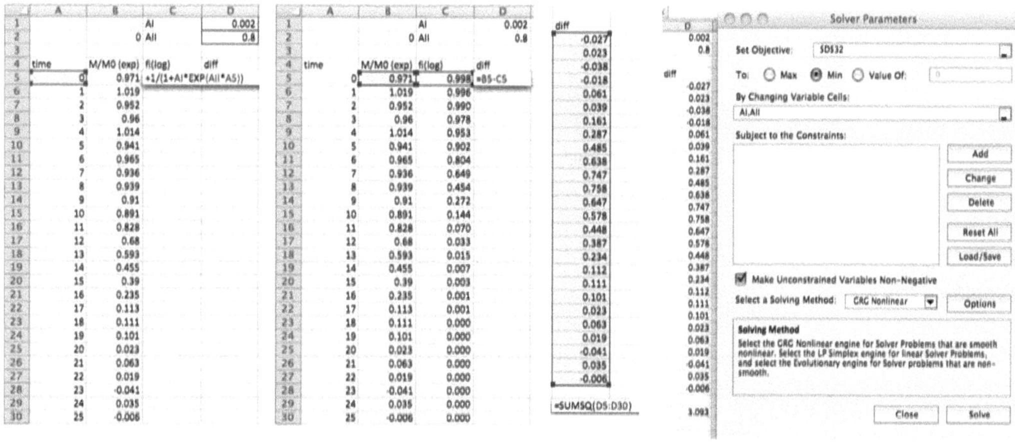

Fig. 1 Screenshots from Excel, illustrating method for fitting parameters

2. Below the cells containing parameter values, input the experimental time points in column A and the experimental data points ($[M]/[M]_0$, or more generally y_i) in column B. In the example in Fig. 1, the first data point goes in row 5 (cell B5).
3. In column C, in the row corresponding to the first data point, type "=" and then the right-hand side of the model equation, using the chosen parameter names and referencing the time points in column A (Fig. 1a). Then use FILL/DOWN on the EDIT menu to apply the equation to each time point. This calculates $[M]/[M]_0$ (or more generally f_i) based on your initial parameter guesses and the chosen model equation.
4. In column D, calculate the residual ($y_i - f_i$) by selecting the cell in the row corresponding to the first data point and then subtracting the entry in column C from that in column B in the same row (Fig. 1b). Then use FILL/DOWN to calculate the residual at each time point. Finally, in the same column just below all the entries, calculate the sum of the squares of the residuals by using the command "=SUMSQ(Dxx:Dyy)," where Dxx and Dyy are the first and last cells, respectively, in the list of residuals (Fig. 1c).
5. Go to the TOOLS menu and select SOLVER. (If it is grayed out, you will need to install it.) Go to the *Set Objective* entry box, then return to your spreadsheet, and click on the cell that calculates SUMSQ. Select *Min* radio button. In the entry box below *By Changing Variable Cells*, type in the chosen names of your parameters, separated by a comma. You may wish to check the box to *Make Unconstrained Variables Non-Negative*. (On some versions of Excel, this will be under Options.) Then *Select a Solving Method* (GRG Nonlinear is the default choice and should work). If you wish, under Options you can adjust the number of iterations, or see intermediate results. Finally, click *Solve* (Fig. 1d).

6. If everything works well, in a few seconds you will see a window that states that *Solver has converged*. Click OK. The converged parameter values that minimize SS will be shown in their cells, and you will also see the minimized SUMSQ, the minimized residuals, and the calculated monomer concentrations for the "best fit" (*see* **Notes 8** and **9**).

2.3 Estimate Parameter Errors

Besides finding the model parameters, it is essential to determine the *error* in the parameter estimates. This is important so that one can place confidence in the values obtained and also to compare parameter values from different sets of data. It can also be very useful to calculate the covariance: This tells you whether the model parameters are independent or correlated. (If the covariance is high, this might mean, for example, that you do not know the parameters well but only their ratio or product.)

To find the errors in the parameter estimates, we first need to calculate the partial derivatives of the model equations with respect to each parameter ($\partial f / \partial a_1, \partial f / \partial a_2, \ldots$) and then evaluate the value of that derivative at each time point, for the optimized values of parameters found in Subheading 2.2. Each of these calculations becomes an element in a Jacobian matrix $J(a^{\min})$, from which we calculate a Hessian matrix $H(a^{\min}) = J^T J$ and invert to find $H^{-1}(a^{\min})$. ($a^{\min}$ is the set of parameter values for $a_1, a_2, \ldots$, found when SS is minimized, in other words, the converged parameter values from Subheading 2.2.) The diagonal entries are used to calculate error estimates of the parameters, and the off-diagonal entries yield the covariances.

Implementing this routine in Excel is straightforward for the analytical model equations under consideration. We will illustrate by estimating errors for Eq. 5:

$$\frac{[M]}{[M]_0} = \left[1 + \left\{\delta \times \exp\left(-k_1 [M]_0 t_{\text{lag}}\right)\right\} \exp\left(k_1 [M]_0 t\right)\right]^{-1}$$

$$= f = \frac{1}{1 + a_1 \exp(a_2 t)}$$

where the two parameters are $a_1 = \delta \times \exp\left(-k_1 [M]_0 t_{\text{lag}}\right)$ and $a_2 = k_1 [M]_0$.

For this model, the partial derivatives of f with respect to each parameter are (*see* **Note 10**)

$$\frac{\partial f_i}{\partial a_1} = \frac{-\exp(a_2 t_i)}{\left(1 + a_1 \exp(a_2 t_i)\right)^2}$$

$$\frac{\partial f_i}{\partial a_2} = \frac{-a_1 t_i \exp(a_2 t_i)}{\left(1 + a_1 \exp(a_2 t_i)\right)^2}$$

Use the same Excel spreadsheet with the minimized sum of squares, and carry out the following steps. Table 1 contains results from completing these calculations for the example illustrated in Fig. 1.

1. In column E row 5 (corresponding to the first data point), type "=" and then the right-hand side of the equation for $\partial f / \partial a_1$, using your chosen parameter names in place of a_1 and a_2 and selecting cell A5 (the time at the earliest data point) in place of t_i. In column F, do the same except use the equation for $\partial f / \partial a_2$. Then FILL/DOWN in both columns to calculate the partial derivatives at each time point. Be sure that the numbers in the cells for the parameter values correspond to the values found when SS is minimized.
2. Below all entries in column E, type "=SUMSQ(Exx:Eyy)" to calculate the sum of squares of all the partial derivatives for a_1, summed over all data points. Below all entries in column F, type "=SUMSQ(Fxx:Fyy)" to calculate the sum of squares of all the partial derivatives for a_2, summed over all data points (*see* **Note 11**).
3. In column G, multiply column E times column F. Below all entries in column G, type "=SUM(Gxx:Gyy)" (*see* **Note 12**).
4. In a separate part of the spreadsheet, set up a two-by-two matrix by inserting the following values into 4 cells, in the pattern shown:

$$\begin{matrix} \text{SUMSQ}(a_1) & \text{SUM}(a_1 a_2) \\ \text{SUM}(a_1 a_2) & \text{SUMSQ}(a_2) \end{matrix}$$

5. This matrix is the Hessian! Now we need to invert this matrix. First, highlight a different two-by-two matrix of cells where you want the inverted matrix entries to go. Then, while highlighting these empty cells, type "=MINVERSE(xx:yy)," where xx is the cell containing SUMSQ(a_1) and yy is the cell containing SUMSQ(a_2). Then, press CTRL and Shift simultaneously, and hit Enter (or Return). You should see 4 values appear in the highlighted cells. (If you do not press Ctrl and Shift simultaneously, you will see only one value.) We will call these values:

$$\begin{matrix} \text{H}^{-1}(a_1\, a_1) & \text{H}^{-1}(a_2\, a_1) \\ \text{H}^{-1}(a_1\, a_2) & \text{H}^{-1}(a_2\, a_2) \end{matrix}$$

6. Calculate the estimates of the errors of the parameter values as

$$\sigma_{a1} = \sqrt{\frac{\text{SS}}{N-2}} \sqrt{H^{-1}(a_1 a_1)}$$

$$\sigma_{a2} = \sqrt{\frac{\text{SS}}{N-2}} \sqrt{H^{-1}(a_2 a_2)}$$

where N = the number of data points, SS is the minimized sum of squares of the residuals, and 2 is the number of fitted parameters (*see* **Note 13**).

Table 1
Error estimates for the example problem solved in Fig. 1

		Converged parameters: $a_1 = 0.00105$; $a_2 = 0.499$				
Time (h)	$[M]/[M]_0$ (data)	$[M]/[M]_0$ (model)	Residual	$\partial f_i / \partial a_1$	$\partial f_i / \partial a_2$	$\left(\partial^2 f_i / \partial a_1 \partial a_2\right)$
0	0.971	0.999	–0.028	–0.998	0.000	0.000
1	1.019	0.998	0.021	–1.642	–0.002	0.003
2	0.952	0.997	–0.045	–2.699	–0.006	0.015
3	0.960	0995	–0.035	–4.432	–0.014	0.062
4	1.014	0.992	0.022	–7.258	–0.030	0.220
5	0.941	0.987	–0.046	–11.841	–0.062	0.733
6	0.965	0.979	–0.015	–19.197	–0.120	2.312
7	0.936	0.967	–0.031	–31.806	–0.226	6.947
8	0.939	0.946	–0.007	–48.636	–0.407	19.790
9	0.910	0.914	–0.004	–74.835	–0.704	52.709
10	0.891	0.866	0.025	–110.692	–1.158	128.134
11	0.828	0.797	0.031	–154.490	–1.777	274.551
12	0.680	0.705	–0.025	–198.918	–2.496	496.543
13	0.593	0.592	0.001	–231.007	–3.140	725.472
14	0.455	0.468	–0.013	–238.086	–3.486	829.898
15	0.390	0.348	0.042	–217.001	–3.404	738.658

16	0.235	0.245	−0.010	−176.769	−2.958	522.828
17	0.113	0.164	−0.051	−131.342	−2.335	306.680
18	0.111	0.107	0.004	−91.107	−1.715	156.243
19	0.101	0.068	0.033	−60.239	−1.197	72.101
20	0.023	0.042	−0.019	−38.582	−0.807	31.134
21	0.063	0.026	0.037	−24.211	−0.532	12.873
22	0.019	0.016	0.003	−14.999	−0.345	5.176
23	−0.041	0.010	−0.051	−9.219	−0.222	2.044
24	0.035	0.006	0.029	−5.638	−0.142	0.798
25	−0.006	0.004	−0.010	−3.438	−0.090	0.309
			0.022 (SUMSQ)	305,189 (SUMSQ)	64.765 (SUMSQ)	4,386.231 (SUM)
Hessian				Inverted Hessian		
305,189		4,386		0.000108		−0.0073
4,386		65		−0.0073		0.508

$$\sigma_{a1} = \sqrt{\frac{0.022}{26-2}}\sqrt{0.000108} = 0.0003,\ \sigma_{a2} = \sqrt{\frac{0.022}{26-2}}\sqrt{0.508} = 0.02$$

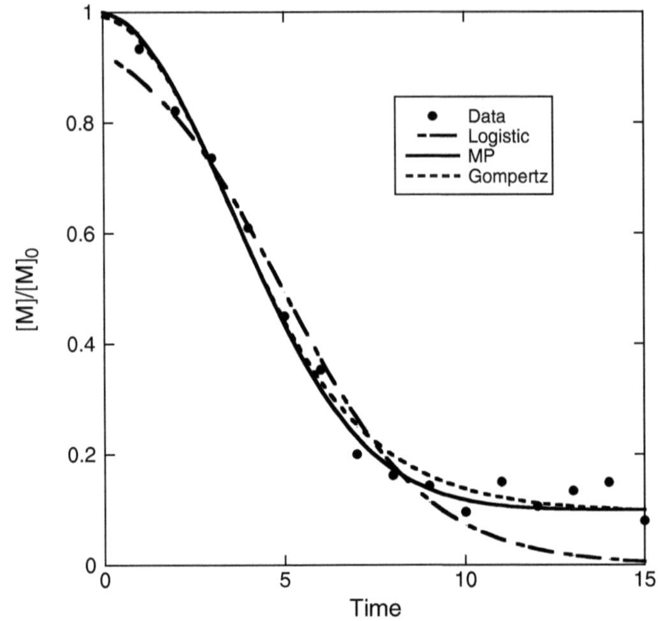

Fig. 2 Experimental data were fitted to three different model equations

7. The off-diagonal entries $H^{-1}(a_1a_2)$ and $H^{-1}(a_2a_1)$ should be the same and should be somewhere between −1 and +1. This is the covariance, and the closer this is to zero, the less correlated the parameters are (*see* **Note 14**).

2.4 Discriminate Among Models

Just because we can find parameter estimates from fitting kinetic data to an equation, we cannot conclude that the mechanism behind the model must be correct! The question must be asked: are there equally good (or even better) models that can also fit the data? A model with more parameters will in general appear to give a better fit to data. We require that any added parameters provide a statistically significant improvement in the fit in order to justify the additional parameters. A straightforward measure of appropriateness of a model is the Akaike Information Criterion (AIC):

$$\mathrm{AIC} = \frac{2K}{N-K} + \ln\left[\frac{\mathrm{SS}}{N-K}\right]$$

where K = the number of fitted parameters, N = the number of experimental data points, and SS is the minimized sum of squares. The "best" model is defined as that with the lowest AIC.

As an example, we created "data" (N = 16) using the monomer partitioning (MP) model and imposing random "errors" and then fit the "data" to the MP, logistic, and Gompertz equations. Reasonable fits were obtained using all three models (*see* Fig. 2).

The minimized sums of squares of the residuals SS were 0.078 for the logistic equation, 0.013 for the monomer partitioning equation, and 0.016 for the Gompertz equation. Even though SS for the logistic equation is worse, it has only two fitted parameters whereas the MP and Gompertz have 3. We calculate that AIC = −4.9, −6.2, and −6.4 for the logistic, Gompertz, and MP models, respectively. So the additional parameter is justified in the improvement in fit, and the MP model is slightly better than the Gompertz (*see* **Note 15**).

3 Notes

1. This equation, like all others in this section, can be readily adapted for experiments such as ThT fluorescence that measure the mass of aggregates rather than the monomer concentration, as long as only monomers and aggregates are present, because

$$\frac{[M]}{[M]_0} = 1 - \frac{[A_{\text{meq}}]}{[M]_0}$$

   where $[A_{\text{meq}}] = \sum_{i=2}^{\infty} i[A_i]$ is the monomer-equivalent concentration in aggregated form and $[Ai]$ is the molar concentration of aggregate containing i monomers.

2. The exponential model is not a good choice for kinetics displaying a clear lag phase nor for evaluating aggregation data over the entire range from start to final stable state. It can capture the initial state of aggregation only.

3. A nucleation-elongation (NE) mechanism is frequently hypothesized for protein aggregation, and Ferrone showed that the NE model reduces to

$$\frac{[M]}{[M]_0} = 1 - k_F [M]_0^{n+1} t^2$$

   at small t [21], where $[M]/[M]_0 \sim 0.9$ or greater. It is sometimes stated in the literature that a t^2 dependence is indicative of an NE mechanism. However note the Taylor series expansion:

$$e^{kt} = 1 + kt + \frac{k^2}{4} t^2 + \ldots$$

   It can be difficult to distinguish exponential versus t^2 dependence at small values of t, within experimental error. Still, the Ferrone equation can be used as an alternative equation for fitting data as long as $[M]/[M]_0 \geq 0.9$.

4. This is a rather disconcerting postulate, because it assumes that the aggregates act as though they are isolated monomers of a different type than M, rather than as true oligomeric species. This postulate states that the rate of monomer loss is proportional to $[M][A_{meq}] = [M]\sum_{i=2}^{\infty} i[A_i]$, which is quite different than the mechanism where fibril elongation occurs by monomer addition to fibrils. In the latter case, the rate of monomer loss is proportional to $[M]\sum_{i=2}^{\infty}[A_i]$. The difference between these two mechanisms is very substantial if i becomes large.
5. The monomer partitioning model, like the NE model but unlike many of the others discussed, makes the more physically realistic assumption that monomer loss by addition to aggregates is proportional to $[M]\sum_{i=2}^{\infty}[A_i]$.
6. Preferably, we have not just one measurement at each time point, but replicates, so we can calculate the standard deviation $\sigma(\sigma_1, \sigma_2 \ldots \sigma_n)$ at each time point. This is particularly important if the experimental technique is such that the error in the measurement at different times (concentrations) is expected to be different. If standard deviations are available, it is best to calculate the weighted sum of squares *w*SS by dividing the residual at each time point, $(yi - fi)$ by the corresponding standard deviation σi, and then squaring and summing:

$$wSS(a) = \sum_{1}^{n}\left[\frac{(y_i - f_i)}{\sigma_i}\right]^2$$

This process assures that the data with the least uncertainty contributes the most to the sum of squares.
7. I recommend simplifying equations by grouping terms if possible and using the group as a parameter. For example, in the example in Fig. 1, we use the bracketed term in Eq. 5 as a single parameter.
8. Convergence to a solution will be obtained more easily if your guessed values are the right order of magnitude. It is also a good idea to repeat the process for different initial guesses, to ensure that the solution converges to the same parameter values. If the solution does not converge, try different initial guesses and check that the calculated values in the spreadsheet change as you manually adjust initial guesses. Check that automatic calculation has not been inadvertently turned off. Try adding the constraint of nonnegative variables (if appropriate) if you have not already done that. Try plotting your data to be sure that it "looks" right. Try a different model equation. If you try

different initial guesses and are able to get convergence to similar values of SS but different parameter values, it means that the solution is not very sensitive to those parameters or that the parameters are highly correlated. This issue is dealt with in the next section.

9. Before we move on, we should check whether the model is reasonable. First, simply plot the experimental data and the calculated model fit on the same graph. Does it "look good"? Second, plot the residuals versus time. The residuals should be randomly scattered about 0; if the residuals are not random, then there is likely some systematic problem with the model equation or a systematic error in the experiment. Third, is the minimized SS reasonably small? There are statistical tests for this, but a simple useful check is to compare $SS/(N-K)$ to the standard deviation of the data. If $SS/(N-K)$ is large compared to the experimental errors, then the model is likely not a good one.
10. For other models, you will need to derive the appropriate partial derivatives or estimate them numerically.
11. This step calculates

$$\mathrm{SUMSQ}(a_1)=\sum_{i=1}^{N}\left(\frac{\partial f_i}{\partial a_1}\right)^2 \text{ and } \mathrm{SUMSQ}(a_2)=\sum_{i=1}^{N}\left(\frac{\partial f_i}{\partial a_2}\right)^2.$$

This procedure is described for two-parameter models but can be easily adapted to three- (or more) parameter models.

12. This step calculates $\mathrm{SUM}(a_1 a_2)=\sum_{i=1}^{N}\left(\frac{\partial f_i}{\partial a_1}\right)\left(\frac{\partial f_i}{\partial a_2}\right)$. Be sure to use SUM and not SUMSQ.
13. There are a number of assumptions about the distribution of errors in the data points and the properties of the model equations that underlie these calculations of the errors in the parameter estimates. Because we are not explicitly checking the validity of these assumptions (or we know that they are not exactly true), then these calculated values can be considered only estimated errors, and not true errors.
14. If the covariance is close to 1 or −1, then the parameters are highly correlated. This means that changing the value of one parameter has a large impact on the value of the other parameter; you may not know the parameter values themselves very well, but rather, e.g., the ratio or the product of the parameters. In this case the confidence interval contour plots are highly elliptical. You may be better off trying to re-parameterize the equation, by, for example, choosing a different way to lump terms.

15. From Fig. 2, it is clear that the Gompertz equation does an excellent job of describing the data, even though these data were generated from the MP model! Even if a model equation is the best of several tested, this does not necessarily mean that the equation is a *mechanistic* description. At a minimum, experimental data at several concentrations should be collected and analyzed. If the same parameter values (within error) are obtained at different concentrations, then confidence in the broader utility of the equation increases. Better yet, orthogonal data such as measurements of fibril size are used along with concentration measurements. It is also notable that the logistic equation fails mostly at long time, when most of the monomer is consumed. If the "data collection" had been truncated at $[M]/[M]_0 > 0.2$, the logistic equation would likely have fared as well as the MP and Gompertz equations in the AIC calculation. This points out how important it is to have data across the entire aggregation profile.

Acknowledgments

This work was supported by grant CBET-085 2278 from the National Science Foundation. The author thanks Dennis Yang for carefully reading a draft of this manuscript.

References

1. Chen SM, Ferrone FA, Wetzel R (2002) Huntington's disease age-of-onset linked to polyglutamine aggregation nucleation. Proc Natl Acad Sci USA 99(18):11884–11889. doi:10.1073/Pnas.182276099
2. Bhattacharyya AM, Thakur AK, Wetzel R (2005) Polyglutamine aggregation nucleation: thermodynamics of a highly unfavorable protein folding reaction. Proc Natl Acad Sci USA 102(43):15400–15405. doi:10.1073/Pnas.0501651102
3. Kar K, Jayaraman M, Sahoo B, Kodali R, Wetzel R (2011) Critical nucleus size for disease-related polyglutamine aggregation is repeat-length dependent. Nat Struct Mol Biol 18(3):328–336. doi:10.1038/Nsmb.1992
4. Ellisdon AM, Pearce MC, Bottomley SP (2007) Mechanisms of ataxin-3 misfolding and fibril formation: kinetic analysis of a disease-associated polyglutamine protein. J Mol Biol 368(2):595–605. doi:10.1016/J.Jmb.2007.02.058
5. Bernacki JP, Murphy RM (2011) Length-dependent aggregation of uninterrupted polyalanine peptides. Biochemistry 50(43): 9200–9211. doi:10.1021/Bi201155g
6. Chen YD, Bjornson K, Redick SD, Erickson HP (2005) A rapid fluorescence assay for FtsZ assembly indicates cooperative assembly with a dimer nucleus. Biophys J 88(1):505–514. doi:10.1529/Biophysj.104.044149
7. Pallitto MM, Murphy RM (2001) A mathematical model of the kinetics of beta-amyloid fibril growth from the denatured state. Biophys J 81(3):1805–1822
8. Ruschak AM, Miranker AD (2007) Fiber-dependent amyloid formation as catalysis of an existing reaction pathway. Proc Natl Acad Sci USA 104(30):12341–12346. doi:10.1073/Pnas.0703306104
9. Brummitt RK, Nesta DP, Chang LQ, Kroetsch AM, Roberts CJ (2011) Nonnative aggregation of an IgG1 antibody in acidic conditions, part 2: nucleation and growth kinetics with competing growth mechanisms. J Pharm Sci 100(6):2104–2119. doi:10.1002/Jps.22447
10. Andrews JM, Roberts CJ (2007) A Lumry-Eyring nucleated polymerization model of

protein aggregation kinetics: 1. Aggregation with pre-equilibrated unfolding. J Phys Chem B 111(27):7897–7913. doi:10.1021/Jp070212j

11. Li Y, Roberts CJ (2009) Lumry-Eyring nucleated-polymerization model of protein aggregation kinetics. 2. Competing growth via condensation and chain polymerization. J Phys Chem B 113(19):7020–7032. doi:10.1021/Jp8083088
12. Bernacki JP, Murphy RM (2009) Model discrimination and mechanistic interpretation of kinetic data in protein aggregation studies. Biophys J 96(7):2871–2887. doi:10.1016/J.Bpj.2008.12.3903
13. Dubay KF, Pawar AP, Chiti F, Zurdo J, Dobson CM, Vendruscolo M (2004) Prediction of the absolute aggregation rates of amyloidogenic polypeptide chains. J Mol Biol 341(5):1317–1326. doi:10.1016/J.Jmb.2004.06.043
14. Colby DW, Cassady JP, Lin GC, Ingram VM, Wittrup KD (2006) Stochastic kinetics of intracellular huntingtin aggregate formation. Nat Chem Biol 2(6):319–323. doi:10.1038/Nchembio792
15. Nielsen L, Khurana R, Coats A, Frokjaer S, Brange J, Vyas S, Uversky VN, Fink AL (2001) Effect of environmental factors on the kinetics of insulin fibril formation: elucidation of the molecular mechanism. Biochemistry 40(20):6036–6046
16. Koo BW, Hebda JA, Miranker AD (2008) Amide inequivalence in the fibrillar assembly of islet amyloid polypeptide. Protein Eng Des Sel 21(3):147–154. doi:10.1093/Protein/Gzm076
17. Pedersen JS, Dikov D, Otzen DE (2006) N- and C-terminal hydrophobic patches are involved in fibrillation of glucagon. Biochemistry 45(48):14503–14512. doi:10.1021/Bi061228n
18. Morris AM, Watzky MA, Agar JN, Finke RG (2008) Fitting neurological protein aggregation kinetic data via a 2-step, Minimal/"Ockham's Razor" model: the Finke-Watzky mechanism of nucleation followed by autocatalytic surface growth. Biochemistry 47(8):2413–2427. doi:10.1021/Bi701899y
19. Necula M, Kuret J (2004) A static laser light scattering assay for surfactant-induced tau fibrillization. Anal Biochem 333(2):205–215. doi:10.1016/J.Ab.2004.05.044
20. Chirita CN, Congdon EE, Yin HS, Kuret J (2005) Triggers of full-length tau aggregation: a role for partially folded intermediates. Bio-chemistry 44(15):5862–5872. doi:10.1021/ Bi0500123
21. Ferrone F (1999) Analysis of protein aggregation kinetics. Meth Enzymol 309:256–274

Chapter 16

A Bioinformatics Method for Identifying Q/N-Rich Prion-Like Domains in Proteins

Eric D. Ross, Kyle S. MacLea, Charles Anderson, and Asa Ben-Hur

Abstract

Numerous proteins contain domains that are enriched in glutamine and asparagine residues, and aggregation of some of these proteins has been linked to both prion formation in yeast and a number of human diseases. Unfortunately, predicting whether a given glutamine/asparagine-rich protein will aggregate has proven difficult. Here we describe a recently developed algorithm designed to predict the aggregation propensity of glutamine/asparagine-rich proteins. We discuss the basis for the algorithm, its limitations, and usage of recently developed online and downloadable versions of the algorithm.

Key words Yeast, Prion, Amyloid, Bioinformatics

1 Introduction

The majority of this book deals with the aggregation and pathogenicity of expanded poly-glutamine tracts. However, many other proteins, while lacking pure poly-glutamine tracts, have regions of high glutamine/asparagine (Q/N) content [1]. Seven different prions (infectious proteins) have been identified in *Saccharomyces cerevisiae* in which prion formation results from conversion of a soluble protein into an insoluble amyloid form [2]. Each of these seven prion proteins contains a Q/N-rich domain that drives aggregation. Recently, four human proteins with domains that compositionally resemble the yeast prion domains have been linked to disease [3]. Cytoplasmic inclusions containing FUS and TDP-43 are seen in both amyotrophic lateral sclerosis (ALS) and some forms of frontotemporal lobar degeneration (FTLD), and mutations in these proteins have been linked to some familial cases of ALS [4]. TAF15 and EWSR1 have separately been connected to ALS and FTLD [5–7]. Each of these four proteins contains a domain that is enriched in Q and N (although not to the same degree as the yeast prion proteins), as well as in other

Danny M. Hatters and Anthony J. Hannan (eds.), *Tandem Repeats in Genes, Proteins, and Disease: Methods and Protocols*, Methods in Molecular Biology, vol. 1017, DOI 10.1007/978-1-62703-438-8_16, © Springer Science+Business Media New York 2013

uncharged disorder-promoting residues such as serine and glycine. Numerous other human proteins contain compositionally similar domains, suggesting that there may be many more disease-associated Q/N-rich proteins yet to be discovered. Unfortunately, predicting the propensity of a given Q/N-rich protein to form aggregates has proven difficult. Here we will examine a recently developed method to predict which of these proteins are likely to aggregate.

Because of the ease of genetic manipulation of yeast, yeast prions have served as a useful model system for defining the sequence features that promote aggregation of Q/N-rich proteins. Scrambling studies of yeast prion domains demonstrated that prion formation is driven by the amino acid composition, not primary sequence, of the prion domain [8, 9]. Early attempts to identify other Q/N-rich proteins with prion-like activity focused on some of the obvious common compositional features of the known yeast prion proteins: high Q/N content and an underrepresentation of charged and hydrophobic residues [1, 10, 11]. However, while such methods were effective at generating long lists of potential prion proteins, these features proved insufficient to predict which of these proteins was likely to show prion-like activity.

Alberti et al. developed a more sophisticated Hidden Markov Model to identify domains of high compositional similarity to known prion proteins [12]. They tested the 100 highest scoring domains from yeast in a series of four assays for prion-like activity. Eighteen of these proteins showed prion-like activity in all four assays [12]. However, among the 100 proteins, there was little correlation between the compositional similarity score and the observed prion-like activity [12, 13]. Therefore, while this method is very effective at identifying potential prion candidates, it is not able to discriminate among these candidates. Importantly, this algorithm, coupled with extensive screening in yeast, was used to identify TAF15 as a candidate ALS-associated protein [6]. Moreover, when the Alberti algorithm is used to scan the human genome, FUS, TDP-43, and EWSR1 are all among the proteins identified as having prion-like domains, demonstrating the ability of this algorithm to identify candidate human disease-causing proteins and highlighting the applicability of algorithms developed in yeast to human proteins.

The downside of such compositional similarity algorithms is simple—they are based on the implicit assumption that any deviations from the composition of yeast prion domains will decrease prion propensity. In reality, because it is unlikely that yeast prion domains have been optimized for maximal prion propensity, some compositional changes will increase prion propensity, and some will decrease it. Therefore, to develop a more accurate scoring method, we used a quantitative mutagenesis method to score the *in vivo* prion propensity of each amino acid [13, 14]. Using these

Table 1
Prion propensity values and prevalence for each amino acid

Amino acid	Prion propensity score	Sup35 PFD composition[a]	Human PrLD composition[b]
Phenylalanine (F)	0.84	1.8	1.4
Isoleucine (I)	0.81	0.0	0.4
Valine (V)	0.81	0.0	0.5
Tyrosine (Y)	0.78	17.5	10.3
Methionine (M)	0.67	0.9	2.6
Tryptophan (W)	0.67	0.0	0.5
Cysteine (C)	0.42	0.0	0.0
Serine (S)	0.13	3.5	18.9
Asparagine (N)	0.080	17.5	6.1
Glutamine (Q)	0.069	28.1	16.3
Glycine (G)	-0.039	16.7	19.9
Leucine (L)	-0.040	0.9	0.6
Threonine (T)	-0.12	0.0	5.0
Histidine (H)	-0.28	0.0	0.6
Alanine (A)	-0.40	4.4	5.7
Arginine (R)	-0.41	1.8	1.3
Glutamic acid (E)	-0.61	0.0	1.1
Proline (P)	-1.20[c]	4.4	5.3
Aspartic acid (D)	-1.28	1.8	2.9
Lysine (K)	-1.58	0.9	0.4

[a]The percentage composition of the prion-forming domain (amino acids 1–114) of Sup35

[b]The average percentage of the compositions of the four identified prion-like domains from EWSR1 (1–280), FUS (1–237), TAF15 (1–152), and TBP-43 (277–414) (as reviewed in King et al. [3])

[c]As described in Toombs et al. [13], when multiple consecutive prolines are separated by no more than one residue, a value of zero is used for the log odds ratio (the prion propensity score) for each proline after the first in that cluster

numbers (Table 1), we developed the first algorithm capable of accurately distinguishing between yeast Q/N-rich domains with and without prion-like activity [13, 15]. Such prion activity in yeast requires proteins not only to be able to form aggregates but also to be able to propagate these aggregates over multiple generations. While the requirements for prion-like propagation may differ between organisms, presumably the basic requirements for amyloid aggregation should be similar. Therefore, human proteins that score highly using this algorithm would seem like good candidates for disease-associated aggregation. Strikingly, when our algorithm is used to scan the human proteome, FUS, EWSR1, and TAF15 are all predicted to have a very high propensity to aggregate (each scoring among the top 1 % of the proteome), while TDP-43 scores right near our previously described cutoff for amyloid aggregation.

A version of this algorithm, called PAPA (Prion Aggregation Prediction Algorithm), is available online either as a downloadable Python script or as a web-based interface, both of which can be accessed at http://combi.cs.colostate.edu/supplements/papa/. While the Python script and website each allow for rapid analysis of proteins, a fuller understanding of the underlying method can allow for more nuanced applications. Therefore, we discuss the method in more detail, examining how our prion propensity values can be used not only to predict overall aggregation propensities of proteins but also to identify key nucleating domains. We also discuss some of the implicit assumptions made in constructing PAPA and the resulting limitations of this method.

2 Materials

1. A FASTA file containing the protein sequence(s) of interest.
2. Python software package installed.
3. PAPA algorithm (downloadable from http://combi.cs.colostate.edu/supplements/papa/).

3 Methods

3.1 The PAPA Algorithm

1. For each amino acid in the protein sequence, assign the appropriate prion propensity score (Table 1).
2. Scan through the protein using a 41-amino-acid window size, scoring each window based on the average prion propensity scores for each amino acid across the window (Fig. 1a; *see* **Notes 1** and **2**). Each window is defined by the central amino acid. Therefore, windows at the termini may contain fewer than 41 amino acids, but should contain at least 21 amino acids.
3. When two or more prolines are separated by no more than one intervening residue, only the first proline in the cluster is counted in the prion propensity score (*see* **Note 3**).
4. Re-scan through the protein, averaging each set of 41 consecutive 41-amino-acid windows (Fig. 1b; *see* **Note 4**). When

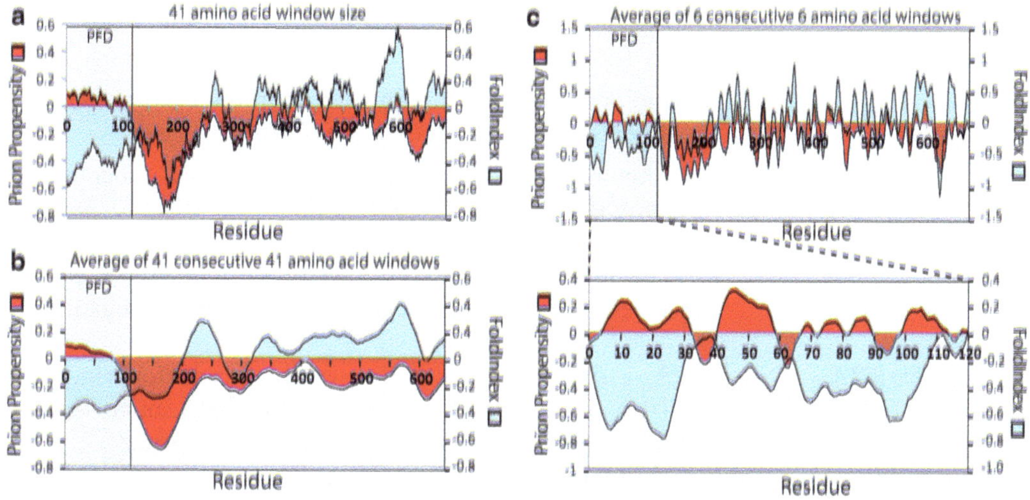

Fig. 1 Prion propensity plots for Sup35, using different window sizes. (**a**) Sup35 was scanned using a 41-amino-acid window size, calculating for each window the order propensity using FoldIndex and the prion propensity, calculated as the average of each amino acid's prion propensity. Each window is defined by its central residue. The prion-forming domain (PFD; amino acids 1–114) is easily recognizable as the only region of the protein with positive prion propensity and negative order propensity. (**b**) Sup35 was re-scanned, averaging the prion propensity and FoldIndex scores for each set of 41 consecutive 41-amino-acid windows. Each set of 41 consecutive 41-amino-acid windows is defined by the central residue of the first 41-amino-acid window in the set. Thus, the score for residue 100 is defined as the average of the forty-one 41-amino-acid windows centered on residues 100–140; therefore, residues 80–160 contribute to this score. (**c**) Prion propensity and FoldIndex analysis of Sup35, averaging each set of six consecutive 6-amino-acid windows. The PFD is enlarged below. Each set of six consecutive 6-amino-acid windows is defined by the central residue of the first 6-amino-acid window. This small window size substantially increases the noise in the analysis, making it more difficult to identify the PFD. However, within a known PFD, this small window size may reveal key nucleating segments

averaging windows near the termini, windows less than 41 amino acids are weighted in proportion to the number of residues in the window.

5. Re-scan through the protein using a 41-amino-acid window size with the prediction algorithm FoldIndex (ref. [16]; http://bip.weizmann.ac.il/fldbin/findex). FoldIndex predicts whether a given window is likely to be natively folded or disordered (*see* **Note 5**).
6. Average the FoldIndex scores for each set of 41 consecutive 41-amino-acid windows (as was done for prion propensity scores in **step 4**).
7. The overall prion propensity of a protein is defined as the 41 consecutive 41-amino-acid windows with the highest prion

propensity score that also has a negative FoldIndex order propensity (i.e., is predicted to be disordered; *see* **Note 6**).

8. A prion propensity score of greater than 0.05 indicates a high likelihood of prion-like activity (*see* **Note 7**).
9. A higher resolution map of a protein can be generated by repeating **step 2** using a smaller window size (Fig. 1c; *see* **Note 8**). Alternatively, once a prion-nucleating segment is experimentally identified, single residue scores can be used to predict which residues within the segment likely drive aggregation.

3.2 Using the PAPA Python Script

1. PAPA, a downloadable version of this prediction algorithm that runs using the Python scripting language, can be found at http://combi.cs.colostate.edu/supplements/papa/.
2. A typical PAPA command (*see* **Note 9**) is entered as:

```
python papa.py fasta_file -o results_file
```

3. In its default mode, PAPA outputs for each sequence the maximum prion propensity score, as calculated in **step** 7 of "scoring overall prion propensity," and the position in the protein sequence where the maximum prion propensity is observed (defined by the central residue of the first 41-amino-acid window of the set). PAPA outputs a score of –1 for proteins with no regions predicted to be disordered.
4. In "verbose" mode, which can be accessed simply by adding "-v" to the command line, PAPA outputs the scores for each set of 41 consecutive 41-amino-acid windows.
5. The requirement for a negative FoldIndex score can be turned off by adding "--ignore_fold_index" to the command line.
6. Window sizes other than the default 41 amino acids can be used by adding to the command line "--window_size X", where "X" represents the desired window size. For example, adding "--window_size 31" would output prion propensity scores for each set of 31 consecutive 31-amino-acid windows.

3.3 Using the Online Version of PAPA

1. A new online version of PAPA allowing for direct entry of protein sequences into a web-based interface can be accessed at http://combi.cs.colostate.edu/supplements/papa/.
2. Paste the sequence to be analyzed into the box at the top of the page.
3. The website returns the maximum prion propensity score, the position in the protein sequence where the maximum prion propensity is observed, and a graph showing the prion propensity and FoldIndex order propensity scores for each set of 41 consecutive 41-amino-acid windows across the length of the protein (as in Fig. 1b).

4 Notes

1. A window size of 41 was chosen both because it approximately equals the minimum-sized fragment required to induce prion formation by the yeast prion protein Ure2 [9] and because it approximates the minimum poly-Q length required for pathological aggregation.
2. This method of averaging the prion propensity scores of each amino acid in a window requires the assumption that there is a linear relationship between the frequency of occurrence of a given amino acid and prion propensity (i.e., if one tyrosine increases prion propensity by a certain amount, a second tyrosine will increase prion propensity by twice this amount). This assumption almost certainly is an oversimplification. Some amino acids may have nonlinear relationships with prion propensity or show a threshold effect. This issue is compounded by the fact that because of sample size limits in the initial library screen, each amino acid's prion propensity score has large confidence intervals—the 95 % confidence interval averages ±0.9 for the twenty amino acids [13]. A consequence of these two issues is that the further a protein's composition deviates from that of Sup35, the less accurate the algorithm is likely to be. Therefore, in screening entire genomes for potential prions, it may prove useful to first prescreen the genome with an algorithm such as the Alberti algorithm [12] to identify proteins with compositional similarity to the yeast prions and then to use PAPA to discriminate among these.
3. Counting only a single proline in a given cluster accounts for the fact that amyloid fibrils are composed of β-sheets, and prolines are known β-sheet breakers [17]. Because a single proline is sufficient to disrupt a β-sheet, additional prolines would be expected to have little additional effect. This theory is experimentally supported by the fact that prolines disproportionately occur in clusters within prion-forming domains and are disproportionately dispersed in compositionally similar domains without prion activity [13].
4. Averaging consecutive windows allows a larger window size to be considered while giving progressively less weight to residues further from the center of the window. Averaging 41 consecutive 41-amino-acid windows effectively means that an 81-amino-acid window size is used but that each residue is weighted in inverse proportion to its distance from the central residue. Therefore, a more direct way to do this calculation is to use an 81-amino-acid window size, but multiply the prion propensity score for each amino acid by 41 minus the distance from the central residue; the sum of these values is then divided

by the total number of individual prion propensity scores that entered into the sum. While this window size appears to be generally effective, clearly it is not perfect for all proteins, as a 37-amino-acid segment from the prion Swi1 is sufficient for prion formation and propagation [18].

5. FoldIndex uses two variables, hydrophobicity and net charge, to predict folding propensity.
6. The requirement for a negative FoldIndex order propensity is based on the observation that the yeast prion domains tend to be characterized not by extremely high prion propensity values, but instead by extended regions that are predicted to be both prion prone and disordered [13]. Our assumption is that the intrinsic disorder of these domains allows them to be more accessible for prion-promoting interactions. However, while FoldIndex has the huge advantage that it is a simple, transparent method for predicting disorder, it is by no means a perfect predictor. Therefore, there is the risk that excluding all regions with positive FoldIndex values incorrectly excludes regions that score just above zero with FoldIndex but are in fact disordered. Thus, to ensure that such candidate regions are not missed, we generally scan proteins both with and without the FoldIndex requirement.
7. This cutoff was defined by comparing the scores of proteins that showed prion-like activity in all four of Alberti's assays versus those that did not show prion-like activity in any assay [12]. Ninety-four percent of the Alberti proteins with no prion-like activity scored below 0.05, while 89 % of proteins with clear prion-like activity score above 0.05. While 0.05 was selected as simple round number, in fact a cutoff of 0.044 performs slightly better, scoring 94 % of the prion proteins correctly, while still scoring 94 % of the non-prion proteins accurately [15]. Importantly, while almost all of the proteins with prion-like activity in all four Alberti assays scored above 0.05, many of the proteins that scored between 0 and 0.05 showed prion-like activity in a subset of the Alberti assays; therefore, proteins scoring just below this cutoff should be considered possible candidates for prion-like aggregation, while those scoring above this cutoff should be considered high-likelihood candidates [13].
8. Smaller window sizes seem to be ineffective at predicting overall prion propensity of proteins; proteins with prion activity appear to be characterized by large regions of modest prion propensity, not short regions of exceptional prion propensity [13]. However, within known prions, the small windows with the highest predicted prion propensity generally seem to agree with experimentally determined key nucleating segments.

Therefore, while we believe that caution should be exercised when interpreting data using small window sizes, they can nonetheless be useful as a starting point for identifying key segments. The presence of short nucleating segments also should not be considered in conflict with the primary sequence independence of yeast prion domains. If specific amino acids strongly promote prion formation, it seems reasonable that moving these residues to a different location in the prion domain would not significantly change overall prion propensity, yet wherever these residues reside will act as a prion nucleation site.

9. Depending on the operating system, the user may need to include the path to the Python executable (see details in the program's readme file). "fasta_file" indicates a path to a FASTA-format file containing the protein sequences. "results_file" is the desired name of the output text file generated containing the results of the PAPA analysis. If no results file is entered, the results will simply be printed on the screen.
10. After this chapter was accepted for publication, mutations in two more human proteins containing prion-like domains were linked to degenerative diseases. Mutations in hnRNPA2B1 and hnRNPA1 were linked to both IBMPFD (inclusion body myopathy associated with Paget's disease of the bone and fronto-temporal dementia) and ALS [19].

Acknowledgements

This work was supported by a National Science Foundation grant (MCB-1023771) to E.D.R.

References

1. Michelitsch MD, Weissman JS (2000) A census of glutamine/asparagine-rich regions: implications for their conserved function and the prediction of novel prions. Proc Natl Acad Sci USA 97(22):11910–11915
2. MacLea KS, Ross ED (2011) Strategies for identifying new prions in yeast. Prion 5(4):263–268
3. King OD, Gitler AD, Shorter J (2012) The tip of the iceberg: RNA-binding proteins with prion-like domains in neurodegenerative disease. Brain Res 1462:61–80. doi:10.1016/j.brainres.2012.01.016
4. Da Cruz S, Cleveland DW (2011) Understanding the role of TDP-43 and FUS/TLS in ALS and beyond. Curr Opin Neurobiol 21(6):904–919. doi:10.1016/j.conb.2011.05.029
5. Couthouis J, Hart MP, Erion R, King OD, Diaz Z, Nakaya T, Ibrahim F, Kim HJ, Mojsilovic-Petrovic J, Panossian S, Kim CE, Frackelton EC, Solski JA, Williams KL, Clay-Falcone D, Elman L, McCluskey L, Greene R, Hakonarson H, Kalb RG, Lee VM, Trojanowski JQ, Nicholson GA, Blair IP, Bonini NM, Van Deerlin VM, Mourelatos Z, Shorter J, Gitler AD (2012) Evaluating the role of the FUS/TLS-related gene EWSR1 in amyotrophic lateral sclerosis. Hum Mol Genet 21(13):2899–2911. doi:10.1093/hmg/dds116
6. Couthouis J, Hart MP, Shorter J, Dejesus-Hernandez M, Erion R, Oristano R, Liu AX, Ramos D, Jethava N, Hosangadi D, Epstein J, Chiang A, Diaz Z, Nakaya T, Ibrahim F, Kim HJ, Solski JA, Williams KL, Mojsilovic-Petrovic J,

Ingre C, Boylan K, Graff-Radford NR, Dickson DW, Clay-Falcone D, Elman L, McCluskey L, Greene R, Kalb RG, Lee VM, Trojanowski JQ, Ludolph A, Robberecht W, Andersen PM, Nicholson GA, Blair IP, King OD, Bonini NM, Van Deerlin V, Rademakers R, Mourelatos Z, Gitler AD (2011) A yeast functional screen predicts new candidate ALS disease genes. Proc Natl Acad Sci USA 108:20881–20890. doi:10.1073/pnas.1109434108

7. Neumann M, Bentmann E, Dormann D, Jawaid A, DeJesus-Hernandez M, Ansorge O, Roeber S, Kretzschmar HA, Munoz DG, Kusaka H, Yokota O, Ang LC, Bilbao J, Rademakers R, Haass C, Mackenzie IR (2011) FET proteins TAF15 and EWS are selective markers that distinguish FTLD with FUS pathology from amyotrophic lateral sclerosis with FUS mutations. Brain 134(Pt 9):2595–2609. doi:10.1093/brain/awr201
8. Ross ED, Baxa U, Wickner RB (2004) Scrambled prion domains form prions and amyloid. Mol Cell Biol 24(16):7206–7213
9. Ross ED, Edskes HK, Terry MJ, Wickner RB (2005) Primary sequence independence for prion formation. Proc Natl Acad Sci USA 102(36):12825–12830
10. Harrison PM, Gerstein M (2003) A method to assess compositional bias in biological sequences and its application to prion-like glutamine/asparagine-rich domains in eukaryotic proteomes. Genome Biol 4(6):R40
11. Sondheimer N, Lindquist S (2000) Rnq1: an epigenetic modifier of protein function in yeast. Mol Cell 5(1):163–172
12. Alberti S, Halfmann R, King O, Kapila A, Lindquist S (2009) A systematic survey identifies prions and illuminates sequence features of prionogenic proteins. Cell 137(1):146–158
13. Toombs JA, McCarty BR, Ross ED (2010) Compositional determinants of prion formation in yeast. Mol Cell Biol 30(1):319–332
14. Ross ED, Toombs JA (2010) The effects of amino acid composition on yeast prion formation and prion domain interactions. Prion 4(2):60–65
15. Toombs JA, Petri M, Paul KR, Kan GY, Ben-Hur A, Ross ED (2012) De novo design of synthetic prion domains. Proc Natl Acad Sci USA 109(17):6519–6524
16. Prilusky J, Felder CE, Zeev-Ben-Mordehai T, Rydberg EH, Man O, Beckmann JS, Silman I, Sussman JL (2005) FoldIndex: a simple tool to predict whether a given protein sequence is intrinsically unfolded. Bioinformatics 21(16):3435–3438
17. Chou PY, Fasman GD (1974) Conformational parameters for amino acids in helical, beta-sheet, and random coil regions calculated from proteins. Biochemistry 13(2):211–222
18. Crow ET, Du Z, Li L (2011) A small, glutamine-free domain propagates the [SWI(+)] prion in budding yeast. Mol Cell Biol 31(16):3436–3444. doi:10.1128/MCB.05338-11
19. Kim HJ, Kim NC, Wang YD, et al (2013) Mutations in prion-like domains in hnRNPA2B1 and hnRNPA1 cause multisystem proteinopathy and ALS. Nature 495:467–473

Chapter 17

Detecting Soluble PolyQ Oligomers and Investigating Their Impact on Living Cells Using Split-GFP

Patrick Lajoie and Erik Lee Snapp

Abstract

Aberrant expansion of the number of polyglutamine (polyQ) repeats in mutant proteins is the hallmark of various diseases. These pathologies include Huntington's disease (HD), a neurological disorder caused by expanded polyQ stretch within the huntingtin (Htt) protein. The expansions increase the propensity of the Htt protein to oligomerize. In the cytoplasm of living cells, the mutant form of Htt (mHtt) is present as soluble monomers and oligomers as well as insoluble aggregates termed inclusion bodies (IBs). Detecting and assessing the relative toxicity of these various forms of mHtt has proven difficult. To enable direct visualization of mHtt soluble oligomers in living cells, we established a split superfolder green fluorescent protein (sfGFP) complementation assay. In this assay, exon 1 variants of Htt (Htt^{ex1}) containing non-pathological or HD-associated polyQ lengths were fused to two different nonfluorescent fragments of sfGFP. If the Htt proteins oligomerize and the sfGFP fragments come into close proximity, they can associate and complement each other to form a complete and fluorescent sfGFP reporter. Importantly, the irreversible nature of the split sfGFP complementation allowed us to trap otherwise transient interactions and artificially increase mHtt oligomerization. When coupled with a fluorescent apoptosis reporter, this assay can correlate soluble mHtt oligomer levels and cell death leading to a better characterization of the toxic potential of various forms of mHtt in living cells.

Key words Split-GFP, Huntingtin exon 1, Oligomers, Cell death

1 Introduction

Huntington's disease occurs due to expansion of CAG repeats in the huntingtin (Htt) gene, which increases the number of polyglutamine (polyQ) repeats and tendency of mutant Htt (mHtt) protein to aggregate in cells. Asymptomatic individuals usually present fewer than 36 CAG repeats [1]. The age of onset of the pathology inversely correlates with the number of repeats [2]. mHtt aggregation leads to the formation of insoluble inclusion bodies (IBs). The role and consequences of IBs remain controversial. While IBs have been proposed to be toxic to neurons [3–5], other studies report evidence for IBs as a cellular coping mechanism to manage the

Danny M. Hatters and Anthony J. Hannan (eds.), *Tandem Repeats in Genes, Proteins, and Disease: Methods and Protocols*, Methods in Molecular Biology, vol. 1017, DOI 10.1007/978-1-62703-438-8_17, © Springer Science+Business Media New York 2013

accumulation of mutant proteins by reducing levels of soluble and potentially toxic monomeric and oligomeric species [6, 7]. Full-length Htt is a large protein (3,144 amino acids) associated with low rates of aggregation in vitro. Indeed, transgenic mice expressing the full-length version of Htt show little impact on the animal longevity with a weak HD phenotype [8]. In contrast, introduction of exon 1 (the first 67 amino acids plus internal variable polyglutamine stretch) of the HD gene in transgenic mice is sufficient to cause a progressive HD phenotype in as little as 3 weeks [9]. In the absence of early biomarkers for HD, the exon 1 model remains a useful alternative to study Htt oligomerization in cell culture.

To determine which form(s) of mHtt is responsible for neuronal cell death in HD, different approaches have been used to correlate formation of soluble oligomers with cell death. In this chapter, we describe a protocol to visualize mHtt oligomers in living cells using split-GFP [10]. In this assay, a fluorescent protein, i.e., superfolder GFP (sfGFP) [11], is split into two nonfluorescent fragments and independently fused to Htt^{ex1} containing either a non-pathological or an HD-associated length of polyQ. When the Htt proteins oligomerize and if the sfGFP fragments come into close enough proximity, molecular complementation between the two nonfluorescent fragments results in the formation of fluorescent reporter [12, 13]. The irreversible nature of this association allows the trapping of transient protein–protein interactions [14]. The trapping leads to an artificial increase in $mHtt^{ex1}$ soluble oligomers [13]. When used in conjunction with cell death reporters, split-sfGFP (or other fluorescent variants) reporters are powerful tools for studying the impact of polyQ oligomers on cell survival.

2 Materials

2.1 Generation of Htt^{ex1} Split-sfGFP Constructs

1. cDNA encoding Htt^{ex1} sequences containing various numbers of polyQ repeats.
2. Vector encoding monomeric superfolder GFP.
3. PCR thermocycler.
4. Reagents for PCR reaction: dNTP mix, Taq polymerase, primers.
5. Restriction enzymes and appropriate buffers.
6. Agarose gel apparatus.
7. Gel and PCR purification kits.
8. T4 ligase and buffer.
9. Competent DH5α bacteria for transformation.
10. DNA purification kit such as a mini prep kit.

2.2 Expression of Htt^{ex1} Split-sfGFP Constructs in Mammalian Cells

1. Laminar flow tissue culture hood and basic tissue culture material.
2. Multi-well Labtek chambers (Thermo Fisher, Pittsburgh, PA).
3. Phenol red-free complete Roswell Park Memorial Institute medium (RPMI) (containing L-glutamine, penicillin/streptomycin, and 10 % fetal bovine serum).
4. Transfection reagent such as Lipofectamine (Life Technologies, Carlsbad, CA).
5. Htt^{ex1} split-sfGFP plasmids and ER-DEVD-tdTomato ODC plasmid.
6. Fluorescence microscope.

2.3 Labeling of Htt^{ex1} Oligomers by Immunofluorescence

1. Formaldehyde solution (37 %), less than 6 months old since opening.
2. 10 % (v/v) fetal bovine serum (FBS) in phosphate buffered saline (PBS).
3. Anti-GFP (raised against the N-terminal part of GFP) and anti-myc primary antibodies (from different hosts, i.e., mouse and rabbit).
4. Fluorescently conjugated secondary antibodies (such as Alexa fluorophore-conjugated antibodies from Life Technologies, Carlsbad, CA).
5. 0.1 % (w/v) Triton X-100 in PBS.

2.4 Quantitation of Htt^{ex1} Split-sfGFP Oligomers in Living or Fixed Cells

1. Image processing software, such as the freely available ImageJ (http://rsbweb.nih.gov/ij/).

3 Methods

3.1 Generation of Htt^{ex1} Plasmids

1. The first step is to design primers to amplify the Htt^{ex1} fragments. In our protocol, we cloned the Htt^{ex1} sequences (Fig. 1a) containing 23, 73, and 145 polyQ (Coriell Institute, Camden, NJ) to the BglII/AgeI sites of pEGFP-N1 vector (Clontech, Mountain View, CA) using primers described in Fig. 1b. Note that we generated the Htt^{ex1} fragments with and without a myc epitope tag (EQKLISEEDL) for future differential recognition by immunofluorescence.
2. Using these primers, prepare the following PCR reaction mix:

   5 μl 10× PCR buffer

   41 μl water

   1 μl Primer mix (dilute at 1 μl 100 μM Forward primer + 1 μl 100 μM Reverse primer + 3 μl dH_2O)

a

Htt[ex1] Q23 sequence:

ATGGCGACCCTGGAAAAGCTGATGAAGGCCTTCGAGTCCCTCAAGTCCTTCCAACAGCAG
CAACAGCAACAACAGCAGCAACAGCAACAACAGCAGCAACAGCAACAACAGCAGCAACAG
CCGCCACCGCCGCCGCCGCCGCCGCCGCCTCCTCAGCTTCCTCAGCCGCCGCCGCAGGCA
CAGCCGCTGCTGCCTCAGCCGCAGCCGCCCCCGCCGCCGCCCCCGCCGCCACCCGGCCCG
GCTGTGGCTGAGGAGCCGCTGCACCGACCA

MATLEKLMKAFESLKSFQQQQQQQQQQQQQQQQQQQQQQQPPPPPPPPPPPQLPQPPPQA
QPLLPQPQPPPPPPPPPPGPAVAEEPLHRP

b

Primer used:
Htt[ex1] fragments can be amplified by PCR using the following primers:
forward primer GATCAGATCTGCCACCATGGCGACCCTGGAAAAG
reverse primer GATCACCGGTCCTGGTCGGTGCAGCGG
and subcloned into the BglII and Age1 site of monomeric
msfGFP-N1 vector.

Split-sfGFP constructs can be generated by PCR using msfGFP-N1
as a template with the following primers:
s157
forward primer 5' GCAAATGGGCGGTAGGCG 3'
reverse primer 5' GATCGCGGCCGCTTACTGCTTGTCGGCCATG 3'
and subcloned into Age1 and Not1 of SFpEGFP-N1.

S238
forward primer 5' GATCACCGGTCCGGGAGCAAGAACGGCATCAAG 3'
reverse primer 5' GGTTCAGGGGGAGGTGT 3'
and subcloned into Age1 and Not1 of msfGFP-N1

c

mSFGFP (s157 in bold, s238 underlined)

MVSKGEELFTGVVPILVELDGDVNGHKFSVRGEGEGDATNGKLTLKFICTTGKLPVPWPT
LVTTLTYGVQCFSRYPDHMKRHDFFKSAMPEGYVQERTISFKDDGTYKTRAEVKFEGDTL
VNRIELKGIDFKEDGNILGHKLEYNFNSHNVYITADKQKNGIKANFKIRHNVEDGSVQLA
DHYQQNTPIGDGPVLLPDNHYLSTQSVLSKDPNEKRDHMVLLERVTAAGITHGMDELYK*

ATGGTGAGCAAGGGCGAGGAGCTGTTCACCGGGGTGGTGCCCATCCTGGTCGAGCTGGAC
GGCGACGTAAACGGCCACAAGTTCAGCGTGCGCGGCGAGGGCGAGGGCGATGCCACCAAC

GGCAAGCTGACCCTGAAGTTCATCTGCACCACCGGCAAGCTGCCCGTGCCCTGGCCCACC
CTCGTGACCACCCTGACCTACGGCGTGCAGTGCTTCAGCCGCTACCCCGACCACATGAAG
CGCCACGACTTCTTCAAGTCCGCCATGCCCGAAGGCTACGTCCAGGAGCGCACCATCAGC
TTCAAGGACGACGGCACCTACAAGACCCGCGCCGAGGTGAAGTTCGAGGGCGACACCCTG
GTGAACCGCATCGAGCTGAAGGGCATCGACTTCAAGGAGGACGGCAACATCCTGGGGCAC
AAGCTGGAGTACAACTTCAACAGCCACAACGTCTATATCACCGCCGACAAGCAGAAGAAC
GGCATCAAGGCCAACTTCAAGATCCGCCACAACGTGGAGGACGGCAGCGTGCAGCTCGCC
GACCACTACCAGCAGAACACCCCCATCGGCGACGGCCCCGTGCTGCTGCCCGACAACCAC
TACCTGAGCACCCAGTCCGTGCTGAGCAAAGACCCCAACGAGAAGCGCGATCACATGGTC
CTGCTGGAACGCGTGACCGCCGCCGGGATCACTCACGGCATGGACGAGCTGTACAAG

Fig. 1 (**a**) Htt[ex1] Q23 nucleotide and amino acid sequences. (**b**) Primers used to generate Htt[ex1]-sfGFP and split-sfGFP plasmids. (**c**) Monomeric sfGFP nucleotide and amino acid sequences. s157 is highlighted in *bold* and s238 is *underlined*

1 μl dNTPs

1 μl template (10 ng/μl stock)

1 μl Pfu polymerase (2.5 U/μl)

50 μl

3. Amplify using standard PCR protocol.
4. Run the resulting PCR product on a 1 % agarose gel.
5. Purify the PCR product using a commercial gel purification kit (such as Qiagen, Valencia, CA).
6. Digest the PCR fragment and the pEGFP-N1 vector using BglII/AgeI enzymes according to manufacturer's instructions.
7. Ligate the digested PCR fragment into the digested vector according to the manufacturer's instructions.
8. Transform into DH5α competent bacteria according to manufacturer's instructions and plate into agar plate containing kanamycin (if using a kanamycin resistance vector).
9. Pick a colony and inoculate into 5 ml of LB broth and grow overnight.
10. Purify the plasmid using commercial kit.
11. Verify successful cloning by diagnostic digestion and sequencing.

3.2 Integrate the sfGFP and Split-sfGFP Fragments into the Htt^{ex1} Plasmids

We used the AgeI/NotI sites of the pEGFP-N1 vector.

1. The first step is to design primers to amplify the different sfGFP fragments. The split-sfGFP first fragment contains the first 157 amino acid of sfGFP (s157) while the second one contains the remaining sequence of the protein (s238) (Fig. 1c). Primers to generate the full-length sfGFP protein should also be designed (Fig. 1).
2. Repeat **steps 2–11** of the previous Subheading 3.1 using appropriate buffers and restriction enzymes.

3.3 Quantitative Imaging of Htt^{ex1} Split-sfGFP Plasmids Oligomerization by Immunofluorescence

First, the ability of both Htt^{ex1} plasmid constructions to complement and produce a fluorescent GFP signal must be verified.

1. Plate Neuro2a cells (N2a) (*see* **Note 1**) into 8-well Labtek imaging chambers.
2. Transfect cells with Htt^{ex1} split-sfGFP plasmids using Lipofectamine 2000 (or your favorite transfection reagent or method) according to manufacturer's instructions. Cells should be transfected with Htt^{ex1} s157 and s238 plasmids separately (a negative control) or in combination. A second set of negative controls is to transfect s157 and s238 sfGFP plasmids that do not have Htt insertions, together or separately or the complementary nonfused version along with the Htt^{ex1} fusion construct. Yet another useful control is to test interactions

with another irrelevant cytoplasmic protein (such as Nalp1b) fused to one of the split-sfGFP fragments. Together these controls will establish the specificity of the polyQ–split-GFP interaction. Finally, it is also important to test whether the position of the split-GFP fragment influences the complementation. This can be determined by generating a plasmid with an Htt^{ex1} protein with a split-sfGFP fragment fused to its NH_2-terminal end.

3. Grow cells for 48 h in complete RPMI media.
4. Fix cells in freshly diluted 3.7 % formaldehyde in PBS containing for 15 min at room temperature.
5. Permeabilize cells using 0.1 % (w/v) Triton X-100 in PBS for 15 min at room temperature.
6. Prevent nonspecific antibody binding by blocking epitopes in cells with 10 % (v/v) FBS in PBS for 1 h at room temperature.
7. Incubate cells with 200 μl of 10 % (v/v) FBS in PBS containing anti-GFP and anti-myc antibodies for 1 h.
8. Wash 3× 10 min with 400 μl of PBS 10 % (v/v) FBS in FBS.
9. Incubate cells with 200 μl of 10 % (v/v) FBS in PBS containing appropriate fluorescently labeled secondary antibodies for 1 h.
10. Wash 3× 10 min with PBS.
11. Observe cells under the fluorescent microscope. No fluorescent signal should be detected when split-GFP fragments are expressed separately. For each fluorescence channel (anti-GFP, anti-myc, and sfGFP fluorescence), set image acquisition settings so the brightest conditions are not saturated (in our case the Q73 s157/Q73 s238) and use the same settings for all different conditions (*see* **Note 2**).
12. Quantitate the fluorescence signal of each channel in ImageJ. This can be done by manually tracing the individual cells as close to the edge as possible for maximum signal and then measuring the average pixel fluorescence intensity for each cell.
13. Plot the average fluorescent intensities for the different conditions (Fig. 2).

3.4 Quantitative Imaging of Htt^{ex1} Split-sfGFP Plasmids Oligomerization and IBs Formation in Living Cells and Impact on Cell Death

1. Plate Neuro2a cells (N2a) (or any other adherent cell type of interest) into 8-well Labtek imaging chambers.
2. Transfect cells with Htt^{ex1} split-sfGFP or intact sfGFP plasmids using Lipofectamine 2000 according to manufacturer's instructions (or favorite transfection method). During the transfection step, a live cell apoptosis reporter such as the ER-DEVD-tdTomato ODC [13, 15] can be incorporated (Fig. 3) (*see* **Note 3**).

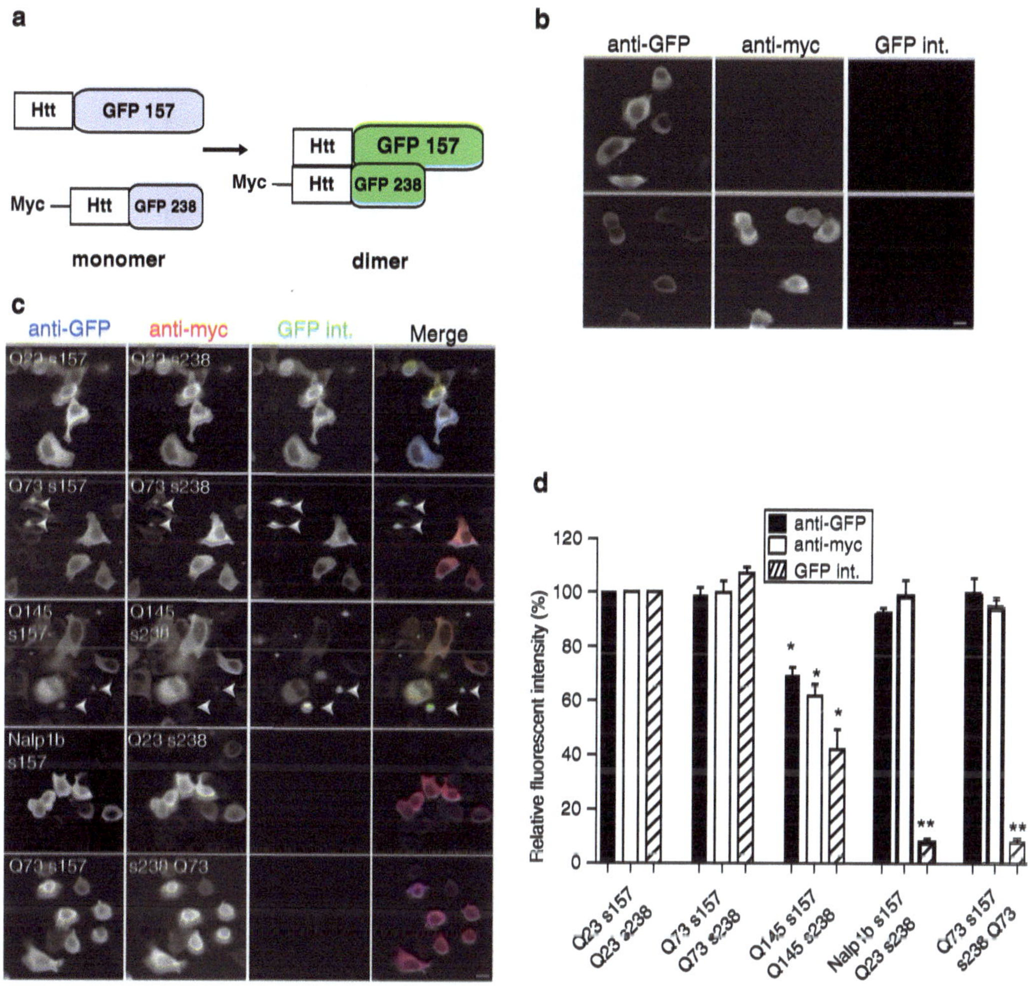

Fig. 2 Visualization of Htt^{ex1}-sfGFP oligomers using split-GFP. (**a**) Illustration of the fusion of wt and mHtt^{ex1} to either 157-GFP or 238-GFP. (**b**) Cells transfected with Q23 s157 or Q23 s238 separately, stained with anti-GFP or anti-myc, and imaged for GFP fluorescence intensity (sfGFP int.). (**c**) Representative images of N2a cells transiently transfected with Htt^{ex1}Q23 157-sfGFP+Htt^{ex1}Q23 238-sfGFP, Htt^{ex1}Q73 157-sfGFP+mHtt^{ex1}Q73 238-sfGFP, mHtt^{ex1}Q145 157-sfGFP+mHtt^{ex1}Q145 238-sfGFP, Nalp1b-157+Htt^{ex1}Q23 238-GFP, or mHtt^{ex1}Q73 157-sfGFP+238-sfGFP mHtt^{ex1}Q73. (**d**) Mean fluorescence intensities in cells without IBs are plotted and compared relative to Q23 means, which were set as 100 % (Reproduced from [13] with the permission of the Public Library of Science)

3. Incubate cells in complete RPMI media.
4. At various time intervals, observe cells under the fluorescent microscope. For the GFP fluorescence channel, set image acquisition settings so the brightest conditions are not saturated and use the same settings for all different conditions (*see* **Note 2**). At the same time, record the percentage of cells expressing IBs.

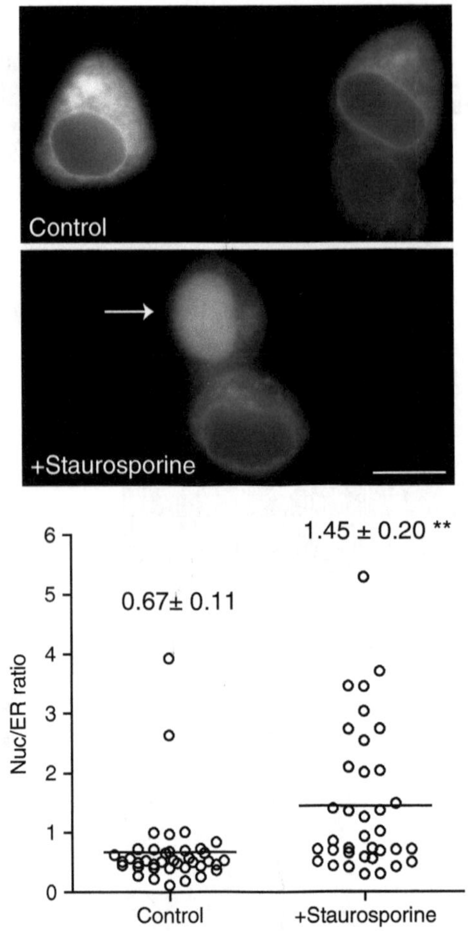

Fig. 3 Validation of the ER-DEVD-tdTomato reporter functionality. N2a cells were transfected with ER-DEVD-tdTomato for 16 h and then treated with or without 5 μM staurosporine for 3 h. The fluorescent intensity ratio of the nucleus over the ER calculated is presented in the plot. ** < 0.0001 compared to untreated cells. Bar = 20 μm (Reproduced from [13] with the permission of the Public Library of Science)

5. Plot the percentage cells expressing IBs for both full sfGFP and split-sfGFP Htt^{ex1} constructs over time (Fig. 4).
6. Quantitate the fluorescent signal of each channel in ImageJ. This can be done by manually tracing the individual cells, as close to the cell edge as possible, and measuring the mean fluorescence intensity (pixel intensity) for each cell.
7. Quantitate DEVDase activity in individual cells by dividing the mean nuclear fluorescence intensity of the ER-DEVD-tdTomato ODC by the ER mean fluorescence intensity.
8. Plot the average fluorescence intensities for the different conditions as a function of DEVDase activity (Fig. 5).

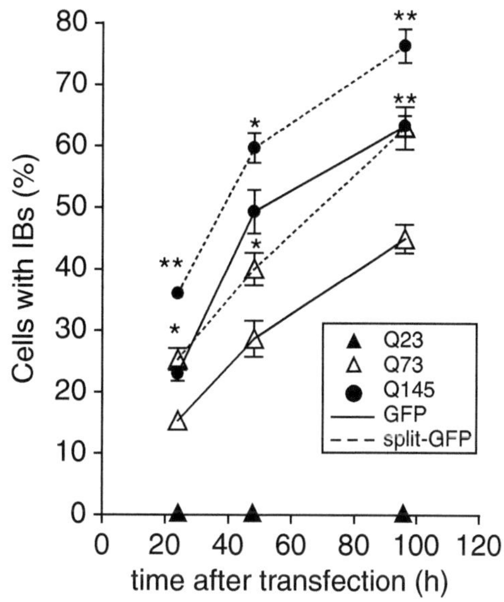

Fig. 4 Quantitation of percentage of cells containing IBs for indicated times post-transfection with Htt^{ex1}Q23, 73, or 145 fused to sfGFP or split-sfGFP constructs. $n > 215$ cells. $^{*}p < 0.05$, $^{**}p < 0.005$ compared to same length of polyQ fused to sfGFP (Reproduced from [13] with the permission of the Public Library of Science)

4 Notes

1. N2a cells can be routinely differentiated into neuron-like cells by incubating the cells with 5 μM dbcAMP (N69, 29-O-dibutyrilaenosine-39:59-cyclic monophosphate sodium salt) (Sigma-Aldrich, St. Louis, MO) for 2 days.
2. When wanting to directly compare fluorescent intensities between two different condition, it is critical that the two images were acquired with the same acquisition settings. To avoid pixel saturation, we routinely acquired the brightest image first and keep the same settings to acquire the other conditions. For the polyQ constructs, because of the very intense signal originating from the IBs, it may be useful to acquire two different images (with high and low exposures) to visualize and measure a nonsaturating cytoplasmic GFP signal.
3. The reporter localizes to the endoplasmic reticulum (ER). Upon induction of apoptosis with staurosporine, activated caspases cleave the DEVD peptide in the reporter, releasing it from the ER membrane and allowing its translocation into the nucleus.

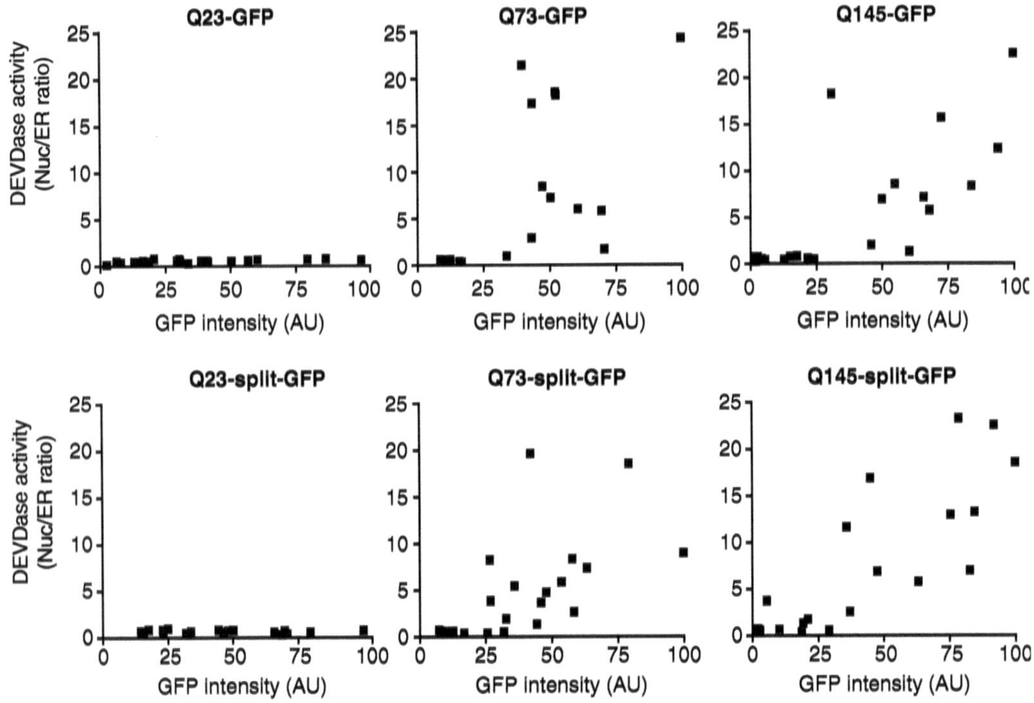

Fig. 5 Levels of soluble cytoplasmic mHtt[ex1] are a negative predictor of cell death. N2a cells were cotransfected with Q23, 73, or 145 Htt[ex1] fused to sfGFP or split-sfGFP and ER-DEVD tdTomato for 48 h. GFP or split-sfGFP intensities and the nuclear/ER ratio of the apoptosis reporter were quantified for individual cells not presenting IBs. In each plot, the cell with the brightest GFP intensity was defined as having a mean intensity of 100 arbitrary units and other cell intensities were converted to this scale. Thus intensities in one plot are not directly comparable to intensities in another plot. Each square represents a single cell. Increasing mHtt[ex1], but not wt Htt[ex1], levels above some threshold were associated with increased cell death. The existence of apparent thresholds correlating with activation of the caspase activity suggests mHtt[ex1] toxicity is sharply concentration dependent and may depend on titration of one or more key cellular factors (Reproduced from [13] with the permission of the Public Library of Science)

References

1. Gusella JF, MacDonald ME (2006) Huntington's disease: seeing the pathogenic process through a genetic lens. Trends Biochem Sci 31(9):533–540
2. Williams AJ, Paulson HL (2008) Polyglutamine neurodegeneration: protein misfolding revisited. Trends Neurosci 31(10):521–528
3. Becher MW, Kotzuk JA, Sharp AH, Davies SW, Bates GP, Price DL, Ross CA (1998) Intranuclear neuronal inclusions in Huntington's disease and dentatorubral and pallidoluysian atrophy: correlation between the density of inclusions and IT15 CAG triplet repeat length. Neurobiol Dis 4(6):387–397
4. Davies SW, Turmaine M, Cozens BA, DiFiglia M, Sharp AH, Ross CA, Scherzinger E, Wanker EE, Mangiarini L, Bates GP (1997) Formation of neuronal intranuclear inclusions underlies the neurological dysfunction in mice transgenic for the HD mutation. Cell 90(3):537–548
5. DiFiglia M, Sapp E, Chase KO, Davies SW, Bates GP, Vonsattel JP, Aronin N (1997) Aggregation of huntingtin in neuronal intranuclear inclusions and dystrophic neurites in brain. Science (New York, NY) 277(5334): 1990–1993
6. Arrasate M, Mitra S, Schweitzer ES, Segal MR, Finkbeiner S (2004) Inclusion body formation

reduces levels of mutant huntingtin and the risk of neuronal death. Nature 431(7010):805–810

7. Takahashi T, Kikuchi S, Katada S, Nagai Y, Nishizawa M, Onodera O (2008) Soluble polyglutamine oligomers formed prior to inclusion body formation are cytotoxic. Hum Mol Genet 17(3):345–356
8. Ehrnhoefer DE, Butland SL, Pouladi MA, Hayden MR (2009) Mouse models of Huntington disease: variations on a theme. Dis Model Mech 2(3–4):123–129
9. Mangiarini L, Sathasivam K, Seller M, Cozens B, Harper A, Hetherington C, Lawton M, Trottier Y, Lehrach H, Davies SW, Bates GP (1996) Exon 1 of the HD gene with an expanded CAG repeat is sufficient to cause a progressive neurological phenotype in transgenic mice. Cell 87(3):493–506
10. Wilson CG, Magliery TJ, Regan L (2004) Detecting protein–protein interactions with GFP-fragment reassembly. Nat Methods 1(3): 255–262
11. Pedelacq JD, Cabantous S, Tran T, Terwilliger TC, Waldo GS (2006) Engineering and characterization of a superfolder green fluorescent protein. Nat Biotechnol 24(1):79–88
12. Herrera F, Tenreiro S, Miller-Fleming L, Outeiro TF (2011) Visualization of cell-to-cell transmission of mutant huntingtin oligomers. PLoS Curr 3:RRN1210. doi: 10.1371/currents.RRN1210k/-/-/2sdo8o1u01fbj/1 [pii]
13. Lajoie P, Snapp EL (2010) Formation and toxicity of soluble polyglutamine oligomers in living cells. PloS One 5(12):e15245. doi:10.1371/journal.pone.0015245
14. Magliery TJ, Wilson CG, Pan W, Mishler D, Ghosh I, Hamilton AD, Regan L (2005) Detecting protein–protein interactions with a green fluorescent protein fragment reassembly trap: scope and mechanism. J Am Chem Soc 127(1):146–157
15. Bhola PD, Simon SM (2009) Determinism and divergence of apoptosis susceptibility in mammalian cells. J Cell Sci 122(Pt 23): 4296–4302

Chapter 18

Cell Biological Approaches to Investigate Polyglutamine-Expanded AR Metabolism

Lori J. Cooper and Diane E. Merry

Abstract

Spinal and bulbar muscular atrophy (SBMA) is a late-onset neurodegenerative disease caused by a polyglutamine expansion in the androgen receptor (AR). In vivo and in vitro studies have suggested that some steps of normal AR function and metabolism, such as hormone binding and nuclear translocation of the AR, are necessary for toxicity and aggregation of the mutant protein. Mutation of discreet functional domains of the AR and sites of posttranslational modification enable the detailed analysis of the role of AR function and metabolism in toxicity and aggregation of polyglutamine-expanded AR. This analysis could potentially lead to the development of targeted therapy for the treatment of SBMA. We have developed a stably transfected, tetracycline-inducible, cell model that replicates many of the hallmarks of disease, including ligand-dependent aggregation and toxicity, and provides a relatively quick system for the reliable expression and analysis of the mutated polyglutamine-expanded AR. Multiple cell lines, each expressing the androgen receptor with a distinct functional mutation, can be created and the dose of tetracycline modulated to produce equal protein expression across lines in order to evaluate the structural and functional requirements of AR toxicity and aggregation. Results from these studies can then be validated in a disease-relevant cell type, spinal motor neurons, using viral delivery of the gene of interest into dissociated spinal cord cultures. Utilization of these cell models provides a relatively rapid, cost-effective experimental pathway to analyze the role of distinct steps in AR metabolism in disease pathogenesis using in vitro systems.

Key words Polyglutamine, CAG repeat, Androgen receptor, Cell models, Nuclear receptor

1 Introduction

Spinal and bulbar muscular atrophy (SBMA) is a late-onset neurodegenerative disease that is characterized by proximal muscle fasciculations and weakness, difficulty speaking and swallowing, and the loss of lower motor and sensory neurons [1]. The disease-causing mutation is an expansion of a polymorphic CAG tract in the androgen receptor gene. In healthy individuals, this tract contains between 6 and 39 CAGs; expansion beyond 39 CAGs causes disease [2–4].

Danny M. Hatters and Anthony J. Hannan (eds.), *Tandem Repeats in Genes, Proteins, and Disease: Methods and Protocols*, Methods in Molecular Biology, vol. 1017, DOI 10.1007/978-1-62703-438-8_18, © Springer Science+Business Media New York 2013

The AR is a ligand-dependent nuclear receptor that, in the absence of its ligand, testosterone or dihydrotestosterone, resides in an aporeceptor complex containing multiple heat shock proteins and accessory proteins in the cytoplasm (reviewed in ref. 5). Upon binding to ligand, the AR dissociates from the complex, undergoes a conformational change, and translocates to the nucleus. In the nucleus, the AR binds to DNA as a dimer and recruits co-regulators to control the expression of androgen-responsive genes. Recent data suggest that some steps of this normal function of the AR are necessary for disease. It has been shown that ligand binding and nuclear localization of the AR are required for disease [6–10]. Additionally, the interdomain interaction between the amino-terminal (N) FxxLF motif and carboxyl-terminal (C) AF2 domain, the N/C interaction, may be necessary for toxicity and aggregation of polyglutamine-expanded AR [11]. Posttranslational modifications of the AR such as acetylation, phosphorylation, and sumoylation have been shown to be important modulators of mutant AR toxicity [12–15]. These findings and the further analysis of steps in normal AR function and metabolism that are necessary for disease could lead to a targeted therapy for patients.

Cell models of SBMA provide an inexpensive, rapid, and effective way to carry out detailed investigations of the role of distinct AR domains that are important for its metabolism and function in the pathogenesis of disease. A stably transfected, undifferentiated, inducible PC12 cell model of SBMA reproduces many key hallmarks of disease, including hormone-dependent toxicity and the formation of nuclear inclusions consisting of proteolyzed polyglutamine-expanded AR protein (Fig. 1), as seen in tissue of patients [16, 17]. An expression plasmid containing the gene encoding the androgen receptor with 112 glutamine repeats can easily be mutated within discrete functional domains to address the role of these domains in polyglutamine-expanded AR toxicity. The inducible nature of the model enables consistent and reliable expression of the mutated polyglutamine-expanded AR. Additionally, the tetracycline concentration can, in most cases, be modulated to produce equal protein expression across cell lines, enabling the role of the mutated functional domain in AR to be determined without differences in protein levels confounding interpretation. Unfortunately, due to variation of transgene insertion site, equal AR expression across cell lines is not always guaranteed. Isogenic FLP-IN HEK293 cell lines can be generated by targeted transgene integration, bypassing this problem and allowing for the rapid generation of stably transfected, inducible cell lines that express comparable levels of polyglutamine-expanded AR with the same doxycycline dose. Such cell lines recapitulate the ligand-dependent toxicity observed in PC12 cells (albeit with a delayed time course); however, relatively few cells exhibit DHT-dependent nuclear inclusions of expanded-polyglutamine AR (our unpublished results),

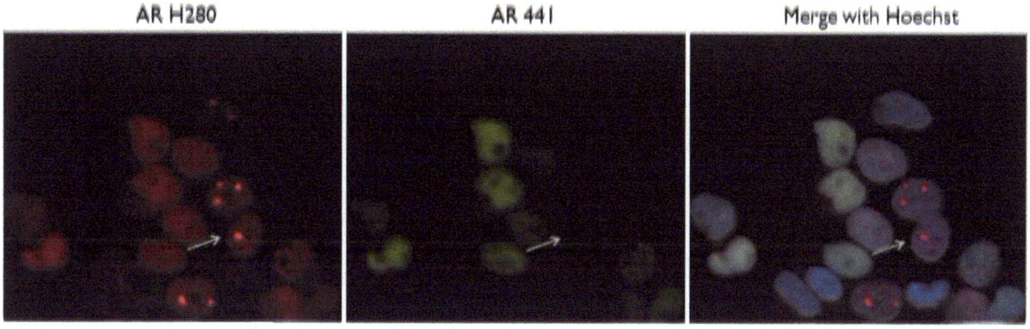

Fig. 1 Stably transfected PC12 cells were induced to express AR with 112 glutamines and treated with 10 nM DHT for 96 h. Cells were fixed with 4 % paraformaldehyde and then immunostained with an antibody to the amino-terminus of the AR (AR H280), an antibody with an internal AR epitope (amino acids 299–315) (AR441), and Hoechst to visualize nuclei. In the presence of DHT, the mutant AR forms intranuclear inclusions comprised primarily of the amino-terminus of polyglutamine-expanded AR

limiting their utility for studies of AR aggregation, and for this reason, this cell model will not be further discussed here.

These simple model systems are excellent starting points for investigating the metabolism of expanded-polyglutamine AR. However, controlled experimentation in a more disease-relevant cell type, spinal motor neurons (using dissociated spinal cord cultures [18]), is necessary for the validation of cell line-derived results before moving to more expensive and complicated in vivo animal models of disease. Polyglutamine-expanded AR can be expressed in motor neurons in dissociated spinal cord cultures using adeno-associated viral delivery. The resulting DHT-dependent toxicity associated with expanded-polyglutamine AR expression can be analyzed through immunohistochemical analysis of the cultures.

Here we describe the development of these cell models of SBMA and explore the unique problems associated with working with polyglutamine-expanded AR in these specific cell types.

2 Materials

2.1 Materials for Site-Directed Mutagenesis and for Determining the Integrity of the Plasmid

1. QuikChange II XL Site-Directed Mutagenesis Kit (Agilent).
2. Agar plates with carbenicillin: Prepare 10.0 g tryptone, 5.0 g yeast extract, 10.0 g NaCl, 15.0 g agar and bring to 1 L with distilled water. Autoclave for 1 h. Let cool at room temperature; when the agar has cooled, add 50 mg of carbenicillin and pour into 50 100 mm × 15 mm polystyrene petri dishes. Store in the dark at 4 °C.
3. LB: Mix 10.0 g tryptone, 5.0 g yeast extract, 10.0 g NaCl with distilled water and bring to 1 L. Autoclave for 1 h. Add selection antibiotics just prior to use.

4. Round-bottom disposable polypropylene tubes.
5. Miniprep DNA plasmid kit.
6. Agarose and agarose gel electrophoresis system.
7. Maxi-prep plasmid kit (e.g., from Invitrogen).

2.2 Cell Culture Components for PC12 Cells

1. PC12 Tet-On cell line (e.g., Clontech, 631137).
2. Cell culture grade vented 25 and 75 mm flasks.
3. PC12 cell media: Add 50 mL of heat-inactivated horse serum (Gibco), 25 mL heat-inactivated fetal bovine serum (Gibco) (*see* **Note 1**), 10 mL glutamine, 100 μg/ml G418 sulfate, DMEM high glucose (with l-glutamine, pyridoxine hydrochloride, without sodium pyruvate) to 500 mL. Filter with Fast PES Filter unit with 0.2 μm pore size (Fisher) and store at 4 °C. Once cells have been cotransfected with pTRE-AR plasmid (*see* Subheading 2.3, **item 1** below) and a hygromycin-resistance-conferring plasmid (pTK-hygro; Clontech, 631750), include 100 mg hygromycin B per 500 mL of media.

2.3 Transfection Components to Create Stable PC12 Cell Line

1. Tetracycline-regulated vector of choice (e.g., from Clontech; 3 generations of vectors). Here we used the pTRE vector (Clontech) (*see* **Note 2**).
2. 100 mm tissue culture dishes.
3. Poly-d-lysine.
4. 96-well tissue culture dishes.
5. 2× HEBS: Combine 3.2 g NaCl, 142 mg KCl, 76 mg $Na_2HPO_4 \cdot 7H_2O$, 540 mg dextrose, 2 g HEPES with 180 mL water and pH to 7.04 with 5 N NaOH. Bring to 200 mL with water and filter-sterilize under cell culture hood.
6. 2.5 M $CaCl_2$.
7. 0.05 % (w/v) trypsin–EDTA.
8. 10× phosphate buffered saline (PBS). Also required is PBS without Ca^{2+} and Mg^{2+}.

2.4 Components to Work with Stable, Inducible PC12 Cell Lines

1. Doxycycline.
2. Triton–DOC lysis buffer: Mix 1 g deoxycholic acid with 70 mL of deionized water, 10 mL 10× PBS, and 0.5 mL Triton X-100. Bring to 100 mL with deionized water.
3. AR H280 antibody (Santa Cruz).
4. 12-well tissue culture dishes.
5. 10 % Tris/glycine SDS-PAGE gels.
6. PVDF or nitrocellulose membranes for Western blotting.

7. 2× sample buffer: Combine 2.5 mL of 0.5 M Tris–HCl pH 6.8, 2.0 mL glycerol, 2.0 mL 20 % w/v sodium dodecyl sulfate (SDS), 0.5 mL 0.1 % w/v Bromophenol Blue and then bring up to 10 mL with deionized water. Add 50 μL of 2-mercapto-ethanol to 950 μL of 2× sample buffer prior to use.

2.5 Materials Needed for Motor Neuron Experiments

1. Pregnant female mouse (E13.5 day).
2. Sterile scissors and forceps.
3. 100 mm petri dish.
4. 12 mm German glass coverslips (Fisher).
5. SMI32 antibody (Covance).
6. Conditioned NFeed medium: Mix 97 mL EMEM (*see* below, **item 9**), 1 mL N3 (100×) (*see* below, **item 8**), 3 mL horse serum (*see* **Note 1**). Divide into 3 dishes of glial cells, initiated from embryonic day 13.5 mouse brains (*see* **Notes 21** and **22**), and incubate for 24 h. Remove and filter with 0.2 μm filter. Store at 4 °C. Add 2.5S nerve growth factor (NGF) (Collaborative Research; stored at −80 °C) to a final concentration of 10 ng/mL prior to use.
7. First feed medium: NFeed, 5 % (v/v) filtered heat-inactivated fetal calf serum, 1 % (v/v) penicillin–streptomycin.
8. N3 (100×): 1 mg/mL insulin, 20 mg/mL apo-transferrin, 1 mg/mL bovine serum albumin, 9.2 mg/mL putrescine, 2.6 μg/mL selenium, 2 μg/mL triiodo-l-thyronine (T3), 910 ng/mL hydrocortisone, 1.3 μg/mL progesterone.
9. Enriched MEM (EMEM): Add powdered MEM (for 1 L) (Invitrogen) with 3.7 g $NaHCO_3$ and 5 g dextrose. Bring volume to 1 L with deionized water.
10. 24-well tissue culture dishes.
11. 60 mm tissue culture dishes.
12. Dissecting medium: Combine 1 g dextrose, 20 g sucrose, 25 mL 20× Colorado serum (0.8 g NaCl, 0.04 g KCl, 0.005 g $Na_2HPO_4{\cdot}7H_2O$, 0.003 g KH_2PO_4), and 14 mL 352 mM HEPES, pH 7.4.
13. Cytosine-B-d-furanoside (CAF).
14. 100 % ethanol.
15. 5α-Dihydrotestosterone (DHT).
16. 4 % paraformaldehyde, pH 7.4: Add 4 g paraformaldehyde to 10 mL 10× PBS and then bring volume to 90 mL with deionized water. Heat to 60 °C while stirring. Add a few drops of 1 N NaOH until solution clears and then pH to 7.4. Bring volume to 100 mL and then filter. Store at 4 °C.

3 Methods

3.1 Site-Directed Mutagenesis and Selecting Plasmid for Transfection

1. Design primers according to QuikChange Mutagenesis Kit protocol to mutate the functional AR domain of interest (*see* **Note 3**).
2. Follow mutagenesis and transformation protocol in kit. Spread transformed bacteria on carbenicillin-containing agar plates and grow at 37 °C overnight (*see* **Note 4**).
3. The next day grow 10–16 colonies from the agar plates in 5 mL LB containing carbenicillin (1 μg/mL) at 30 °C for 6–12 h (*see* **Note 5**).
4. Isolate plasmid DNA from each culture using miniprep kit. Check DNA concentration and quality using a spectrophotometer.
5. Digest 1 μg of the plasmid DNA from each culture with restriction enzymes to check the integrity of the plasmid. Digest with a single-site restriction enzyme to assay the size of the plasmid and with restriction enzymes to determine the size and instability of the CAG tract within the AR cDNA (*see* **Note 6**).
6. Plasmids that contain the expanded CAG tract with few contracted species as determined by restriction digest should be sequenced to confirm the mutation, the CAG tract length, and the absence of additional mutations (*see* **Note 7**).
7. Inoculate 300 mL of LB (containing 1 μg/mL carbenicillin) with 500 μL of the culture containing the expanded CAG tract containing plasmid and grow at 30 °C overnight. Isolate plasmid DNA using a maxi-prep kit. Confirm integrity of CAG repeat tract by repeating restriction digest used in **step 5**.

3.2 Culturing and Transfecting PC12 Cells

1. Thaw frozen, early-passage, Tet-activator-containing PC12 cells in a 37 °C water bath and then quickly transfer them to a 15 mL conical tube containing parental PC12 cell media and pellet at low speed (*see* **Note 8**).
2. Resuspend cells in media, then transfer to a tissue culture flask, and place in a humidified 37 °C incubator with 10 % atmospheric CO_2.
3. Coat 100 mm tissue culture dishes with poly-d-lysine (0.005 mg/mL) and let sit in a laminar flow hood for 20 min at room temperature.
4. Remove poly-d-lysine and let dry under UV light for 20–30 min.
5. In laminar flow hood using sterile technique, aspirate media from flask and wash cells with 3 mL of sterile PBS without Ca^{2+} and Mg^{2+}. Aspirate PBS and add 1 mL 0.05 % trypsin–EDTA; incubate at 37 °C until the cells detach.

6. Using a serological pipette, transfer cells to 15 mL conical tube and then spin at 160 × *g* for 5 min at room temperature to pellet cells. Aspirate trypsin–EDTA and resuspend cells in PC12 cell media (*see* **Note 9**).
7. Plate sufficient cells on poly-d-lysine-coated 100 mm dishes to ensure cell confluency of approximately 90 % on the following day (*see* **Note 10**).
8. The next day, combine 10 μg pTRE-AR112Q with a plasmid conferring hygromycin resistance (pTK-hygro plasmid) in a 4:1 molar ratio (pTRE:pTK-hygro). Add 50 μL of 2.5 M $CaCl_2$ and then bring volume to 500 μL with sterile cell culture grade water.
9. Add solution dropwise to 500 μL of 2× HEBS, pH 7.04, and then mix and incubate at room temperature for 5 min or until cloudy.
10. Add mixture dropwise to the plate of PC12 cells with 10 mL of fresh PC12 cell media. Incubate at 37 °C for 5–7 h, and then refresh media. On the next day, split cells onto 6 poly-d-lysine-coated plates and let settle overnight.
11. The next day, change to media containing hygromycin in order to select for cells expressing the hygromycin-resistance gene (*see* **Note 11**).
12. Change the media once every 2 days for the first week to eliminate dying, untransfected cells. The surviving cells should be spatially separated on the plate, eventually allowing for the isolation of individual colonies.
13. The media should be changed once a week for 3–4 weeks until individual colonies emerge that are visible to the naked eye. At this point, each colony should be transferred to an individual well of a 96-well plate (*see* **Note 12**).
14. Once the clonal cells are sufficiently confluent, transfer the clonal culture to a 24-well plate and eventually move the cells to a 6-well dish (*see* **Note 13**).

3.3 Testing Cell Lines for Transgene Expression

1. Each clone from the transfection must be assayed for AR expression (*see* **Note 14**) as well as polyglutamine tract length (*see* **Note 15**).
2. Transfer approximately 200,000 cells to each well of a 12-well plate.
3. On the next day, treat the cells with media containing 1 μg/mL doxycycline (*see* **Note 16**) for 48 h.
4. After 48 h, aspirate media, wash cells with PBS, and then lyse cells with Triton–DOC lysis buffer.

5. Run 20 μg of protein in 1× sample buffer (*see* **Note 17**) on a 10 % Tris/glycine gel at 120 V for 90 min and then transfer protein to PVDF or nitrocellulose at 100 V for 60 min.
6. Probe membrane for AR with AR H280 antibody (Santa Cruz) and identify clones that express polyglutamine-expanded AR of correct size and appropriate amount. CAG repeat length is then confirmed by sequencing (*see* **Note 18**).
7. Upon identification of candidate clonal cell lines (*see* **Note 19**), a doxycycline dose–response experiment is initiated to determine the amount of doxycycline needed to produce comparable AR expression across PC12 cell lines.

3.4 Embryonic Spinal Cord Dissection

1. Euthanize pregnant female (E13.5 day) by CO_2 inhalation and decapitate as a secondary measure of euthanasia. Remove paired uterine horns using sterile scissors and forceps and place in 100 mm dish on ice containing 5 mL dissecting medium.
2. Remove amniotic sacs around embryos and transfer embryos to new 100 mm dish with 5 mL dissecting medium (*see* **Note 20**).
3. Use forceps to cut the spinal cord along with the dorsal root ganglia, starting at the base of the tail, and transfer cords to 60 mm dish with 1 mL dissecting medium (for glial cultures, *see* **Note 21**).
4. Cut cords into small pieces, add 60 μL 2.5 % trypsin and 2 mL dissecting medium, and then incubate at 37 °C for 15 min.
5. Transfer cords to a 15 mL tube containing 3 mL of first feed, triturate lightly (approximately 10 times) with sterile transfer pipette, and let settle.
6. Transfer the large settled pieces to another 15 mL tube containing 2 mL of first feed and let settle. Transfer supernatant to an empty 15 mL tube, then add 1.5 mL first feed to the undissociated cords, and triturate again.
7. Repeat until most of the cords are dissociated, then increase volume to 0.5 mL/cord, count cells, and plate 265,000 cells in 24-well plates that contain poly-d-lysine-coated 12 mm German glass coverslips in 0.5 mL first feed media.
8. On the next day, add 0.5 mL conditioned media. Perform half change of media two days later (*see* **Note 23**).
9. Add 1.4 μg/mL CAF to the NFeed on day 4–5 (to kill dividing cells) and incubate for 1–3 days.

3.5 Viral Infection of Motor Neurons

1. Culture motor neurons in conditioned media for 3 weeks. Feed 3 times per week by performing a half change of conditioned media.
2. Infect each well of dissociated spinal cord cultures with AAV-AR (serotype 1; cis-plasmid for AAV-AR production created by

inserting full-length AR cDNA with the desired mutation and CAG repeat length) to ensure ≥90 % infectivity (*see* **Note 24**). Plasmids for AAV1-AR production are available from the authors upon request.

3. Perform a half change of conditioned media the next day.
4. 5 days after infection, treat cultures with either ethanol or 10 μM DHT in conditioned media. Retreat every 2 days performing half changes of media (*see* **Note 25**).
5. After 7 days of treatment, fix cultures with 4 % paraformaldehyde and immunostain with SMI32 to visualize motor neurons (*see* **Note 26**).
6. Count motor neurons (*see* **Note 27**) in 10 fields per coverslip (3–4 coverslips per condition) at 20,000× magnification while blinded to the condition.
7. Total number of motor neurons is reported.

4 Notes

1. Toxicity of polyglutamine-expanded AR is ligand dependent; therefore, serum used in all experimental conditions must be charcoal-stripped of hormone.
2. For inducible expression of the mutated polyglutamine-expanded AR in PC12 cells, the AR gene is under the control of a Tet-response element promoter (pTRE; Tet-On, Clontech). The full-length AR cDNA with 112 CAG repeats was cloned into the pTRE vector as described [17]. This plasmid is available from the authors upon request. Additional tetracycline-regulated expression vectors are available from Clontech.
3. Primers used for mutagenesis should be between 25 and 45 bases in length, have a 40 % GC content, and end in one or more C or G bases. The mutation should be in the middle of the primers.
4. The pTRE plasmid also contains the β-lactamase gene, which confers ampicillin resistance.
5. The expanded CAG tract has a high propensity to contract in bacteria when grown at 37 °C for longer periods of time. Growing the bacteria at a lower temperature (and for a shorter time, if possible) will minimize the CAG tract contraction.
6. When evaluating the CAG tract for contraction using restriction enzyme digestion, enzymes that produce a CAG-containing fragment less than 1 kb in size should be used. SmaI/AflII double digestion of the AR cDNA produces a fragment of 679 bp when the CAG tract numbers 112 repeats; this fragment size allows for easy separation of contracted species

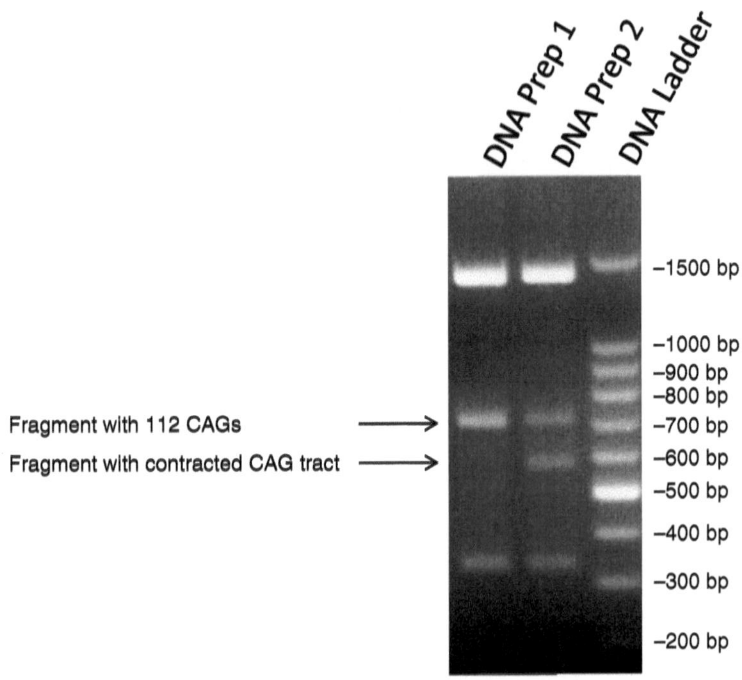

Fig. 2 Plasmid DNA containing AR cDNA with 112 CAGs was extracted from two different bacterial cultures and digested with SmaI and AflII. The DNA prep from culture 1 is relatively homogeneous in CAG length, whereas the DNA prep from culture 2 has a subspecies with a contracted CAG tract

on a 2 % w/v agarose gel, creating either a smear or a distinct band that migrates faster than the major band (Fig. 2). Contracted species are more difficult to resolve from larger DNA fragments.

7. It is possible that additional mutations could be introduced during the mutagenesis process. To ensure the final plasmid that will be transfected into cells only contains the mutation introduced by the primers, clone the desired mutation into a new plasmid containing the AR gene. Perform appropriate cuts to select plasmids that are expanded and confirm by sequencing the fidelity of sequence in the amplified region.
8. PC12 cell freezing medium contains 10 % v/v DMSO and no antibiotics. Quickly move cells out of this media into fresh media. For best results, culture the cells in antibiotic-free media for 12–24 h after thawing.
9. Make sure to break up the PC12 cells by pipetting vigorously up and down.
10. PC12 cells show optimal transfection efficiency when they are 85–90 % confluent and distributed evenly across the plate. If the cells are in clumps, trypsinize and replate them.

11. The hygromycin-resistance gene is not part of the pTRE-AR plasmid that we have used, but is instead on a cotransfected plasmid (however, newer generation vectors contain the selectable marker). The pTK-hygro plasmid is transfected at a 1:4 molar ratio with the pTRE-AR plasmid, increasing the likelihood that a hygromycin-resistant cell will contain the pTRE-AR plasmid.
12. To transfer colonies to the 96-well plate, aspirate media and wash cells with 2 mL of PBS containing Ca^{2+} and Mg^{2+}. Aspirate the PBS; then, with the plate at a 20° angle, pipette 2–5 μL of trypsin onto the first colony, pipetting up and down to loosen and collect the cells. Transfer cells and trypsin to the 96-well plate. Maintaining the same plate orientation, continue to pick colonies, quickly working from the bottom to the top of the plate and using a fresh tip and well for each colony. Add 200 μL of media to each cell-containing well of the 96-well plate. Fill wells without cells with 200 μL of PBS to prevent evaporation.
13. To quickly and efficiently transfer many PC12 clones from 96-well to 24-well plates, once they have grown to near confluency, pipette the existing media up and down to dissociate and dislodge the cells. Transfer cells and media to wells of a new 24-well plate. Add fresh media to cells. (Transfer can be similarly accomplished using trypsin–EDTA; however, the relatively loose adherence of PC12 cells makes the use of trypsin–EDTA for this time-consuming step unnecessary.)
14. Not all clones will express AR; some will only have been transfected with the pTK-hygro plasmid. Therefore each clone needs to be checked for expression of the plasmid interest.
15. Although the pTRE-AR plasmid is selected based on the minimal presence of contracted species, it is nonetheless likely that the plasmid prep used for transfection contains some DNA molecules with shorter CAG tracts.
16. Doxycycline will induce the expression of AR driven by the TRE.
17. Include on the gel a control cell line that is known to express polyglutamine-expanded AR for a comparison of polyglutamine tract size.
18. To determine AR CAG repeat length in clonal cell lines, first isolate DNA using DNA isolation kit (Qiagen). Use appropriate primers to amplify the CAG repeat of the transgene (forward primer in the promoter and reverse primer 3′ to the CAG repeat); electrophorese PCR product on 2 % w/v agarose gel. Extract PCR product from gel using QIAquick Gel Extraction Kit (Qiagen) and sequence the product to determine CAG repeat length. The introduced mutation should be confirmed in similar fashion.

19. Be sure to freeze multiple vials of clones expressing AR with an expanded-polyglutamine tract in freezing media.
20. If culturing transgenic motor neurons, genotype tail clips of embryos [6] while holding spinal cords on ice and subsequently, separately pool transgenic and non-transgenic spinal cords.
21. For glial culture preparation, remove brains from E13.5 embryonic mice and place in a 60 mm dish with 1 mL dissecting medium. Carry out **steps 4–6** of the dissociated spinal cord culture protocol. Plate dissociated cells on a poly-d-lysine-coated 100 mm plate (2 brains/3 plates) and feed with MEM containing 10 % FBS and antibiotics (50 units/mL penicillin and 50 μg/mL streptomycin). Change media after 3 days to remove dead cells. Glial cells should completely cover the plates after 4–7 days in culture and be ready to use for conditioning NFeed.
22. The original protocol for culturing mouse dissociated spinal cords [18] did not call for the use of glial-conditioned media. However, the requirement for using charcoal-stripped serum in the growth of these AR-expressing cultures (due to the hormone-dependent toxicity of polyglutamine-expanded AR) leads to the diminished health of the cultures, likely due to the loss of other hormones and growth factors (despite the addition of N2 in this protocol). We have found that the use of glial-conditioned media allows the full differentiation of motor neurons in cultures derived from transgenic mice and maintains the health of motor neurons in non-transgenic cultures after infection with AAV-AR, with no difference in motor neuron number prior to the addition of DHT [8, 12].
23. Perform half change of media on dissociated spinal cord cultures every 2 days. Full media changes will kill cultures.
24. The polyglutamine-expanded AR-encoding cDNA in the cis-plasmid contains sequences encoding 3 copies of the SV40 NLS (NLSx3) fused to the amino-terminus of the AR. This signal ensures nuclear localization of the overexpressed protein, which is critical for toxicity of polyglutamine-expanded AR. In the absence of this exogenous NLS, the polyglutamine-expanded AR will, upon hormone binding, form cytoplasmic aggregates composed of full-length AR protein, a feature not observed in SBMA patients or in mouse models of SBMA.
25. The final concentration of DHT should be 10 μM after carrying out the half feed.
26. Motor neurons cultured from mice expressing AR with 112 glutamines show significant ligand-dependent toxicity when treated with DHT for 7 days, compared to normal AR-overexpressing or non-transgenic motor neurons [8, 11, 12].
27. Motor neurons are identified by both SMI32 immunoreactivity and morphology [18].

Acknowledgement

This work was supported by NIH grant NS32214.

References

1. Sobue G, Hashizume Y, Mukai E, Hirayama M, Mitsuma T, Takahashi A (1989) X-linked recessive bulbospinal neuronopathy: a clinico-pathological study. Brain 112:209–232
2. Edwards A, Hammond HA, Jin L, Caskey CT, Chakraborty R (1992) Genetic variation at five trimeric and tetrameric tandem repeat loci in four human population groups. Genomics 12:241–253
3. Giovannucci E, Stampfer MJ, Krithivas K, Brown M, Dahl D, Brufsky A, Talcott J, Hennekens CH, Kantoff PW (1997) The CAG repeat within the androgen receptor gene and its relationship to prostate cancer. Proc Natl Acad Sci USA 94:3320–3323
4. La Spada AR, Wilson EM, Lubahn DB, Harding AE, Fischbeck KH (1991) Androgen receptor gene mutations in X-linked spinal and bulbar muscular atrophy. Nature 353:77–79
5. Pratt WB, Toft DO (1997) Steroid receptor interactions with heat shock protein and immunophilin chaperones. Endocr Rev 18:306–360
6. Chevalier-Larsen ES, O'Brien CJ, Wang H, Jenkins SC, Holder L, Lieberman AP, Merry DE (2004) Castration restores function and neurofilament alterations of aged symptomatic males in a transgenic mouse model of spinal and bulbar muscular atrophy. J Neurosci 24:4778–4786
7. Katsuno M, Adachi H, Kume A, Li M, Nakagomi Y, Niwa H, Sang C, Kobayashi Y, Doyu M, Sobue G (2002) Testosterone reduction prevents phenotypic expression in a transgenic mouse model of spinal and bulbar muscular atrophy. Neuron 35(5):843–854
8. Montie HL, Cho MS, Holder L, Liu Y, Tsvetkov AS, Finkbeiner S, Merry DE (2009) Cytoplasmic retention of polyglutamine-expanded androgen receptor ameliorates disease via autophagy in a mouse model of spinal and bulbar muscular atrophy. Hum Mol Genet 18(11):1937–1950
9. Nedelsky NB, Pennuto M, Smith RB, Palazzolo I, Moore J, Nie Z, Neale G, Taylor JP (2010) Native functions of the androgen receptor are essential to pathogenesis in a Drosophila model of spinal and bulbar muscular atrophy. Neuron 67:936–952
10. Takeyama K, Ito S, Yamamoto A, Tanimoto H, Furutani T, Kanuka H, Miura M, Tabata T, Kato S (2002) Androgen-dependent neurodegeneration by polyglutamine-expanded human androgen receptor in Drosophila. Neuron 35(5):855–864
11. Orr CR, Montie HL, Liu Y, Bolzoni E, Jenkins SC, Wilson EM, Joseph JD, McDonnell DP, Merry DE (2010) An interdomain interaction of the androgen receptor is required for its aggregation and toxicity in spinal and bulbar muscular atrophy. J Biol Chem 285: 35567–35577
12. Montie HL, Pestell RG, Merry DE (2011) SIRT1 modulates aggregation and toxicity through deacetylation of the androgen receptor in cell models of SBMA. J Neurosci 31:17425–17436
13. Mukherjee S, Thomas M, Dadgar N, Lieberman AP, Iniguez Lluhi JA (2009) Small ubiquitin-like modifier (SUMO) modification of the androgen receptor attenuates polyglutamine-mediated aggregation. J Biol Chem 284(32):21296–21306
14. Palazzolo I, Burnett BG, Young JE, Brenne PL, La Spada AR, Fischbeck KH, Howell BW, Pennuto M (2007) Akt blocks ligand binding and protects against expanded polyglutamine androgen receptor toxicity. Hum Mol Genet 16(13):1593–1603
15. Palazzolo I, Stack C, Kong L, Musaro A, Adachi H, Katsuno M, Sobue G, Taylor JP, Sumner CJ, Fischbeck KH, Pennuto M (2009) Overexpression of IGF-1 in muscle attenuates disease in a mouse model of spinal and bulbar muscular atrophy. Neuron 63(3): 316–328
16. Li M, Miwa S, Kobayashi Y, Merry DE, Yamamoto M, Tanaka F, Doyu M, Hashizume Y, Fischbeck KH, Sobue G (1998) Nuclear inclusions of the androgen receptor protein in spinal and bulbar muscular atrophy. Ann Neurol 44:249–254
17. Walcott JL, Merry DE (2002) Ligand promotes intranuclear inclusions in a novel cell model of spinal and bulbar muscular atrophy. J Biol Chem 277(52):50855–50859
18. Roy J, Minotti S, Dong L, Figlewicz DA, Durham HD (1998) Glutamate potentiates the toxicity of mutant Cu/Zn-superoxide dismutase in motor neurons by postsynaptic calcium-dependent mechanisms. J Neurosci 18:9673–9684

INDEX

Danny M. Hatters and Anthony J. Hannan (eds.), *Tandem Repeats in Genes, Proteins, and Disease: Methods and Protocols*, Methods in Molecular Biology, vol. 1017, DOI 10.1007/978-1-62703-438-8, © Springer Science+Business Media New York 2013

M

N

O

P

R

MIX
Papier aus verantwortungsvollen Quellen
Paper from responsible sources
FSC® C105338

If you have any concerns about our products,
you can contact us on
ProductSafety@springernature.com

In case Publisher is established outside the EU,
the EU authorized representative is:
Springer Nature Customer Service Center GmbH
Europaplatz 3, 69115 Heidelberg, Germany

Printed by Libri Plureos GmbH
in Hamburg, Germany